# XFEM FRACTURE ANALYSIS OF COMPOSITES

# XFEM FRACTURE ANALYSIS OF COMPOSITES

**Soheil Mohammadi**

*University of Tehran, Iran*

A John Wiley & Sons, Ltd., Publication

*Library of Congress Cataloging-in-Publication Data*

Mohammadi, S. (Soheil)
    XFEM fracture analysis of composites / Soheil Mohammadi.
       pages cm
    Includes bibliographical references and index.
    ISBN 978-1-119-97406-2
    1. Composite materials–Fracture.   2. Composite materials–Fatigue.   3. Fracture mechanics.   4. Finite element method.   I. Title.
    TA418.9.C6M64 2012
    620.1′186–dc23

                                                                        2012016776

A catalogue record for this book is available from the British Library.

ISBN: 978-1-119-97406-2

Typeset in 10/12pt Times by Aptara Inc., New Delhi, India
Printed and bound in Singapore by Markono Print Media Pte Ltd

*To: Mansoureh, Sogol & Soroush*

# Contents

# Preface

A decade after its introduction, the extended finite element method (XFEM) has now become the primary numerical approach for analysis of a wide range of discontinuity applications, including crack propagation problems. The simplicity of the idea of enrichment for reproducing a singular/discontinuous nature of the field variable, the flexibility in handling several cracks and crack propagation patterns on a fixed mesh, and the level of accuracy with minimum additional degrees of freedom (DOFs) have transformed XFEM into the most efficient computational approach for handling various complex discontinuous problems. Concepts of XFEM are now even taught in a number of postgraduate courses, for instance advanced fracture mechanics and meshless methods, in major engineering departments, such as Civil, Mechanical, Material, Aerospace and so on, all over the world.

On the other hand, the highly flexible design of composites allows attractive prescribed tailoring of material properties, fitted to the engineering requirements for a wide range of engineering and industrial applications; from advanced aerospace and defence systems to traditional structural strengthening techniques, and from large scale turbines and power plants to nanoscale carbon nanotubes (CNTs) applications. Despite excellent characteristics, composites suffer from a number of shortcomings, mainly in the form of unstable cracking which can be initiated and propagated under different production imperfections and service circumstances. Therefore, the study of the crack stability and load bearing capacity of these types of structures, which directly affect the safety and economics of many important industries, has become one of the important topics of research for the computational mechanics community.

This text is dedicated to discussing various aspects of the application of the extended finite element method for fracture analysis of composites on the macroscopic scale. Nevertheless, most of the discussed subjects can be similarly used for fracture analysis of other materials, even on microscopic scales. The book is designed as a textbook, which provides all the necessary theoretical bases before discussing the numerical issues.

The book can be classified into four parts. The first part is dedicated to the basics. The introduction chapter provides a general overview of the problem in hand and summarily reviews available analytical and numerical techniques for fracture analysis of composites. The second chapter deals with the basics of the theory of elasticity, and is followed by discussions on asymptotic solutions for displacement and stress fields in different fracture modes, basic concepts of stress intensity factors, energy release rate, various forms of the $J$ contour integral and mixed mode fracture criteria.

The second part, Chapter 3, is a redesigned and completed edition of the same chapter in my previous book, and presents a detailed discussion on the extended finite element method.

After presenting the basic formulation, the chapter continues with three sections on available options for strong discontinuity enrichment functions, weak discontinuity enrichments for material interfaces, and a collection of several crack-tip enrichments for various engineering applications. It concludes with sample simulations of a wide range of problems, including classical in-plane mixed mode fracture mechanics, cracking in plates and shells, simulation of shear band creation and propagation, self-similar fault rupture, sliding contact, hydraulic fracture, and dislocation dynamics, to assess the accuracy, performance, robustness and efficiency of XFEM.

The main part of the book includes four comprehensive chapters dealing with various aspects of fracture in composite structures. Static crack analysis in orthotropic materials, dynamic fracture mechanics for stationary and moving cracks, inhomogeneous functionally graded materials and bimaterial delamination analysis are discussed in detail. After a review of anisotropic and orthotropic elasticity, all chapters begin with a complete discussion of available analytical solutions for near crack-tip fields in the corresponding orthotropic problem, followed by orthotropic mixed mode fracture mechanics and associated forms of the interaction integral. XFEM enrichment functions for each class of orthotropic materials are obtained and numerical issues related to XFEM discretization are addressed. A number of illustrative numerical simulations are presented and discussed at the end of each chapter to assess the performance of XFEM compared to alternative analytical and numerical techniques.

The final part reviews a number of ongoing research topics for orthotropic materials. First, the orthotropic version of the extended isogeometric analysis (XIGA) is presented by briefly explaining the basic concepts of NURBS and IGA methodology and discussing a number of simple isotropic and orthotropic simulations. Then, the newly developed idea of plane strain anisotropic dislocation dynamics is briefly presented and related XFEM formulation and necessary enrichment functions for the self-stress state of edge dislocations are explained. The book concludes with two brief introductory sections on orthotropic biomaterial applications of XFEM and the piezoelectric problems.

I would like to express my sincere gratitude to Prof. T. Belytschko, for his valuable friendly comments and encouraging message after the publication of the first book on XFEM, and to Prof. A.R. Khoei, as a friend, a colleague and a referee with excellent comments and discussions on various subjects of computational mechanics, especially XFEM. In preparing the present book, particularly the first two parts, I have used the available contributions from brilliant research works by many others, and I have done my best to properly and explicitly acknowledge their achievements within the text, relevant figures, tables and formulae. I am much indebted to their outstanding works, and I apologize sincerely for any unintentional failure in appropriately acknowledging them.

My special thanks to many of my former and present M.Sc. and Ph.D. students who have endeavored many aspects of XFEM over the past decade. First, Dr A. Asadpoure, with whom we started to explore XFEM and new orthotropic enrichment functions in 2002. The results for the dynamic fracture of stationary and moving cracks were obtained by Dr D. Motamedi, and Ms S. Esna Ashari developed the orthotropic bimaterial enrichment functions and simulated all the delamination and interlaminar crack problems. Most of the results for FGM problems were prepared by Mr H. Bayesteh, who actively contributed in many other parts of the book. Mr S.S. Ghorashi and Mr N. Valizadeh skillfully developed the XIGA methodology and Mr S. Malekafzali implemented XFEM for anisotropic dislocation dynamics. Other results were obtained by my students: S.H. Ebrahimi, A. Daneshyar, M. Parchei, M. Goodarzi, S.N. Rezaei, S.N. Mahmoudi and M.M.R. Kabiri.

My acknowledgement is extended to John Wiley & Sons, Ltd., Publication for the excellent professional work that facilitated the whole process of publication of the book; in particular to Dr E.F. Kirkwood and E. Willner, D. Cox, R. Davies, L. Wingett, A. Hunt, C. Lim, S. Sharma and Dr R. Whitelock.

The inspiration and power for completing this work have been the love and understanding of my family, as they had to comply with all my commitments. After a life-time engagement in mathematics, physics and engineering, satisfaction is not obtained just in academic or professional progress, novelty and innovation; it should perhaps be sought in ethics, responsibility, love and freedom. This book has been completed at the twilight of a long hard winter, with a hope for a bright flourishing spring of prosperity and freedom to come. I would like to proudly dedicate this work to all spirited noble Iranian students who accomplish academic achievements while challenging for more DOFs!

Soheil Mohammadi
*Spring 2012*
*Tehran, IRAN*

# Nomenclature

Parameters not shown in this nomenclature are temporary variables or known constants, defined immediately when cited in the text.

| | |
|---|---|
| $\alpha$ | Curvilinear coordinate |
| $\alpha$ | First Dundurs parameter |
| $\alpha, \beta$ | Newmark parameters |
| $\alpha, \beta, \gamma$ | FGM constants |
| $\alpha_0$ | Curvilinear coordinate $\alpha$ of an ellipse |
| $\alpha_{lk}$ | Components of coordinate transformation tensor |
| $\beta$ | Curvilinear coordinate |
| $\beta$ | Second Dundurs parameter |
| $\beta_0$ | Curvilinear coordinate $\beta$ of an ellipse |
| $\beta_d, \beta_s$ | Dilatational and shear wave functions |
| $\gamma$ | Wedge angle |
| $\gamma_s$ | Surface energy density |
| $\gamma_d, \gamma_s$ | Dilatational and shear wave functions |
| $\gamma_{ij}$ | Engineering shear strain |
| $\delta$ | Plastic crack tip zone |
| $\delta$ | Variation of a function |
| $\delta(\xi)$ | Dirac delta function |
| $\delta_{ij}$ | Kronecker delta function |
| $\delta_{\mathrm{I}}, \delta_{\mathrm{II}}$ | Local displacements of crack edges |
| $\boldsymbol{\varepsilon}$ | Strain tensor |
| $\varepsilon$ | Oscillation index |
| $\varepsilon_{ij}, \varepsilon_i$ | Strain components |
| $\bar{\varepsilon}_{ij}$ | Dimensionless angular geometric function |
| $\varepsilon_{ij}^{\mathrm{aux}}$ | Auxiliary strain components |
| $\varepsilon_{\mathrm{o}}$ | Applied displacement loading |
| $\varepsilon_{\mathrm{yld}}$ | Yield strain |
| $\xi$ | Local curvilinear (mapping) coordinate system |

| | |
|---|---|
| $\xi_i$ | Knot $i$ |
| $\xi_{\text{tip}}$ | Crack-tip position |
| $\xi(\mathbf{x})$ | Distance function |
| $\xi_{\text{g}}, \eta_{\text{g}}$ | Gauss point position along the contour $J$ |
| $\eta$ | Local curvilinear (mapping) coordinate system |
| $\eta$ | Equivalent inelastic strain |
| $\theta$ | Crack propagation angle with respect to initial crack |
| $\theta$ | Angular polar coordinate |
| $\theta_0$ | Crack angle |
| $\theta_k, \bar{\theta}_k$ | Orthotropic angular functions |
| $\theta_{\text{d}}, \theta_{\text{s}}$ | Dynamic distance functions |
| $\kappa, \kappa'$ | Material parameters |
| $\kappa_0$ | Effective material parameter |
| $\lambda$ | Lame modulus |
| $\lambda$ | Power of radial enrichment |
| $\lambda$ | Ratio of orthotropic Young modules $E_2/E_1$ |
| $\lambda, \lambda_n$ | Roots of the characteristic equation |
| $\mu, \mu_{ij}$ | Isotropic and orthotropic shear modulus |
| $\nu, \nu_{ij}$ | Isotropic and orthotropic Poisson's ratios |
| $\bar{\nu}$ | Average orthotropic Poisson's ratios |
| $\rho$ | Radius of curvature |
| $\rho$ | Density |
| $\boldsymbol{\sigma}$ | Stress tensor |
| $\sigma_0$ | Applied normal traction |
| $\sigma_{\text{cr}}$ | Critical stress for cracking |
| $\sigma_{\text{eff}}$ | von Mises effective stress |
| $\sigma_{ij}, \sigma_i$ | Stress components |
| $\bar{\sigma}_{ij}$ | Dimensionless angular geometric function |
| $\sigma_{ij}^{\text{aux}}$ | Auxiliary stress components |
| $\sigma_{\text{yld}}$ | Yield stress |
| $\sigma_{\theta\theta}$ | Hoop stress |
| $\tau_0$ | Applied tangential traction |
| $\tau_n$ | Decohesive shear stress |
| $\phi(\mathbf{x})$ | Level set function |
| $\phi(z)$ | Complex stress function |
| $\varphi$ | Angle of orthotropic axes |
| $\varphi$ | Crack angle |
| $\varphi$ | Ramp function for transition domain |
| $\varphi$ | Electric potential |

| | |
|---|---|
| $\chi_m(\mathbf{x})$ | Enrichment function for weak discontinuities |
| $\chi(z)$ | Complex stress function |
| $\psi$ | Friction coefficient |
| $\psi$ | Phase angle |
| $\psi(\mathbf{x})$ | Enrichment function |
| $\psi(\mathbf{x}, t)$ | Level set function |
| $\psi(z)$ | Complex stress function |
| $\Gamma$ | Boundary |
| $\Gamma_1$ | Infinitesimally small internal contour |
| $\Gamma_c$ | Crack boundary |
| $\Gamma_t$ | Traction (natural) boundary |
| $\Gamma_u$ | Displacement (essential) boundary |
| $\Delta$ | Finite variation of a function |
| $\Delta t$ | Time-step |
| $\Delta a$ | Crack length increment |
| $\Lambda_i(t)$ | Time interval shape functions |
| $\Xi$ | Knot vector |
| $\Pi$ | Potential energy |
| $\Phi$ | Airy stress function |
| $\Phi_j(\mathbf{x})$ | MLS shape functions |
| $\Phi_i(z_i)$ | Complex functions |
| $\Omega$ | Domain |
| $\Omega_1, \Omega_2$ | Non-overlapping subdomains |
| $\Omega_{pu}$ | Domain associated with the partition of unity |
| $\Psi_H^\alpha$ | Dislocation glide enrichment |
| $(1, 2)$ | Material axes |
| $a$ | Crack length/half length |
| $a$ | Semi-major axis of ellipse |
| $\bar{a}$ | Effective crack length |
| $\mathbf{a}(\mathbf{x})$ | Vector of unknown coefficients |
| $\mathbf{a}, \mathbf{a}_h$ | Heaviside enrichment degrees of freedom |
| $\mathbf{a}_i, \mathbf{a}_k$ | Enrichment degrees of freedom |
| $\mathbf{a}^{enr}$ | Enrichment degrees of freedom |
| $A$ | Area associated with the domain $J$ integral |
| $A_1$ | Area inside the infinitesimally small internal contour $\Gamma_1$ |
| $A^+, A^-$ | Area of the influence domain above and below the crack |
| $A_i, A_{ij}$ | Coefficients |
| $b$ | Width of a plate |
| $b$ | Semi-minor axis of ellipse |

| | |
|---|---|
| $\mathbf{b}_k, \mathbf{b}_k^l$ | Crack tip enrichment degrees of freedom |
| $\mathbf{b}^\alpha$ | Burgers vector for dislocation $\alpha$ |
| $b_\alpha$ | Magnitude of the Burgers vector for dislocation $\alpha$ |
| $b_n$ | Series coefficients |
| $\mathbf{B}$ | Matrix of derivatives of shape functions |
| $B_{12}, B_{66}$ | Coefficients of characteristic equation |
| $B_{i,p}(\xi)$ | B-spline basis function of order $p$ |
| $\mathbf{B}^h$ | Matrix of derivatives of final shape functions |
| $\mathbf{B}_i^r$ | Strain-displacement matrix (derivatives of shape functions) |
| $\mathbf{B}_i^\mathbf{u}$ | Matrix of derivatives of classical FE shape functions |
| $\mathbf{B}_i^\mathbf{a}$ | Matrix of derivatives of Heaviside enrichment shape functions |
| $\mathbf{B}_i^\mathbf{b}$ | Matrix of derivatives of crack tip enrichment shape functions |
| $\mathbf{B}_i^\mathbf{c}$ | Matrix of derivatives of weak discontinuity enrichment shape functions |
| $\mathbf{B}_i^\mathbf{c}$ | Matrix of derivatives of transition shape functions |
| $c$ | Dugdale effective crack length |
| $c_J$ | Size of crack tip contour for $J$ integral |
| $c_{ij}$ | Material compliance constants |
| $\mathbf{c}_\mathrm{m}$ | Degrees of freedom for weak discontinuity enrichment |
| $\mathbf{c}_\mathrm{m}$ | Degrees of freedom for transitional enrichment |
| $c_\mathrm{d}$ | Dilatational wave speed |
| $c_\mathrm{L}$ | Wave speed along the loading axis |
| $c_\mathrm{R}$ | Rayleigh speed |
| $c_\mathrm{s}$ | Shear wave speed |
| $C$ | Material constitutive matrix |
| $\mathsf{C}$ | 4th order material compliance tensor |
| $\mathsf{C}_{ijkl}$ | Cartesian components of $\mathsf{C}$ |
| $C_n$ | Coefficient |
| $\mathbf{C}(\xi)$ | NURBS curve |
| $d_{ij}$ | Material modulus constants |
| $d_{ij}^d$ | Dynamic material modulus constants |
| $\mathbf{d}_\mathrm{m}$ | Degrees of freedom for transitional enrichment |
| $D$ | Dynamic function |
| $\mathbf{D}$ | Two dimensional Material modulus matrix |
| $\mathsf{D}$ | 4th order material elasticity modulus tensor |
| $\mathsf{D}_{ijkl}$ | Cartesian components of $\mathbf{D}$ |
| $\mathbf{D}_{\alpha\beta}$ | Components of $\mathbf{D}$ |
| $D_i, D_x, D_y$ | Elastic displacement vector |
| $E, E_i$ | Isotropic and orthotropic Young's modules |
| $E_i, E_x, E_y$ | Electric field |

| | |
|---|---|
| $E'$ | Effective material parameter |
| $E^0$ | Reference Young modulus |
| $E^{12}$ | Equivalent bimaterial elastic modulus |
| $\bar{E}$ | Average orthotropic Young modules |
| $f_k(\mathbf{x})$ | Set of PU functions |
| $\mathbf{f}$ | Nodal force vector |
| $\mathbf{f}_i^r$ | Nodal force components (classic and enriched) |
| $\mathbf{f}^b$ | Body force vector |
| $\mathbf{f}^t$ | External traction vector |
| $\mathbf{f}^c$ | Cohesive crack traction vector |
| $\mathbf{f}^{ext}$ | External force vector |
| $f_I, f_{II}$ | Functions of the crack-tip speed |
| $f_I^d, f_{II}^d$ | Universal functions of the crack-tip speed |
| $f_{ij}^I, f_{ij}^{II}, f_{ij}^{III}$ | Mode I, II and II angular functions |
| $f_k^{pu}$ | Set of PU functions |
| $F_l(\mathbf{x}), F_\alpha(r, \theta)$ | Crack tip enrichment functions |
| $F(\sigma, \alpha)$ | Delamination function |
| $\mathsf{F}_{ij}$ | Deformation gradient |
| $g(\theta)$ | Angular function for a crack-tip kink problem |
| $g_j(\theta), \bar{g}_k(\theta)$ | Orthotropic crack-tip enrichment functions |
| $\mathsf{G}$ | Shear modulus |
| $G, G(\theta)$ | Fracture energy release rate |
| $G_c$ | Critical fracture energy release rate |
| $G_1, G_2$ | Mode I and II fracture energy release rates |
| $h$ | Intrinsic shear band thickness |
| $h_t$ | Characteristic thickness of the bonding layer |
| $H$ | Slope of linear softening curve |
| $\bar{H}$ | Intrinsic hardening coefficient |
| $H(\xi), H(\mathbf{x})$ | Heaviside function |
| i | Complex number $i^2 = -1$ |
| $\mathbf{I}$ | 2$^{nd}$ order identity tensor |
| $\mathsf{I}$ | 4$^{th}$ order symmetric identity tensor |
| $I(t)$ | Corresponding creep compliance |
| $\mathbf{J}$ | Jacobian matrix |
| $J, J^s$ | $J$ integral |
| $J^{act}$ | Actual $J$ integral |
| $J^{aux}$ | Auxiliary $J$ integral |
| $J_1, J_2$ | Components of the $J$ vector |
| $J_k^d$ | Dynamic $J$ integral |

| $k_0$ | Dimensionless constant for the power hardening law |
|---|---|
| $K$ | Bulk modulus |
| $\mathbf{K}$ | Stiffness matrix |
| $\mathbf{K}_{ij}^{rs}$ | Components of stiffness matrix |
| $K$ | Stress intensity factor |
| $K, \bar{K}$ | Complex stress intensity factor |
| $K_c$ | Critical stress intensity factor |
| $K_0$ | Reference stress intensity factor |
| $K_I, K_{II}, K_{III}$ | Mode I, II and III stress intensity factors |
| $\bar{K}_I, \bar{K}_{II}$ | Normalized mode I and mode II stress intensity factors |
| $K_I^{aux}, K_{II}^{aux}$ | Auxiliary mode I and mode II stress intensity factors |
| $K_{Ic}, K_{IIc}$ | Critical mode I and mode II stress intensity factors |
| $K_{Ic}^1, K_{Ic}^2$ | Fracture toughnesses along the principal planes of elastic symmetry |
| $K_{Ic}^{\theta}$ | Fracture toughness at propagation |
| $K_{Ic}^d$ | Dynamic crack initiation toughness |
| $K_{Ic}^D$ | Dynamic crack growth (propagation) toughness |
| $K_{\theta\theta}, K_{tt}$ | Hoop stress intensity factor |
| $l_{ij}$ | Coefficient |
| $L$ | Length of the singular element |
| $L(v_c)$ | Dynamic matrix for orthotropic materials |
| $m$ | Number of enrichment functions |
| $m$t | Number of nodes to be enriched by crack-tip enrichment functions |
| $m$h | Number of nodes to be enriched by Heaviside enrichment functions |
| $m$f | Number of crack-tip enrichment functions |
| $m$m | Number of weak discontinuity enrichment functions |
| $m$st | Number of transition enrichment functions 1 |
| $m$ sh | Number of transition enrichment functions 2 |
| $m_k$ | Roots of characteristic equation $m_k = m_{kx} + im_{ky}$ |
| $M$ | Concentrated bending moment |
| $M$ | Interaction integral |
| $M^{(1)}, M^{(2)}$ | Interaction integral associated with two modes I and II |
| $M^d$ | Dynamic interaction integral |
| $\mathbf{M}$ | Mass matrix |
| $\mathbf{M}_{ij}$ | Components of mass matrix |
| $n$ | Power number for the HKK plastic model |
| $n$ | Number of nodes for each finite element |
| $ng^A$ | Number of gauss points inside contour area $A$ |
| $ng^\Gamma$ | Number of gauss points on contour $\Gamma$ |
| $n$p | Number of independent domains of partition of unity |

| | |
|---|---|
| $n_n$, $n_{nodes}$ | Number of nodes in a finite element |
| $n_e$, $n_{elem.}$ | Number of finite elements |
| $n_{cp}$ | Number of control points |
| $n_{cells}$ | Number of background cells of EFG |
| $n_{DOFs}$ | Number of degrees of freedom |
| $\mathbf{n}$ | Normal vector |
| $\mathbf{N}_j$ | Matrix of shape functions |
| $N_j$ | Shape function |
| $N_{elements}$ | Number of finite elements |
| $N_{nodes}$ | Number of nodes |
| $N_{enrich.}$ | Number of enrichment functions |
| $N_{DOFs}$ | Number of degrees of freedom |
| $\bar{N}_i$ | Hierarchical shape functions for the transition domain |
| $\bar{N}_j$ | New set of GFEM shape functions |
| $p(\mathbf{x})$ | Basis function |
| $p$ | A point on curvilinear coordinate system $p = \alpha + i\beta$ |
| $p$, $p_k$, $\bar{p}_k$ | Orthotropic parameters |
| $\boldsymbol{p}$ | Basis function |
| $\boldsymbol{p}^{enr}$ | Enrichment basis function |
| $\boldsymbol{p}^{lin}$ | Linear basis function |
| $p_k$ | $k$-th basis function |
| $p^l(\mathbf{x})$ | $l$-order polynomial function |
| $P$ | Concentrated force |
| $\boldsymbol{P}$ | External load vector |
| $P_{cr}$ | Critical load |
| $q$ | Arbitrary smoothing function |
| $q$, $q_k$, $\bar{q}_k$ | Orthotropic parameters |
| $q_i$ | Nodal values of the arbitrary smoothing function |
| $(r, \theta)$ | Local crack tip polar coordinates |
| $r_J$ | Radius of $J$ integral contour |
| $r_d$, $r_s$ | Dilatational and shear distance functions |
| $r_1$, $r_s$ | Orthotropic distance functions |
| $r_p$, $r_{p1}$, $r_{p2}$ | Crack tip plastic zone |
| $R$ | Ramp function |
| $R_K$ | Ratio of dynamic stress intensity factors |
| $R_\delta$ | Ratio of opening to sliding displacements |
| $R_i^p(\xi)$ | NURBS function of order $p$ |
| $s$ | Roots of characteristic equation $s = s_1 + is_2$ |
| $s_k$ | Roots of characteristic equation $s_k = s_{kx} + is_{ky}$ |

| | |
|---|---|
| $\bar{s}_k$ | Roots of characteristic equation $\bar{s}_k = \bar{s}_{kx} + i\bar{s}_{ky}$ |
| $s_m$ | Roots of characteristic equation $s_m = s_{m1} + is_{m2}$ |
| $\bar{S}_n$ | Slope of softening curve |
| $S_{ij}$ | Material constants |
| $\mathbf{S}(\xi_1, \xi_2)$ | NURBS surface |
| $t$ | Time |
| $\mathbf{t}$ | Traction |
| $\mathbf{t}_h$ | Unit vector for tangential direction |
| $t_{ij}$ | Material function |
| $t_0$ | Time for the wave to reach the crack tip |
| $t_\mathrm{p}$ | FRP thickness |
| $T_i(t)$ | Time shape functions |
| $T_j$ | Enriched time interval |
| $T_j$ | Transformation matrix |
| $\mathbf{T}_i$ | Control points |
| $\mathbf{u}$ | Displacement vector |
| $\dot{\mathbf{u}}$ | Velocity vector |
| $\bar{\mathbf{u}}$ | Prescribed displacement |
| $\bar{\dot{\mathbf{u}}}$ | Prescribed velocity |
| $\ddot{\mathbf{u}}$ | Acceleration vector |
| $\mathbf{u}_i$ | Displacement field component |
| $\dot{\mathbf{u}}_i$ | Velocity field component |
| $\ddot{\mathbf{u}}_i$ | Acceleration field component |
| $\mathbf{u}_i^{\mathrm{aux}}$ | Auxiliary displacement field component |
| $\dot{\mathbf{u}}_i^{\mathrm{aux}}$ | Auxiliary velocity field component |
| $\mathbf{u}^{\mathrm{enr}}$ | Enriched displacement field |
| $\mathbf{u}^{\mathrm{FE}}$ | Classical finite element displacement field |
| $\mathbf{u}^{\mathrm{XFEM}}$ | XFEM displacement field |
| $\mathbf{u}^{\mathrm{tra}}$ | Transition enrichment part of the displacement field |
| $\mathbf{u}^{\mathrm{H}}$ | Heaviside enrichment part of the displacement field |
| $\mathbf{u}^{\mathrm{tip}}$ | Crack-tip enrichment part of the displacement field |
| $\mathbf{u}^{\mathrm{mat}}$ | Weak discontinuity enrichment part of the displacement field |
| $\mathbf{u}_{\mathrm{tip}}^{\mathrm{Enr}}(\mathbf{x}, t)$ | Crack-tip part of the approximation |
| $u_n^h$ | Displacement at time $n$ |
| $\dot{u}_n^h$ | Velocity at time $n$ |
| $\ddot{u}_n^h$ | Acceleration at time $n$ |
| $\mathbf{u}^h, \mathbf{u}^h(\mathbf{x})$ | Approximated displacement field |
| $\dot{\mathbf{u}}^h$ | Approximated velocity field |
| $\ddot{\mathbf{u}}^h$ | Approximated acceleration field |

| | |
|---|---|
| $\bar{\mathbf{u}}_j$ | Nodal displacement vector |
| $\bar{\mathbf{u}}_{ij}$ | Displacement angular functions |
| $u_x, u_y, u_z$ | $x$, $y$ and $z$ displacement components |
| $u_{x'}, u_{y'}$ | Local displacements of the nodes along the crack in the singular element |
| $U^I, U^{II}$ | Symmetric and antisymmetric crack tip displacements |
| $U_s$ | Strain energy |
| $U_s^e, U_s^p$ | Elastic and plastic strain energies |
| $U_\Gamma$ | Surface energy |
| $v_c$ | Crack-tip velocity |
| $\mathbf{v}$ | Velocity vector |
| $\mathbf{v}^c$ | Classical velocity DOFs |
| $\mathbf{v}^e$ | Additional velocity DOFs |
| $\mathbf{V}(t)$ | Vector of approximated velocity degrees of freedom |
| $V$ | Volume |
| $W$ | External work |
| $w^{aux}$ | Auxiliary work |
| $W^{ext}$ | Virtual work of the external loading |
| $W_g$ | Gauss weight factor |
| $W_i$ | Weights associated with each control point $i$ |
| $W_g^\Gamma$ | Gauss weighting factor for contour $\Gamma$ |
| $W_g^A$ | Gauss weighting factor for area inside contour integral $J$ |
| $W^{int}$ | Internal virtual work |
| $w^M$ | Interaction work |
| $w_d$ | Kinetic energy density |
| $w_s$ | Strain energy density |
| $W_t(t)$ | Time weight function |
| $x, y, z$ | Cartesian coordinates |
| $(X, Y)$ | Global coordinate system |
| $(X_1, X_2)$ | Global coordinate system |
| $\mathbf{x}$ | Position vector |
| $\mathbf{x}_c$ | Position of crack or discontinuity |
| $\mathbf{x}_{tip}$ | Position of the crack tip |
| $\mathbf{x}_\Gamma$ | Position of the projection point on an interface |
| $(x_1, x_2)$ | 2D coordinate system |
| $(x_1, x_2)$ | Material axes |
| $(x', y')$ | Local crack tip coordinate axes |
| $z$ | Complex variable $z = x + iy$ |
| $\bar{z}$ | Conjugate complex variable $\bar{z} = x - iy$ |
| $z_i$ | Complex parameters |

| | |
|---|---|
| $\frac{d}{dt}$ | Time derivative |
| $\frac{D}{Dt}$ | Material time derivative |
| $\dot{f}, \ddot{f}$ | The first and second temporal derivatives of a function |
| $f', f''$ | The first and second spatial derivatives of a function |
| $\bar{f}, \bar{\bar{f}}$ | The first and second integrals of a function |
| $\nabla = \partial/\partial x$ | Nabla operator |
| $\langle \rangle$ | Jump operator across an interface |
| $:$ | Inner product of two second order tensors |
| $\otimes$ | Tensor product of two vectors |
| Re | Real part of a complex number |
| Im | Imaginary part of a complex number |
| $\propto$ | Proportional |
| BEM | Boundary element method |
| CAD | Computer aided design |
| CNT | Composite nanotube |
| COD | Crack opening displacement |
| CTOD | Crack-tip opening displacement |
| DCT | Displacement correlation technique |
| DEM | Discrete element method |
| DOF | Degree of freedom |
| EDI | Equivalent domain integral |
| EFG | Element free Galerkin |
| ELM | Equilibrium on line |
| EPFM | Elastic plastic fracture mechanics |
| FDM | Finite difference method |
| FE | Finite element |
| FEM | Finite element method |
| FGM | Functionally graded materials |
| FMM | Fast marching method |
| FPM | Finite point method |
| FRP | Fibre reinforced polymer |
| GFEM | Generalized finite element method |
| GNpj | Generalized Newmark approximation of degree $p$ for equations of order $j$ |
| HRR | Hutchinson-Rice-Rosengren |
| IGA | Isogeometric analysis |
| LEFM | Linear elastic fracture mechanics |
| LSM | Level set method |
| MCC | Modified crack closure |
| MLPG | Meshless local Petrov Galerkin |

| | |
|---|---|
| MLS | Moving least squares |
| NURBS | Non-uniform rational B-splines |
| OUM | Ordered upwind method |
| PU | Partition of unity |
| PUFEM | Partition of unity finite element Method |
| RKPM | Reproducing kernel particle method |
| SAR | Statically admissible stress recovery |
| SIF | Stress intensity factor |
| SPH | Smoothed particle hydrodynamics |
| SSpj | Single step approximation of degree $p$ for equations of order $j$ |
| STXFEM | Space time extended finite element method |
| TXFEM | Time extended finite element method |
| WLS | Weighted least squares |
| XFEM, X-FEM | Extended finite element method |
| XIGA | Extended isogeometric analysis |

# 1

# Introduction

## 1.1 Composite Structures

Composite materials are used extensively in engineering and industrial applications; from traditional structural strengthening to advanced aerospace and defence systems, and from nanoscale carbon nanotube (CNT) applications to large scale turbines and power plants and so on. Their highly flexible design allows prescribed tailoring of material properties, fitted to the engineering requirements. These include a wide variety of properties across various length scales, including nano and micro-mechanical structural needs, thermo-mechanical specifications and even electro-magneto-mechanical characteristics. In addition, multilayer and orthotropic functionally graded materials (FGMs) have been increasingly used in advanced material systems in high-tech industries to withstand hostile operating conditions, where conventional homogeneous composites may fail.

Composite materials are created by the combination of two or more materials to form a new material with enhanced properties compared to those of the individual constituents. By this definition, reinforced concrete, as a mixture of stone, sand, cement and steel, wood comprised of cellulose and lignin, and bone consisting of collagen and apatite can be regarded as special types of composites. The conventional forms of composites, however, are made of two main ingredients: fibres and matrix. Fibres are required to have a number of specifications, such as high elasticity modulus and ultimate strength, and must retain their geometrical and mechanical properties during fabrication and handling. The matrix constituent must be chemically and thermally compatible with the fibres over a long period of time, and is meant to bind together the fibres, protect their surfaces, and transfer stresses to the fibres efficiently.

High specific strength, excellent fatigue durability, significant corrosion, chemical and environmental resistances, especially important in food and chemical processing plants, cooling towers, offshore platforms and so on, designable mechanical properties, electromagnetic transparency or electrical insulation, together with relatively fast deployment and low maintenance have made composites attractive materials for almost all engineering applications.

Despite excellent characteristics, composites suffer from a number of shortcomings, such as brittleness, high thermal and residual stresses, poor interfacial bonding strength and low toughness, which may facilitate the process of unstable cracking under different conditions, such as imperfection in material strength, fatigue, yielding and production faults. These failures

*XFEM Fracture Analysis of Composites*, First Edition. Soheil Mohammadi.
© 2012 John Wiley & Sons, Ltd. Published 2012 by John Wiley & Sons, Ltd.

can cause extensive damage accompanied by substantial reduction in stiffness and load bearing capacity, decreased ductility and the possibility of abrupt collapse mechanisms. The problem becomes even more important in intensive concentrated loading conditions, such as moving and dynamic loadings, high velocity impact and explosion.

Moreover, composite materials are utilized in thin forms which are susceptible to various types of defects. Cracking, the most likely type of defect in these structures, can be initiated and propagated under different production imperfections and service circumstances, such as initial weakness in material strength, fatigue and yielding. Therefore, the study of the crack stability and load bearing capacity of these types of structures, which directly affect the safety and economics of many important industries, has become an important topic of research for the computational mechanics community.

## 1.2 Failures of Composites

Layered, orthotropic, sometimes inhomogeneous and multi-material characteristics of composites allow the possibility for occurrence of various failure modes under different loading conditions. In general, however, the failure modes of composite plies can be categorized into four classes: fibre failure, ply delamination, matrix cracking and fibre/matrix deboning. These failure modes, or any combination of them, reduce and may ultimately eliminate the composite action altogether.

### 1.2.1  Matrix Cracking

The matrix material is the lowest strength component in a composite action to withstand a specific loading. The brittle nature of matrix cracking is the main source of failure in composites and may initiate other modes of failure, such as delamination and debonding.

### 1.2.2  Delamination

Delamination, also called interlaminar debonding or interface cracking, is among the most commonly encountered failure modes in composite laminates and may become a major source of concern in the performance and safety of composites by reducing the ductility, stiffness and strength of the composite specimen and even cause sudden brittle fracture mechanisms.

Delaminations can be initiated or extended from high stress concentrations that originate from mechanical effects, such as manufacturing, transportation and service effects, such as temperature, moisture, matrix shrinkage, or from general loading conditions, especially sudden concentrated loadings such as impact and explosion.

These effects may become more severe around curved sections, sudden changes of cross sections, and free edges. One important aspect of delamination failure is that substantial internal damage may exist in the interface adjacent plies without any apparent external destruction.

### 1.2.3  Fibre/Matrix Debonding

A perfect bonding between the fibre and matrix is necessary to ensure the composite action. Any debonding, or even local sliding, may substantially affect the overall strength of the

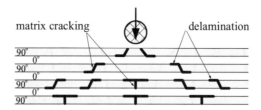

**Figure 1.1**    Main modes of cracking in composites.

composite specimen. It is generally accepted that the composite material should be designed and manufactured in such a way that fibre/matrix debonding never occurs before matrix cracking and delamination.

### 1.2.4    Fibre Breakage

Fibre breakage is probably the last mode of failure of a composite specimen prior to its collapse. Once the fibres are broken, the load bearing capacity of the specimen suddenly drops to almost zero.

### 1.2.5    Macro Models of Cracking in Composites

Homogeneous composites are primarily assumed to behave in an orthotropic linear elastic state. Equivalent homogeneous orthotropic material properties are determined based on the assumption of an equivalent smeared fibre/matrix mixture. This is certainly the case for most numerical solutions at the macroscopic level. As a result, fracture is assumed to occur only in an in-plane cracking mode (matrix/fibre cracking) or in an interlaminar cracking state (delamination), as depicted in Figure 1.1.

## 1.3    Crack Analysis

In this section, a brief review of the main available theoretical approaches for analysis of crack stability and propagation is presented. There are different classifications in crack analysis. For example, from the geometrical point of view, a crack may be represented as an internal discontinuity or external boundary (discrete crack model), characterizing a strong displacement discontinuity (Figure 1.2a), or its equivalent continuum mechanical effects (in terms of stiffness and strength reduction) can be considered within the numerical model in a distributed fashion without explicitly defining its geometry (smeared crack model) (see Figure 1.2b).

### 1.3.1    Local and Non-Local Formulations

Early attempts to simulate crack problems by numerical methods adopted a simple rule to check the stress state at any sampling point against a material strength criterion. The constitutive

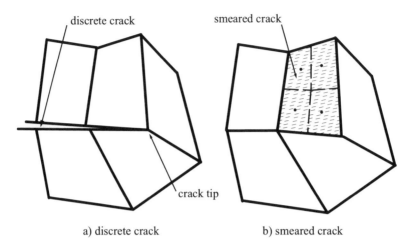

**Figure 1.2**   Discrete and smeared crack models in a typical finite element mesh.

behaviour of the point was only affected by its own local stress–strain state (point 1 in Figure 1.3). Soon it was realized that cracking could not be regarded solely as a local point-wise stress-based criterion, and such a local approach for fracture analysis may become size or mesh dependent and unreliable.

The remedy was the introduction of non-local formulations based on characteristic length scales (Bazant and Planas, 1997), defined for the material constitutive law as an intrinsic material property, or for a numerical model based on the geometrical requirements. To clarify the basic idea, consider a very simplified case (point 2 in Figure 1.3), where the fracture behaviour of this point is determined from a non-local criterion expressed in terms of the state

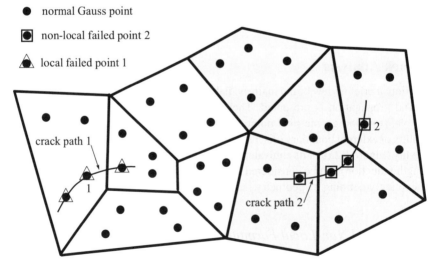

**Figure 1.3**   Local and non-local evaluation of the cracking state.

variables (including the length scale) at that point and a number of surrounding points in its support domain.

## 1.3.2 Theoretical Methods for Failure Analysis

Three fundamental approaches are available for discussion of the effects of defects and failures: continuum-based plasticity and damage mechanics and the crack-based approach of fracture mechanics. All three approaches can be implemented within different numerical methods. These methods, however, are applied to fundamentally different classes of failure problems. While the theory of plasticity and damage mechanics are basically designed for problems where the displacement field, and usually the strain field, remain continuous everywhere, fracture mechanics is essentially formulated to deal with strong discontinuities (cracks) where both the displacement and strain fields are discontinuous across a crack surface (Mohammadi, 2003a, 2008).

In practice, however, damage mechanics and the theory of plasticity have been modified and adapted for failure/fracture analysis of structures with strong discontinuities and fracture mechanics is sometimes used for weak discontinuity problems. It is, therefore, difficult to distinguish between the practical engineering applications exclusively associated with each class.

### 1.3.2.1 Plasticity

Plasticity theory is well developed to deal with plastic deformations and is based on various local-failure criteria, written in terms of local (point-wise) state variables, such as stress tensor and elastic and plastic strain components. Most of the plasticity based crack analyses are based on softening plasticity models of smeared cracking, which may become mesh or size dependent, if higher order formulations (such as the Cosserate or gradient theories) are not adopted. Plasticity models are capable of predicting the initiation of crack as well as predicting its growth, and can be readily implemented in different numerical techniques.

### 1.3.2.2 Fracture Mechanics

In contrast, the theory of linear elastic fracture mechanics (LEFM) is based on the existence of an initial crack or flaw and adopts the laws of thermodynamics to formulate an energy-based criterion for analysis of the existing crack. Such an approach ensures the size-independence of the solution. Clearly, the method is based on an explicit discrete definition of crack in the form of internal or external boundaries. Another important aspect of LEFM is its capability in deriving the singular stress field, predicted by the analytical solution at a crack tip. These two specifications have substantially complicated the numerical techniques designed for fracture mechanics analysis.

In addition to original linear elastic formulation, fracture mechanics has been extended to limited nonlinear behaviour and plasticity around the crack tip, forming the theory of elastoplastic fracture mechanics (EPFM).

### 1.3.2.3  Damage Mechanics

Damage mechanics has been increasingly adopted to analyze failure in various engineering application involving concrete, rock, metals, composites, and so on. Damage mechanics is a non-local approach (similar to fracture mechanics) but with a formulation apparently similar to the softening theory of plasticity. In damage mechanics, both the strength and stiffness of a material point are decreased if it experiences some level of damage. This is in contrast to the classical theory of plasticity, where the stiffness remains unchanged and only the strength is updated according to the hardening/softening behaviour.

Thermodynamics principles are adopted to derive the necessary formulation based on the micro-cracking state of the material and one of the fundamental assumptions of equivalent strain or equivalent strain energy principles to relate the equivalent undamaged model with the real damaged one. Such an equivalent undamaged model holds the continuity of the model intact.

## 1.4  Analytical Solutions for Composites

### 1.4.1  Continuum Models

Conventional continuum lamination models for analysis of composites are based on a composite element, which considers the fibre/matrix mixture as an equivalent homogeneous orthotropic continuum laminate, with perfect bond between the constituents in each single lamina, no strain discontinuity across the interface, and a regular arrangement of fibres. Equilibrium equations, compatibility conditions and the linear elastic Hook's law determine the elasticity constants and govern the stress–strain constitutive law or its generalized form. The classical lamination theory formulates the multilayered laminate based on variations of fibre orientation, stacking sequence and ply-level material properties. More advanced models assume a viscoelasticity model for the matrix. A number of analytical models have also been developed to account for a number of failure modes in each ply, but they are, unfortunately, limited to very simplified geometries, and specific orientations, stacking and loading conditions.

### 1.4.2  Fracture Mechanics of Composites

Linear elastic fracture mechanics (LEFM) is based on the existence of a crack or a flaw and determines the state of its stability and possible propagation. Its non-local nature guarantees the size or mesh (in case of a finite element analysis) independency of the solution. Definitions of non-local concepts such as the stress intensity factor, energy release rate and energy-based criteria allow the classical fracture mechanics to be extended to nonlinear problems. Most of the research in this field, however, can be classified into four categories: static cracking in a single orthotropic material, dynamic orthotropic cracking, orthotropic bimaterial interface cracks and fracture in orthotropic FGMs. All categories include topics on definition and evaluation of stress intensity factors, associated $J$ and interaction integrals, deriving the asymptotic solutions, crack propagation criteria, and so on.

The fracture mechanics of composite structures has been studied by many researchers. Beginning with the pioneering work by Muskelishvili (1953), several others such as Sih, Paris and Irwin (1965), Bogy (1972), Bowie and Freese (1972), Barnett and Asaro (1972), Kuo and Bogy (1974), Tupholme (1974), Atluri, Kobayashi and Nakagaki (1975a), Forschi and Barret (1976), Boone, Wawrzynek and Ingraffea (1987), Viola, Piva and Radi (1989) and, more recently, Lim, Choi and Sankar (2001), Carloni and Nobile (2002), Carloni, Piva and Viola (2003) and Nobile and Carloni (2005) have proposed solutions for various anisotropic static and quasi-static crack problems.

Simultaneously, several researchers have contributed to finding the elastodynamic fields around a propagating crack within an anisotropic medium, including Achenbach and Bazant (1975), Arcisz and Sih (1984), Piva and Viola (1988), Viola, Piva and Radi (1989), Shindo and Hiroaki (1990), De and Patra (1992), Gentilini, Piva and Viola (2004), Kasmalkar (1996) and Chen and Erdogan (1996). Lee, Hawong and Choi (1996) derived the dynamic stress and displacement components around the crack tip of a steady state propagating crack in an orthotropic material. The same subject was then followed by Gu and Asaro (1997), Rubio-Gonzales and Mason (1998), Broberg (1999), Lim, Choi and Sankar (2001), Federici *et al.* (2001), Nobile and Carloni (2005), Piva, Viola and Tornabene (2005) Sethi *et al.* (2011), and Abd-Alla *et al.* (2011), among others.

The research has not been limited to single layer orthotropic homogeneous composites. Analytical solutions for delamination in multilayer composites have also been investigated comprehensively. The first attempt was probably by Williams (1959) who discovered the oscillatory near-tip behaviour for a traction-free interface crack between two dissimilar isotropic elastic materials, followed by several others such as Erdogan (1963), Rice and Sih (1965), Malysev and Salganik (1965), England (1965), Comninou (1977), Comninou and Schmuser (1979), Sun and Jih (1987), Hutchinson, Mear and Rice (1987) and Rice (1988), among others. The study of interface cracks between two anisotropic materials was performed by Gotoh (1967), Clements (1971) and Willis (1971), followed by Wang and Choi (1983a, 1983b), Ting (1986), Tewary, Wagoner and Hirth (1989), Wu (1990), Gao, Abbudi and Barnett (1992) and Hwu (1993a, 1993b), Bassani and Qu (1989), Sun and Manoharan (1989), Suo (1990), Yang, Sou and Shih (1991), Hwu (1993b), Qian and Sun (1998), Lee (2000) and Hemanth *et al.* (2005).

Fracture mechanics of FGMs has similarly been an active topic for analytical research. For instance, Yamanouchi *et al.* (1990), Holt *et al.* (1993), Ilschner and Cherradi (1995), Nadeau and Ferrari (1999), Takahashi *et al.* (1993), Pipes and Pagano (1970, 1974), Pagano (1974), Kurihara, Sasaki and Kawarada (1990), Niino and Maeda (1990), Sampath *et al.* (1995), Kaysser and Ilschner (1995), Erdogan (1995) and Lee and Erdogan (1995) have studied various aspects of FGM properties. Despite material inhomogeneity, Sih and Chen (1980), Eischen (1983) and Delale and Erdogan (1983) have shown that the asymptotic crack-tip stress and displacement fields for certain classes of FGMs follow the general form of homogeneous materials and Ozturk and Erdogan (1997) and Konda and Erdogan (1994) analytically solved for crack-tip fields in inhomogeneous orthotropic infinite FGM problems. Evaluation of the $J$ integral for determining the mixed-mode stress intensity factors in general FGM problems was studied by Gu and Asaro (1997), Gu, Dao and Asaro (1999), Anlas, Santare and Lambros (2000) and, in particular, Kim and Paulino (2002a, 2002b, 2000c, 2003a, 2003b, 2005) who examined and developed three independent formulations: non-equilibrium, incompatibility and constant-constitutive-tensor, for the $J$ integral.

## 1.5   Numerical Techniques

Due to the limitations and inflexible nature of analytical methods in handling arbitrary complex geometries and boundary conditions and general crack propagations, several numerical techniques have been developed for solving composite fracture mechanics problems.

The finite element method has been widely used for fracture analysis of structures for many years and is probably the first choice of analysis for general engineering problems, including fracture, unless a better solution is proposed. Despite outstanding advantages, alternative methods are also available, including the adaptive finite/discrete element method (DEM), the boundary element method (BEM), a variety of meshless methods, the extended finite element method (XFEM), the extended isogeometric analysis (XIGA) and, more recently, advanced multiscale techniques. In the following, a brief review of a number of studies on fracture analysis of composites for each class of numerical methods is presented.

### 1.5.1   Boundary Element Method

Cruse (1988), Aliabadi and Sollero (1998) and García-Sánchez, Zhang and Sáez (2008) developed boundary element solutions for quasi-static crack propagation and dynamic analysis of cracks in orthotropic media. In the boundary element method, a number of elements are used to discretize the boundary of the problem domain, and the domain itself is analytically represented in the governing equations. The boundary element method, regardless of all the benefits, cannot be readily extended to nonlinear systems and is not suited to general crack propagation problems (Figure 1.4c).

### 1.5.2   Finite Element Method

Most of the performed numerical analyses of structures prior to the end of the twentieth century were related to the finite element method. The finite element method can be easily adapted to complex geometries and general boundary conditions and is well developed into almost every possible engineering application, including nonlinear, inhomogeneous, anisotropic, multilayer, large deformation, fracture and dynamic problems. Several general purpose finite element softwares have been developed, verified and calibrated over the years and are now available to almost anyone who asks (and pays) for them. Furthermore, concepts of FEM are now offered by all engineering departments in the form of postgraduate and even undergraduate courses.

Introduction and fast development of the finite element method drastically changed the extent of application of LEFM from classical idealized models to complex practical engineering problems. After earlier application of FEM to fracture analysis of composites (for instance by Swenson and Ingraffea, 1988) a large number of studies adopted FEM to simply obtain the displacement, strain and stress fields required for numerical evaluation of fracture mechanics parameters such as the stress intensity factors, the energy release rate or the $J$ integral to assess the stability of crack (Rabinovitch and Frosting, 2001; Pesic and Pilakoutas, 2003; Rabinovitch, 2004; Colombi, 2006; Lu *et al.*, 2006; Yang, Peng and Kwan, 2006; Bruno, Carpino and Greco, 2007; Greco, Lonetti and Blasi, 2007).

Later, various techniques were developed within the FEM framework to allow reproduction of a singular stress field at a crack tip and to facilitate simulation of arbitrary crack propagations.

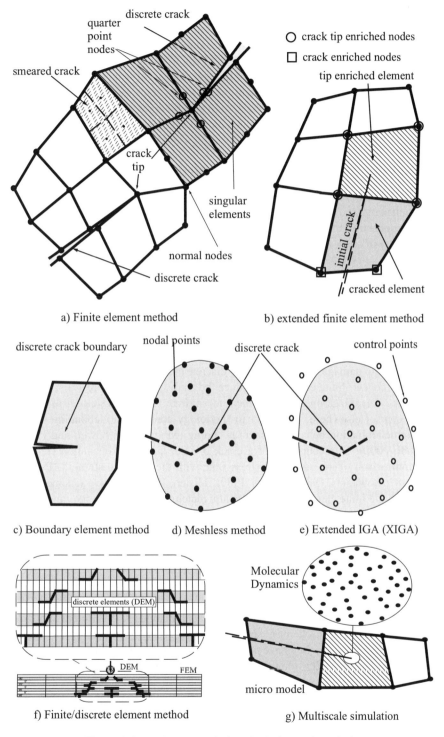

**Figure 1.4** Various numerical methods for crack analysis.

Singular finite elements, developed to accurately represent crack-tip singular fields (Owen and Fawkes, 1983), provide the major advantage of simple construction of the model by simply moving the nearby midside nodes to the quarter points with no other changes in the finite element formulation being required (Figure 1.4a). Prior to the development of XFEM, singular elements were the most popular approach for fracture analysis of structures. Singular elements, however, have to be used in a finite element mesh, where crack faces have to match element boundaries. This largely limits their application to general crack propagation problems, unless combined with at least a local adaptive finite element scheme.

## 1.5.3 Adaptive Finite/Discrete Element Method

The adaptive finite element method, combined with the concepts of contact mechanics of the discrete element method (DEM), has been adopted in several studies for simulation of progressive crack propagation in composites under quasi-static and dynamic/impact loadings. They include a variety of crack models, such as smeared crack, discrete inter-element crack models, cracked interface elements and the discrete contact element, which may use general contact mechanics algorithms to simulate progressive delamination and fracture problems (Mohammadi, Owen and Peric, 1997; Sprenger, Gruttmann and Wagner, 2000; Wu, Yuan and Niu, 2002; Mohammadi and Forouzan-sepehr, 2003; Wong and Vecchio, 2003; Wu and Yin, 2003; Mohammadi, 2003b; Lu *et al.*, 2005; Wang, 2006; Teng, Yuan and Chen, 2006; Mohammadi and Mousavi, 2007; Moosavi and Mohammadi, 2007; Mohammadi, 2008; Rabinovitch, 2008).

In a composite progressive fracture analysis, part of the composite model, which is potentially susceptible to damage, is represented by discrete elements and the rest of the specimen is modelled with coarser finite elements to reduce the analysis time (Figure 1.4f). Each discrete element can be discretized by a finite element mesh; finer for the plies closer to the damaged region and coarser elsewhere. A contact methodology controls the debonding mechanisms and all post delamination behaviour, such as sliding (Mohammadi, Owen and Peric, 1997; Mohammadi, 2008). On occurrence of a crack or after a crack propagation step based on a simple comparison of the computed stress state with the adhesive strength (Parker, 1981; O'Brien, 1985; Rowlands, 1985; Roberts, 1989; Taljsten, 1997), adaptive schemes are used to locally remesh the finite element model to ensure matching of element edges and crack faces. Nonlinear material properties and geometric nonlinearities can be considered in the basic FE formulation. The method, however, is numerically expensive and the nodal alignments may cause numerical difficulties and mesh dependency to some extent in propagation problems.

## 1.5.4 Meshless Methods

Meshless methods include a wide variety of numerical methods with different approximation techniques, diverse solution schemes, variable levels of accuracy and dissimilar applications. Ironically, in a general view, they have nothing in common but being different from the finite element method. The main idea of meshless methods is to avoid a predefined fixed connectivity between the nodal points which are used to define the geometry and to set necessary degrees of freedom to discretize the governing equation. As a result of such a connection-free style of nodal discretization, any existing cracks or crack propagation paths can be efficiently embedded geometrically within the numerical model (Figure 1.4d).

Meshless methods have been used extensively for analysis of various engineering problems (Belytschko *et al.*, 2002). They include the frequently used classes of the element-free Galerkin method (EFG) (Belytschko, Lu and Gu, 1994; Belytschko, Organ and Krongauz, 1995; Ghorashi, Valizadeh and Mohammadi, 2011), the meshless local Petrov–Galerkin (MLPG) (Atluri and Shen, 2002), smoothed particle hydrodynamics (SPH) (Belytschko *et al.*, 2000; Madani and Mohammadi, 2011; Ostad and Mohammadi, 2012), the finite point method (FPM) (Onate *et al.*, 1995; Bitaraf and Mohammadi, 2010), isogeometric analysis (IGA) (Hughes, Cottrell and Bazilevs, 2005) and its extended version (XIGA) (Ghorashi, Valizadeh and Mohammadi, 2012), and other approaches including the reproducing kernel particle method (RKPM) (Liu *et al.*, 1996), HP-clouds (Duarte and Oden, 1995), the equilibrium on line method (ELM) (Sadeghirad and Mohammadi, 2007) and the smoothed finite element method (Liu and Trung, 2010), among others. It is noted that some of the mentioned methods are in fact a combination of finite element and meshless concepts. By this definition, the extended finite element method (XFEM) may somehow be included.

Despite higher accuracy and flexible adaptive schemes, the majority of meshless methods are yet to be user friendly because of weak versatility to deal with arbitrary boundary conditions and geometries, complicated theoretical bases, high numerical expense and the need for sensitivity analysis, calibration and difficult stabilization schemes in many of them.

## *1.5.5   Extended Finite Element Method*

Since the introduction of the extended finite element method (XFEM) for fracture analysis by Belytschko and Black (1999), Moës, Dolbow and Belytschko (1999) and Dolbow (1999), based on the mathematical foundation of the partition of unity finite element method (PUFEM), proposed earlier by Melenk and Babuska (1996) and Duarte and Oden (1996), XFEM methodology has been rapidly extended to a vastly wide range of applications that somehow include a local discontinuity or singularity within the solution.

The natural extension of FEM into XFEM allows new capabilities while preserving the finite element original advantages. The two main superiorities of XFEM are its capability in reproducing the singular stress state at a crack tip, and allowing several cracks or arbitrary crack propagation paths to be simulated on an independent unaltered mesh (Figure 1.4b). In fact, while the presence of the crack is not geometrically modelled and the mesh does not need to conform to the virtual crack path, the exact analytical solutions for singular stress and discontinuous displacement fields around the crack (tip) are reproduced by inclusion of a special set of enriched shape functions that are extracted from the asymptotic analytical solutions.

Apart from earlier works that were directed towards the development of the extended finite element method for linear elastic fracture mechanics (LEFM), simulation of failure and localization has been the target of several studies including, Jirásek and Zimmermann (2001a, 2001b), Sukumar *et al.* (2003), Dumstorff and Meschke (2003), Patzak and Jirásek (2003), Ventura, Moran and Belytschko ( 2005), Samaniego and Belytschko (2005), Areias and Belytschko (2005a, 2005b, 2006) and Song, Areias and Belytschko (2006). In addition, various contact problems have been simulated by XFEM. For example, Dolbow, Moes and Belytschko (2000c, 2001), Shamloo, Azami and Khoei (2005), Belytschko, Daniel and Ventura (2002), Khoei and Nikbakht (2006), Khoei, Shamloo and Azami (2006) and Khoei *et al.* (2006) have used simple Heaviside enrichments to deal with contact discontinuities in different problems, while Ebrahimi, Mohammadi and Kani (2012) have recently proposed a numerical approach within the partition of unity finite element method to determine the order of singularity at a sliding contact corner.

Moreover, several dynamic fracture problems have been studied by XFEM. Among them, Peerlings *et al.* (2002), Belytschko *et al.* (2003), Oliver *et al.* (2003), Ventura, Budyn and Belytschko (2003), Chessa and Belytschko (2004, 2006), Belytschko and Chen (2004), Zi *et al.* (2005), Rethore, Gravouil and Combescure (2005a), Rethore *et al.* (2005), Menouillard *et al.* (2006), Gregoire *et al.* (2007), Nistor, Pantale and Caperaa (2008), Prabel, Marie and Combescure (2008), Combescure *et al.* (2008), Gregoire, Maigre and Combescure (2008), Motamedi (2008), Kabiri (2009), Gravouil, Elguedj and Maigre (2009a, 2009b) and Rezaei (2010) have discussed various aspects of general dynamic fracture problems.

In the past decade, development of XFEM has substantially contributed to new studies of fracture analysis of various types of composite materials. Dolbow and Nadeau (2002) employed XFEM to simulate fracture behaviour of micro-structured materials with a focus on functionally graded materials. Then, Dolbow and Gosz (2002) described a new interaction energy integral method for the computation of mixed mode stress intensity factors in functionally graded materials. In a related contribution, Remmers, Wells and de Borst (2003) presented a new formulation for delamination in thin-layered composite structures. A study of bimaterial interface cracks was performed by Sukumar *et al.* (2004) by developing new bimaterial enrichment functions. Nagashima, Omoto and Tani (2003) and Nagashima and Suemasu (2004, 2006) described the application of XFEM to stress analyses of structures containing interface cracks between dissimilar materials and concluded the need for orthotropic enrichment functions to represent the asymptotic solution for a crack in an orthotropic material. Hettich and Ramm (2006) simulated the delamination crack as a jump in the displacement field without using any crack-tip enrichment. Also, a number of XFEM simulations have focused on thermo-mechanical analysis of orthotropic FGMs (Dag, Yildirim and Sarikaya, 2007), 3D isotropic FGMs (Ayhan, 2009; Zhang *et al.*, 2011; Moghaddam, Ghajar and Alfano, 2011) and frequency analysis of cracked isotropic FGMs (Natarajan *et al.*, 2011). Recently, Bayesteh and Mohammadi (2012) have used XFEM with orthotropic crack-tip enrichment functions to analyze several FGM crack stability and propagation problems.

Development of independent orthotropic crack-tip enrichment functions was reported in a series of papers by Asadpoure, Mohammadi and Vafai (2006, 2007), Asadpoure and Mohammadi (2007) and Mohammadi and Asadpoure (2006). Later, Motamedi (2008) and Motamedi and Mohammadi (2010a, 2010b, 2012) studied the dynamic crack stability and propagation in composites based on static and dynamic orthotropic enrichment functions and Esna Ashari (2009) and Esna Ashari and Mohammadi (2009, 2010b, 2011a, 2012) have further extended the method for orthotropic bimaterial interfaces.

## 1.5.6 Extended Isogeometric Analysis

Since the introduction of isogeometric analysis (IGA) by Hughes, Cottrell and Bazilevs (2005) a new and fast growing chapter has been opened in unifying computer aided design (CAD) and numerical solutions by using the non-uniform rational B-splines (NURBS) functions (Figure 1.4e). The method, in fact, can be categorized with other meshless methods, but it is briefly introduced due to its fast growing state and excellent potentials.

IGA has been successfully adopted in several engineering problems, including structural dynamics (Cottrell, Hughes and Bazilevs, 2009; Hassani, Moghaddam and Tavakkoli, 2009), Navier–Stokes flow (Nielsen *et al.*, 2011), fluid–solid interaction (Bazilevs *et al.*, 2009), shells

(Benson *et al.*, 2010a; Benson *et al.*, 2011; Kiendl *et al.*, 2010; Kiendl *et al.*, 2009), damage mechanics (Verhoosel *et al.*, 2010a), cohesive zone simulations (Verhoosel *et al.*, 2010b), heat transfer (Anders, Weinberg and Reichart, 2012), large deformation (Benson *et al.*, 2011), electromagnetic (Buffa, Sangalli and Vazquez, 2010), strain localization (Elguedj, Rethore and Buteri, 2011), contact mechanics (Lu, 2011; Temizer, Wriggers and Hughes, 2011), topology optimization (Hassani, Khanzadi and Tavakkoli, 2012) and crack propagation (Verhoosel *et al.*, 2010b, Benson *et al.*, 2010b; De Luycker *et al.*, 2011; Haasemann *et al.*, 2011; Ghorashi, Valizadeh and Mohammadi, 2012).

The first attempt at enhancing IGA for crack problems was reported by Verhoosel *et al.* (2010b) and followed by Benson *et al.* (2010c), De Luycker *et al.* (2011) and Haasemann *et al.* (2011). Recently, a full combination of XFEM and IGA methodologies has been developed for general mixed mode crack propagation problems by the introduction of extended isogeometric analysis (XIGA) by Ghorashi, Valizadeh and Mohammadi (2012). XIGA uses the superior concepts of XFEM to extrinsically enrich the versatile IGA control points with Heaviside and crack-tip enrichment functions.

## 1.5.7   Multiscale Analysis

The traditional borders between simulation of mechanics, physics and even biology problems have been removed by recent computational advances in the form of multiscale simulations. Most problems in science involve many scales in time and space. An example is the overall stress state in a cracked solid which can well be described by macroscopic continuum equations, but requires details on a microscale at the tip of the crack. In a multiscale method, a part of the model which requires a more accurate theoretical basis or numerical approximation, due to lack of theoretical bases or existing inconsistencies of many conventional models, is simulated by a finer modelling scale, which can better represent details of the material behaviour and the interacting effects of material constituents (Figure 1.4g).

A common difficulty with the simulation of these problems and many others in physics, chemistry, engineering and biology is that an attempt to represent all fine scales will lead to an enormous computational model with unacceptably long computation times and huge memory requirements, even for state of the art modern supercomputers. On the other hand, if the discretization at a coarse level ignores the fine scale information then the solution may not be physically meaningful. Therefore, the challenging task of incorporating the influence of the fine scales into the coarse model must logically be determined.

Several multiscale techniques have been developed for different applications including the bridging scale method, the bridging domain method, homogenization techniques, the quasi continuum method, the heterogeneous multiscale method, and so on. These methods can be incorporated into multiscale FEM, meshless or XFEM methodologies to analyze various multiscale problems at macro/micro, micro/nano, macro/nano, macro/micro/nano and bio/nano interface levels (Mohammadi, 2012).

## 1.6   Scope of the Book

This text is dedicated to discussing various aspects of application of the extended finite element method for fracture analysis of composites on the macroscopic scale. Nevertheless, many subjects can be similarly used for fracture analysis of other materials, even in microscopic scales.

The book is designed as a textbook, which provides all the necessary theoretical bases before discussing the numerical issues. This preliminary chapter briefly introduced the subject of fracture in composite structures, and summarily reviewed the existing classes of analytical and numerical techniques. In each case, a short description and a number of reference works were presented. The aim was to provide a general overview of the wide extent of applications without going into detail.

Chapter 2 provides a review of classical isotropic fracture mechanics, which quickly examines the basic concepts and fundamental formulations but does not provide proofs or in depth detail. It is, in fact, a slightly modified and corrected edition of a similar chapter in Mohammadi (2008). The chapter begins with an introduction to the basics of the theory of elasticity, and is followed by discussions on classical problems of LEFM. Asymptotic solutions for displacement and stress fields in different fracture modes are presented and basic concepts of stress intensity factors, energy release rate and the $J$ contour integral and its modifications, such as the equivalent domain integral and the interaction integral, are explained. The methodology of extracting mixed mode stress intensity factors is addressed and various mixed mode fracture criteria are explained. The chapter briefly reviews issues related to the basic finite element models of fracture mechanics and describes the basic formulation of popular singular finite elements. The chapter ends by extending some of the basic ideas of LEFM to elastoplastic fracture mechanics (EPFM).

Chapter 3, which is a redesigned and completed edition of the XFEM chapter in Mohammadi (2008), begins with the concepts of the partition of unity and the extended finite element method. It is followed by a detailed XFEM formulation for analysis of cracks in isotropic materials. This section constitutes the basic relations for modelling displacement discontinuity across a crack and the stress singularity at the crack tip. The chapter continues with three sections on available options for strong discontinuity enrichment functions, weak discontinuity enrichments for material interfaces, and a collection of several crack-tip enrichments for various engineering applications. A review of the level set method for tracking moving boundaries is provided before the concluding section which includes simulations and discussion of a wide range of problems to assess the accuracy, performance and efficiency of XFEM in dealing with various discontinuity and singularity problems, which include classical in-plane mixed mode fracture mechanics problems, cracking in plates and shells, simulation of shear band creation and propagation, self-similar fault rupture, sliding contact, hydraulic fracture, and dislocation dynamics.

Static orthotropic fracture analysis is the comprehensive subject of Chapter 4. It begins with a review of anisotropic and orthotropic elasticity, followed by a complete discussion on the available analytical solutions for near crack-tip fields in orthotropic materials. These fields are then used to develop the orthotropic enrichment functions for the XFEM formulation. A section is then devoted to orthotropic mixed mode fracture mechanics, which discusses orthotropic mixed mode criteria, the $J$ integral, crack propagation criteria for orthotropic media and other related issues. Finally, a number of numerical simulations are provided to illustrate the validity, robustness and efficiency of the proposed approach for evaluation of mixed mode stress intensity factors in homogeneous orthotropic materials.

Chapter 5 is devoted to dynamic fracture analysis of composites. After a comprehensive literature review, analytical solutions for near crack-tip in dynamic states are provided for isotropic and orthotropic media. Then, a section on dynamic stress intensity factors discusses the dynamic fracture criteria for different cases of stationary and moving cracks. It also includes details of the dynamic $J$ and interaction integrals and explains the existing techniques for determining the dynamic stress intensity factors. The next section provides the dynamic XFEM formulation and discusses various options for the dynamic crack-tip enrichment functions. A separate section is dedicated to time integration techniques and reviews existing developments in the fields of the time and time–space extended finite element methods (TXFEM and STXFEM, respectively). The chapter concludes with numerical simulations and comprehensive discussions of several dynamic fracture problems, including stationary and propagating cracks in orthotropic media. The results, usually in the form of time histories, complicate the process of comparisons with available reference data.

Orthotropic functionally graded materials are comprehensively discussed in Chapter 6. Again, after a literature review, analytical asymptotic solutions are presented for a crack in an inhomogeneous orthotropic medium. Then, fracture mechanics concepts, such as the toughness, stress intensity factors and crack propagation criteria, are explained. A comprehensive discussion is dedicated to various options for deriving the interaction integral for an inhomogeneous medium. It is followed by details of inhomogeneous XFEM formulation which also includes necessary enrichment functions and the numerical requirement for a transition domain. Several numerical examples from simple tensile FGM plates to crack propagation in an orthotropic FGM bending beam are presented and discussed.

Delamination or interlaminar crack analysis by XFEM is discussed in Chapter 7 with a comprehensive review of the existing techniques, followed by the concepts of fracture mechanics for bimaterial interface cracks. This includes the analytical solutions for the displacement and stress fields near an interlaminar crack tip in an orthotropic bimaterial problem. The simplified case of isotropic bimaterial is also discussed. Then, the interaction integral and the method of computing stress intensity factors are explained. The chapter also includes a section on delamination propagation criteria, before presenting the details of bimaterial XFM. Several numerical issues are addressed and the necessary strong and weak discontinuity enrichment functions and bimaterial crack-tip enrichment functions are explained. The chapter ends with a broad section on numerical simulations, which include conventional composite bimaterial applications and a number of FRP-strengthening problems from the retrofitting industry.

The final chapter presents a number of on-going research topics based on new engineering applications for orthotropic materials or new numerical tools for more efficiently tackling the cracking in composites. It begins by introducing the orthotropic version of the extended isogeometric analysis (XIGA) for fracture analysis of composites. This section briefly reviews the basic concepts of NURBS and IGA methodology and explains the enrichment techniques within an extended IGA framework (XIGA). Numerical issues are addressed and a number of simple simulations, both isotropic and orthotropic, are presented and discussed. The next section is dedicated to the newly developed idea of plane strain anisotropic dislocation dynamics. Related XFEM formulation and necessary enrichment functions for the self-stress state of edge dislocations are explained. A number of numerical simulations are presented without any available reference results. The chapter concludes with two brief sections on orthotropic biomaterial applications of XFEM and the piezoelectric problems.

# 2

# Fracture Mechanics, A Review

## 2.1 Introduction

The cracking and failure of tools and constructions have always been of concern for human societies. A broken tool, or a cracked windshield may be frustrating or a little costly, but nowadays problems in highly complex technological systems, such as an unfortunate airliner crash, are far more serious than in previous centuries, both from safety concerns and from the potential costs. Fortunately, advances in the field of fracture mechanics have helped to offset some of the potential dangers, but much remains to be learned and applied when appropriate (Anderson, 1995).

The field of fracture mechanics has gradually evolved over almost a century from a scientific curiosity of a few specialists in 1920 into a mature widespread engineering discipline. Most engineering universities offer the fracture mechanics course at graduate level, and a number of important concepts of fracture mechanics are now parts of engineering design courses, even at the undergraduate level.

The first reported fracture-related experiment goes back to Leonardo da Vinci, several centuries ago, who provided some clues as to the root cause of fracture by measuring the strength of iron wires. He found that the strength varied inversely with the wire length. A qualitative conclusion was that for a longer wire, there was a higher probability of inclusion of internal flaws which could trigger the failure (Anderson, 1995).

Analytically, Kirsh (1898) and Kolosov (1909) were the first to analyse the effect of a circular flaw in an infinite tensile plate and discuss the biaxial stress state around the flaw with a stress concentration factor of 3. Then, in 1913 Inglis published a stress analysis for an elliptical hole in an infinite linear elastic tensile plate and discussed a sharp crack by degeneration of the elliptical hole into a line (Inglis, 1913).

The major quantitative connection between the fracture state and flaw size came from the work of Griffith in 1920 who was studying the effects of scratches and similar flaws on aircraft engine components (Griffith, 1921, 1924). He transformed the stress analysis of an elliptical hole, performed earlier by Inglis, to calculate the effect of the crack on the strain energy stored in an infinite cracked plate and studied the unstable propagation of a crack. He used the first law of thermodynamics to formulate his fracture theory: a flaw becomes unstable, and thus

---

*XFEM Fracture Analysis of Composites*, First Edition. Soheil Mohammadi.
© 2012 John Wiley & Sons, Ltd. Published 2012 by John Wiley & Sons, Ltd.

fracture occurs, when the change in strain energy from a potential crack growth increment is sufficient to overcome the finite surface energy of the material. While the Griffith model could correctly predict the relationship between strength and flaw size in glass specimens and ideally brittle solids, it was unsuccessful for metals, and it took until 1948 for a modification that made it applicable to metals. Other major related analytical studies were published by Westergaard (1939) and Williams (1952, 1957) for stress and displacement solutions near a crack tip.

Development of fracture mechanics is closely related to some well-known catastrophes in recent history. Many experts believe that fracture mechanics progressed from being a scientific interest to an engineering discipline, primarily because of what happened to the several hundred damaged Liberty ships during World War II, (approximately 20 ships were totally ruptured and some even broke in two). The failures occurred primarily because of the combination of poor weld properties with stress concentrations, and poor choice of brittle materials in construction.

The field of fracture mechanics, as we know it today, was probably born in the US Naval Research Laboratory (NRL), beginning with the study of Liberty failures. The first contribution of Irwin, who used to lead NRL, was to extend the Griffith approach to metals by including the energy dissipated by local plastic flow (Irwin, 1948). Orowan independently proposed a similar modification to the Griffith theory (Orowan, 1948). The concept of energy release rate $G$, which is related to the Griffith theory but in a form that is more useful for solving engineering problems, was proposed by Irwin in 1957, soon to be presented as a means for showing that the stresses and displacements near the crack tip could be described by a single parameter, known as the stress intensity factor (SIF) (Irwin, 1957, 1958, 1961; Irwin, Kies and Smith, 1958).

The two accidents of the Comet, the world's first jetliner (Svensson, 2012), in 1954 sparked an extensive investigation of the causes and led to significant progress in the understanding of fracture and fatigue. Tests and studies of fragments showed that a crack had developed due to metal fatigue near a window that eventually grew into an effectively very large crack and subsequent crash. In 1955, Wells used fracture mechanics to show that the fuselage failures resulted from fatigue cracks reaching a critical size. These cracks were caused by insufficient local strength combined with severe stress concentration around square corners (Wells, 1955). A major study on general fatigue analysis was later presented by Paris, Gomez and Anderson (1961) who provided experimental and theoretical arguments for their approach of applying fracture mechanics principles to fatigue crack growth.

Another early application of fracture mechanics was the analysis of steam turbines of General Electric by Winnie and Wundt (1958) who applied the Irwin's energy release rate approach to predict the bursting behaviour of large disks extracted from rotor forgings (Anderson, 1995).

Around 1960, when the fundamentals of LEFM were fairly well established, researchers turned their attention to different views of crack tip plasticity, including the first and second order models by Irwin (1961), the strip yield model by Dugdale (1960), the linear model of Barenblatt (1959a, 1959b) and the crack-tip opening displacement (CTOD) by Wells (1961). In 1968, Rice developed the contour independent $J$ integral to characterize the nonlinear material behaviour ahead of a crack, by idealizing plastic deformation as a nonlinear elastic response (Rice, 1968a, 1988; Rice and Rosengren, 1968; Rice and Levy, 1972). The same concept had been previously published by Eshelby (1956) in the form of so-called conservation integrals for continuous domains. Then, Hutchinson (1968) and Rice and Rosengren (1968) related the *J* integral to crack-tip stress fields in nonlinear materials (Anderson, 1995; Barsom and Rolfe, 1987).

The safety requirements of the nuclear power industry in the 1970s led to the extensive support of crack stability studies. Begley and Landes (1972) at Westinghouse, used the *J* integral to characterize the fracture toughness of the extremely tough steels of nuclear pressure vessels, and developed a standard procedure for *J* testing of such metals. In the UK, the same industry developed its own fracture design analysis (Harrison *et al.*, 1980) based on the strip yield model of Dugdale (1960) and Barenblatt (1959a, 1959b).

The finite element method (FEM) was originally used as a simple analytical tool for obtaining the continuum based displacement and stress fields. Later, sophisticated singular elements were proposed by Barsoum (1974, 1975, 1976a, 1976b, 1977, 1981) and Henshell and Shaw (1975) and efficiently implemented by Fawkes, Owen and Luxmoore (1979) and Owen and Fawkes (1983) to simulate the singularity condition at crack tips. Then, it was extensively adopted as a major improvement to the already available numerical techniques in LEFM. Major numerical progress, however, includes development of the meshless methods, such as the element-free Galerkin method (EFG) (Belytschko, Lu and Gu, 1994), the meshless local Petrov–Galerkin (MLPG) method (Atluri and Shen, 2002), smoothed particle hydrodynamics (SPH) (Belytschko *et al.*, 1996) and the extended finite element method (XFEM) (Dolbow, Moës and Belytschko, 2000b; Mohammadi, 2008). Related literature reviews of the numerical issues will be presented in the relevant chapters.

Though, comprehensively studied, documented and included in design codes, crack analysis methods have continuously been challenged by a number of now and then accidents. These accidents, in fact, have challenged the notion that fracture was well understood and under control in modern structures. In 1988, multiple fatigue cracks linked up to form a large, catastrophic crack in the roof of a Boeing 737 during flight, killing one and injuring many passengers (Malnic and Meyer, 1988; Chandler, 1989). On November 1994, a magnitude 6.7 earthquake in Southern California damaged many welded beam-column joints in steel framed buildings; the joints were designed to absorb energy by plastic deformation, but fractured in an almost brittle fashion (Mahin, 1997). In another example, in 1998, the German InterCity Express (ICE) train, travelling at a speed of 280 km/h crashed, resulting in hundreds of casualties and massive destruction. The cause of the accident was a tyre detachment due to a fatigue crack, as described by Esslinger *et al.* (2004). More recent studies include the study of early cracking of Girth-Gear of an industrial Ball-Mill (up to 12 meters in diameter and over 90 tonnes in weight, with a manufacturing cost of about half a million dollars), which was expected to have a fatigue life of 20 years and more, but cracks initiated and propagated within the first two years of operation (Mirzaei, Razmjoo and Pourkamali, 2001). More or less similar trends were reported in the failure of exploded gas cylinders containing hydrogen (Mirzaei, 2008).

This chapter only briefly reviews the basic theoretical concepts of fracture mechanics for linear and nonlinear analyses of isotropic materials. Concepts that are related to specific subjects such as orthotropic materials, dynamic materials and the effect of inhomogeneity will be dealt with in the corresponding chapters. Little or no originality is claimed for this chapter, nor is there any claim of completeness. The sections are selected, organized and presented with regard to the needs of subsequent sections and chapters as a basis of LEFM for isotropic materials, and a precursor to the main subject of the book, XFEM fracture analysis of composites.

## 2.2   Basics of Elasticity

### 2.2.1   Stress–Strain Relations

Beginning with the definition of stress and strain tensors, $\sigma$ and $\varepsilon$, respectively,

$$\sigma = \begin{bmatrix} \sigma_{11} & \sigma_{12} & \sigma_{13} \\ \sigma_{12} & \sigma_{22} & \sigma_{23} \\ \sigma_{13} & \sigma_{23} & \sigma_{33} \end{bmatrix} \tag{2.1}$$

$$\varepsilon = \begin{bmatrix} \varepsilon_{11} & \varepsilon_{12} & \varepsilon_{13} \\ \varepsilon_{12} & \varepsilon_{22} & \varepsilon_{23} \\ \varepsilon_{13} & \varepsilon_{23} & \varepsilon_{33} \end{bmatrix} \tag{2.2}$$

or in a vector (array) notation,

$$\sigma = \begin{bmatrix} \sigma_{11} \\ \sigma_{22} \\ \sigma_{33} \\ \sigma_{12} \\ \sigma_{23} \\ \sigma_{13} \end{bmatrix} \tag{2.3}$$

$$\varepsilon = \begin{bmatrix} \varepsilon_{11} \\ \varepsilon_{22} \\ \varepsilon_{33} \\ 2\varepsilon_{12} \\ 2\varepsilon_{23} \\ 2\varepsilon_{13} \end{bmatrix} = \begin{bmatrix} \varepsilon_{11} \\ \varepsilon_{22} \\ \varepsilon_{33} \\ \gamma_{12} \\ \gamma_{23} \\ \gamma_{13} \end{bmatrix} \tag{2.4}$$

where $\gamma_{ij} = 2\varepsilon_{ij}$ is the engineering shear strain component. The generalized Hooke's law for linear elastic materials can be defined as:

$$\sigma = \mathbf{D} : \varepsilon \tag{2.5}$$

where $\sigma$ and $\varepsilon$ are the second order stress and strain tensors, respectively, and $\mathbf{D}$ is the fourth order elasticity modulus tensor with Cartesian components $D_{ijkl}$. For general three-dimensional problems, the fourth order elasticity tensor $\mathbf{D}$ can be represented by a two-dimensional matrix $\mathbf{D}$ with components $\mathbf{D}_{\alpha\beta}$, $\alpha, \beta = 1 - 6$.

Equation (2.1) can now be rewritten as,

$$\sigma_{ij} = D_{ijkl}\varepsilon_{kl} \tag{2.6}$$

Similarly, the strain tensor can be defined in terms of the stress tensor,

$$\varepsilon = \mathbf{C} : \sigma \tag{2.7}$$

or in component form

$$\varepsilon_{ij} = C_{ijkl}\sigma_{kl} \tag{2.8}$$

For isotropic materials, the elasticity modulus can be determined from the bulk and shear moduli, $K$ and $G$, respectively, (Neto, Peric and Owen, 2008)

$$\mathbf{D} = 2G\bar{\mathbf{I}} + A\left(K - \frac{2}{3}G\right)\mathbf{I} \otimes \mathbf{I} = 2\mu\mathbf{I} + A\lambda\mathbf{I} \otimes \mathbf{I} \tag{2.9}$$

where $\mathbf{I}$ is the second-order identity tensor and $\mathbf{I}$ is the fourth-order symmetric identity tensor, and $\lambda$ and $\mu$ are the Lame and shear modules

$$\lambda = \frac{\nu E}{(1+\nu)(1-2\nu)} \tag{2.10}$$

$$\mu = \frac{E}{2(1+\nu)} = G \tag{2.11}$$

and

$$A = \begin{cases} 1 & \text{3D, plane strain, axisymmetric} \\ \dfrac{2G}{K + \dfrac{4}{3}G} = \dfrac{1-2\nu}{1-\nu} & \text{plane stress} \end{cases} \tag{2.12}$$

The expanded form for 3D isotropic materials in Cartesian coordinates $(x, y, z)$ can then be written as:

$$\begin{Bmatrix} \sigma_{xx} \\ \sigma_{yy} \\ \sigma_{zz} \\ \sigma_{xy} \\ \sigma_{yz} \\ \sigma_{zx} \end{Bmatrix} = \begin{bmatrix} \lambda+2\mu & \lambda & \lambda & 0 & 0 & 0 \\ \lambda & \lambda+2\mu & \lambda & 0 & 0 & 0 \\ \lambda & \lambda & \lambda+2\mu & 0 & 0 & 0 \\ 0 & 0 & 0 & \mu & 0 & 0 \\ 0 & 0 & 0 & 0 & \mu & 0 \\ 0 & 0 & 0 & 0 & 0 & \mu \end{bmatrix} \begin{Bmatrix} \varepsilon_{xx} \\ \varepsilon_{yy} \\ \varepsilon_{zz} \\ \gamma_{xy} = 2\varepsilon_{xy} \\ \gamma_{yz} = 2\varepsilon_{yz} \\ \gamma_{zx} = 2\varepsilon_{zx} \end{Bmatrix} \tag{2.13}$$

Equation (2.13) is further simplified for the plane stress, plane strain and two-dimensional axisymmetric cases. For the plane stress case, the dimension in one direction (thickness) is neglected compared to the two others, and

$$\begin{Bmatrix} \sigma_{xx} \\ \sigma_{yy} \\ \sigma_{xy} \end{Bmatrix} = \frac{E}{1-\nu^2} \begin{bmatrix} 1 & \nu & 0 \\ \nu & 1 & 0 \\ 0 & 0 & \dfrac{1-\nu}{2} \end{bmatrix} \begin{Bmatrix} \varepsilon_{xx} \\ \varepsilon_{yy} \\ \gamma_{xy} \end{Bmatrix} \tag{2.14}$$

The strain component along the thickness direction, $\varepsilon_{zz}$, can also be derived from the Poisson's effect:

$$\varepsilon_{zz} = -\frac{\nu}{1-\nu}(\varepsilon_{xx} + \varepsilon_{yy}) \tag{2.15}$$

A plane strain case resembles a long body undergoing no variation in load or geometry in the longitudinal direction. As a result,

$$
\begin{Bmatrix} \sigma_{xx} \\ \sigma_{yy} \\ \sigma_{zz} \\ \sigma_{xy} \end{Bmatrix} = \frac{E}{(1+\nu)(1-2\nu)} \begin{bmatrix} 1-\nu & \nu & 0 \\ \nu & 1-\nu & 0 \\ \nu & \nu & 0 \\ 0 & 0 & \dfrac{1-2\nu}{2} \end{bmatrix} \begin{Bmatrix} \varepsilon_{xx} \\ \varepsilon_{yy} \\ \gamma_{xy} \end{Bmatrix} \tag{2.16}
$$

For an axisymmetric solid of revolution, a form almost similar to the plane strain relation is derived in terms of polar coordinate systems:

$$
\begin{Bmatrix} \sigma_{zz} \\ \sigma_{rr} \\ \sigma_{\theta\theta} \\ \sigma_{rz} \end{Bmatrix} = \frac{E}{(1+\nu)(1-2\nu)} \begin{bmatrix} 1-\nu & \nu & \nu & 0 \\ \nu & 1-\nu & \nu & 0 \\ \nu & \nu & 1-\nu & 0 \\ 0 & 0 & 0 & \dfrac{1-2\nu}{2} \end{bmatrix} \begin{Bmatrix} \varepsilon_{zz} \\ \varepsilon_{rr} \\ \varepsilon_{\theta\theta} \\ \gamma_{rz} \end{Bmatrix} \tag{2.17}
$$

### 2.2.2 Airy Stress Function

Airy developed the idea of a stress function which can satisfy both the equilibrium and the compatibility conditions (Sharifabadi, 1990). He showed that, in the absence of body forces, such a function $\Phi$ must satisfy the Laplace equation:

$$
\nabla^4 \Phi = \nabla^2 (\nabla^2 \Phi) = \left( \frac{\partial^2}{\partial x^2} + \frac{\partial^2}{\partial y^2} \right) \left( \frac{\partial^2 \Phi}{\partial x^2} + \frac{\partial^2 \Phi}{\partial y^2} \right) = 0 \tag{2.18}
$$

and

$$
\sigma_{xx} = \frac{\partial^2 \Phi}{\partial y^2} \tag{2.19}
$$

$$
\sigma_{yy} = \frac{\partial^2 \Phi}{\partial x^2} \tag{2.20}
$$

$$
\sigma_{xy} = -\frac{\partial^2 \Phi}{\partial x \, \partial y} \tag{2.21}
$$

The Airy function must also satisfy the natural (stress) boundary conditions, which further limits its application for complex problems.

### 2.2.3 Complex Stress Functions

Kolosov (1909) and Muskheshvili (1953) developed the idea of complex stress functions, which enables one to find solutions for more general problems including sharp corners, cracks and openings. Assuming $\psi(z)$ and $\chi(z)$ to be two harmonic analytic functions of $z$,

$$
\nabla^2 (\psi) = \nabla^2 (\chi) = 0 \tag{2.22}
$$

$$
z = x + iy = re^{i\theta} \tag{2.23}
$$

Any stress function $\Phi$ can be expressed as:

$$\Phi = \mathrm{Re}\left[\bar{z}\psi(z) + \chi(z)\right] \tag{2.24}$$

where $\bar{z} = x - iy$:

The three stress components $\sigma_{xx}$, $\sigma_{yy}$ and $\sigma_{xy}$ can be solved by substituting Eq. (2.24) into Eqs. (2.19)–(2.21),

$$\sigma_{xx} + \sigma_{yy} = 4\,\mathrm{Re}\left[\psi'(z)\right] \tag{2.25}$$

$$-\sigma_{xx} + \sigma_{yy} + 2i\sigma_{xy} = 2[\bar{z}\psi''(z) + \chi''(z)] \tag{2.26}$$

where $'$ and $''$ denote the first and second derivatives of a function, respectively.

Cartesian displacements $u_x$ and $u_y$ can also be expressed in terms of the complex functions:

$$u_x + iu_y = \frac{3-\nu}{E}\psi(z) - \frac{1+\nu}{E}[\bar{z}\psi'(\bar{z}) + \chi'(\bar{z})] \tag{2.27}$$

In a general curvilinear coordinate system any point can be represented by $(\alpha, \beta)$, where $\alpha$ and $\beta$ are functions of coordinates $(x, y)$. The stress components in terms of curvilinear coordinates can then be defined as:

$$\sigma_{\alpha\alpha} + \sigma_{\beta\beta} = 4\,\mathrm{Re}\left[\psi'(z)\right] \tag{2.28}$$

$$-\sigma_{\alpha\alpha} + \sigma_{\beta\beta} + 2i\sigma_{\alpha\beta} = 2[\bar{z}\psi''(z) + \chi''(z)]e^{2i\theta} \tag{2.29}$$

and for displacements

$$2\mu(u_\alpha - iu_\beta) = \left[\frac{3-\nu}{1+\nu}\psi(z) - z\psi'(\bar{z}) - \chi'(\bar{z})\right]e^{-i\theta} \tag{2.30}$$

The special case of polar coordinates is obtained by setting $\alpha \to r$ and $\beta \to \theta$.

## 2.3 Basics of LEFM

### 2.3.1 Fracture Mechanics

Various kinds of geometric discontinuity, such as a sharp change in geometry, opening, hole, notch, crack, and so on are known to be the main source of failure in a large number of catastrophic failures of structures. Such discontinuities generate substantial stress concentrations which reduce the overall strength of a material.

In order to explain the fundamental differences of fracture mechanics and the conventional theory of the strength of a material, consider a simple infinite tensile plate, as depicted in Figure 2.1. The plate is assumed to include a number of isolated defects, such as a tiny hole and a microcrack and the rest of the specimen is presumed flawless.

The average stress field far from the flaws (region A) is equal to the applied tensile stress, $\sigma_0$. Limiting the internal stress field to the material yield stress results in determination of the maximum allowable traction from the following simple material strength criterion:

$$\sigma_0 = \sigma_{\mathrm{yld}} \tag{2.31}$$

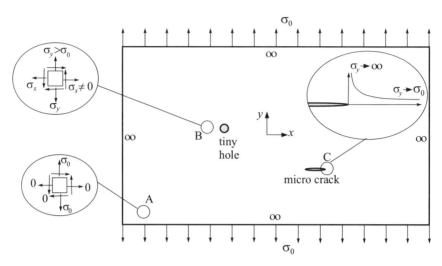

**Figure 2.1**  An infinite tensile plate and the stress states around an intact point A, tiny hole B and microcrack-tip C.

In contrast, the elasticity solution for an infinite plate with a circular defect/hole (as will be discussed in Section 2.3.2) predicts a biaxial non-uniform stress field with a stress concentration factor of 3 at the centre of the plate, regardless of the size of the hole. In a limiting case of a line crack, the solution from a degenerated elliptical hole shows an infinite stress state at the crack tip (point C). While no material can withstand such an infinite stress state, real life observations show that many materials with internal cracks remain stable. As a result, instead of comparing the existing maximum stress field with a critical strength value, a different approach based on finite values is required. The theory of fracture mechanics uses a local stress intensity factor or a global fracture energy release rate in comparison with their critical values to assess the stability of a flawed specimen.

These fracture-based parameters define the way a crack affects the behaviour of a cracked specimen. This is primarily achieved by classifying the general complex behaviour of a crack into three independent categories, as depicted in Figure 2.2. In the opening mode I, crack surfaces are pulled apart in the normal direction ($y$) but remain symmetric about the $xz$ and $xy$ planes. The shearing mode II represents the sliding mode of movement of crack surfaces in the $x$ direction, while remaining symmetric about the $xy$ plane and skew symmetric about the $xz$ plane. Finally, in the tearing mode III, the crack surfaces slide over each other in the $z$ direction, while remaining skew symmetric about the $xy$ and $xz$ planes.

In the following, a number of classical problems of fracture mechanics are reviewed, and then the fundamental formulations of fracture mechanics are briefly discussed.

## 2.3.2   Infinite Tensile Plate with a Circular Hole

In 1898, Kirsch analysed the problem of an infinite plate with a circular hole under uniform tensile stress, as depicted in Figure 2.3a.

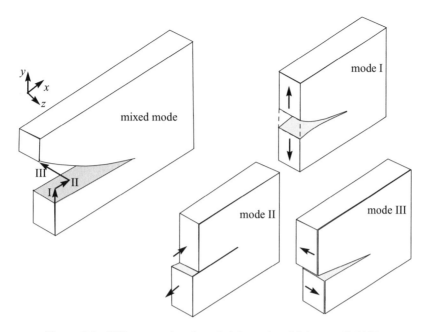

**Figure 2.2**   Different modes of crack deformation (Mohammadi, 2008).

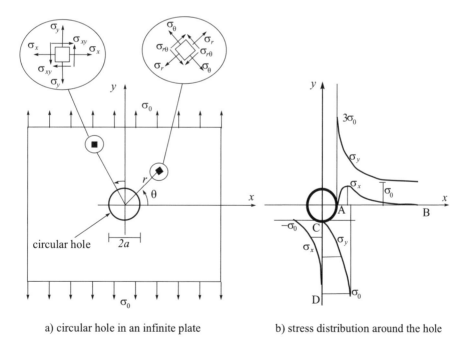

a) circular hole in an infinite plate          b) stress distribution around the hole

**Figure 2.3**   Infinite tensile plate with a circular hole. (a) geometry, (b) distribution of stress components (Mohammadi, 2008). (Reproduced by permission of John Wiley & Sons, Ltd.)

The problem is difficult to analyse in a single coordinate system. The boundary conditions around the circular hole can naturally be expressed in polar coordinate systems, whereas the far field Cartesian boundary conditions are better expressed in the $xy$ coordinate system.

Recalling the definition of the stress, $\sigma_{yy}$, in terms of the Airy stress function $\sigma_{yy} = \partial^2\Phi/\partial x^2$ would suggest a stress function of the form $\Phi = \sigma_0 x^2$ to represent the far field boundary condition $\sigma_{yy} = \sigma_0$. Alternatively, a polar representation of $\Phi$ is required to suit the circular hole with $x = r\cos\theta$:

$$\Phi = \sigma_0 r^2 \cos^2\theta \quad \text{or} \quad \Phi = \sigma_0 f(r)\cos 2\theta \tag{2.32}$$

After some manipulations, the following solutions are obtained (Meguid, 1989):

$$\sigma_{rr} = \frac{\sigma_0}{2}\left(1 - \frac{a^2}{r^2}\right)\left[1 - \left(1 - 3\frac{a^2}{r^2}\right)\cos 2\theta\right] \tag{2.33}$$

$$\sigma_{\theta\theta} = \frac{\sigma_0}{2}\left[\left(1 + \frac{a^2}{r^2}\right) + \left(1 + 3\frac{a^4}{r^4}\right)\cos 2\theta\right] \tag{2.34}$$

$$\sigma_{r\theta} = \frac{\sigma_0}{2}\left[\left(1 - \frac{a^2}{r^2}\right)\left(1 + 3\frac{a^2}{r^2}\right)\sin 2\theta\right] \tag{2.35}$$

It is of interest to examine the stress values at the edge of the hole, $r = a$:

$$\begin{cases} \sigma_{rr} = 0 \\ \sigma_{\theta\theta} = \sigma_0(1 + 2\cos 2\theta) \\ \sigma_{r\theta} = 0 \end{cases} \tag{2.36}$$

which shows that despite application of a uniform unidirectional tensile traction, a non-uniform biaxial stress state is generated around the hole, which may even become compressive at $\theta = \pi/2, 3\pi/2$ ($\sigma_{\theta\theta} = \sigma_{xx} = -\sigma_0$). The stress concentration factor for $\sigma_{\theta\theta}$ is 3 at $\theta = 0, \pi$ ($\sigma_{\theta\theta} = \sigma_{yy} = 3\sigma_0$). Figure 2.3b illustrates the distribution of stress components along the major axes of the plate.

### 2.3.3  Infinite Tensile Plate with an Elliptical Hole

Inglis (1913), a professor of naval architecture, independently of an earlier work by Kolosov, solved the problem of stress concentration around an elliptical hole in an infinite plate subjected to uniform stress loading, as depicted in Figure 2.4.

The following complex stress potential functions, proposed by Inglis in the curvilinear coordinate system $\alpha$ and $\beta$, satisfy the boundary conditions and are periodic in $\beta$:

$$\psi(z) = \frac{1}{4}\sigma_0 c[(1 + e^{2\alpha_0})\sinh p - e^{2\alpha_0}\cosh p] \tag{2.37}$$

$$\chi(z) = -\frac{1}{4}\sigma_0 c^2\left[(\cosh 2\alpha_0 - \cos\pi)p + \frac{1}{2}e^{2\alpha_0}\cosh 2\left(p - \alpha_0 - i\frac{\pi}{2}\right)\right] \tag{2.38}$$

where $p = \alpha + i\beta$ and $c = a/\cosh\alpha_0 = b/\sinh\alpha_0$. The solution at $\alpha = \alpha_0$ is:

$$(\sigma_{\beta\beta})_{\alpha=\alpha_0} = \sigma_0 e^{2\alpha_0}\left[\frac{\sinh 2\alpha_0(1 + e^{-2\alpha_0})}{\cosh 2\alpha_0 - \cos 2\beta} - 1\right] \tag{2.39}$$

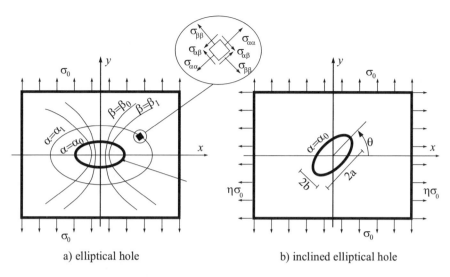

a) elliptical hole                                    b) inclined elliptical hole

**Figure 2.4**   An elliptical hole in an infinite plate (Mohammadi, 2008). (Reproduced by permission of John Wiley & Sons, Ltd.)

and for points located at the end of the ellipse in terms of $a$ and $b$,

$$(\sigma_{\beta\beta})_{\substack{\alpha=\alpha_0 \\ \beta=0,\pi}} = \sigma_0\left(1 + 2\frac{a}{b}\right) \tag{2.40}$$

Equation (2.40) shows that for a circular hole ($a = b$) the stress concentration factor becomes 3, similar to the conclusion made from the circular hole problem. Another extreme case is where the ellipse is degenerated into a crack ($b = 0$), generating an infinite stress. For a crack propagating along the applied tensile stress ($a = 0$), however, the stress at the crack tip remains at the finite value $\sigma_0$.

It has been shown that Eq. (2.40) can be rewritten in terms of the radius of curvature $\rho$:

$$(\sigma_{\beta\beta})_{\substack{\alpha=\alpha_0 \\ \beta=0,\pi}} = \sigma_0\left(1 + 2\sqrt{\frac{a}{\rho}}\right) \tag{2.41}$$

which shows that the stress concentration factor is proportional to $\rho^{-1/2}$.

Figure 2.5 illustrates the $\sigma_{xx}$ and $\sigma_{yy}$ stress contours around two elliptical holes with aspect ratios of $a/b = 2$ and $a/b = 5$ within an infinite tensile plate. It is noted that as the ellipse changes towards a sharp crack, the stress concentration factor becomes very large and tends to infinity for a sharp crack. The second observation is the change of the uniaxial stress state to a biaxial one, which also includes compressive regions.

For a more general form of the problem, which is an inclined elliptical hole in an infinite biaxial ($\sigma_0$, $\eta\sigma_0$) tensile plate (Figure 2.4b), the solution at $\alpha = \alpha_0$ is obtained as (Murrel, 1964):

$$(\sigma_{\beta\beta})_{\alpha=\alpha_0} = \sigma_0\left[\frac{(1+\eta)\sinh 2\alpha_0 - (1-\eta)(\cos 2\theta - e^{2\alpha_0}\cos 2(\beta - \theta))}{\cosh 2\alpha_0 - \cos 2\beta}\right] \tag{2.42}$$

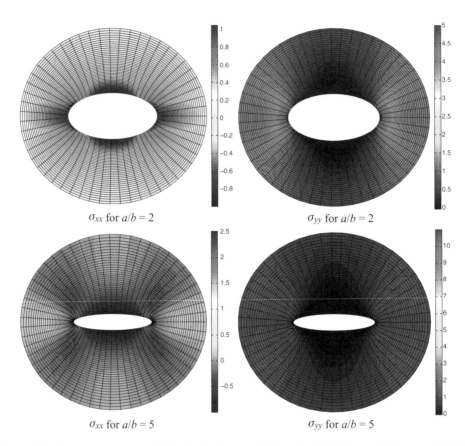

$\sigma_{xx}$ for $a/b = 2$    $\sigma_{yy}$ for $a/b = 2$

$\sigma_{xx}$ for $a/b = 5$    $\sigma_{yy}$ for $a/b = 5$

**Figure 2.5** Stress contours around elliptical holes in an infinite plate (Mahdavi and Mohammadi, 2012).

### 2.3.4 Westergaard Analysis of a Line Crack

Consider an infinite plate with a central traction-free crack of length $2a$ subjected to uniform biaxial stress $\sigma_0$, as depicted in Figure 2.6a.

One solution is to superimpose Inglis solutions in the two cases of $a = 0$ and $b = 0$. As an alternative approach, Westergaard (1939) proposed the following biharmonic stress function $\Phi$ as a solution to the crack problem:

$$\Phi = \mathrm{Re}\,\bar{\bar{\phi}}(z) + y\,\mathrm{Im}\,\bar{\phi}(z) \qquad (2.43)$$

where $\bar{\phi}$ and $\bar{\bar{\phi}}$ are the first and second integrals of $\phi(z)$, respectively. The stress components then become:

$$\sigma_{xx} = \mathrm{Re}\,\phi(z) - y\,\mathrm{Im}\,\phi'(z) \qquad (2.44)$$

$$\sigma_{yy} = \mathrm{Re}\,\phi(z) + y\,\mathrm{Im}\,\phi'(z) \qquad (2.45)$$

$$\sigma_{xy} = -y\,\mathrm{Re}\,\phi'(z) \qquad (2.46)$$

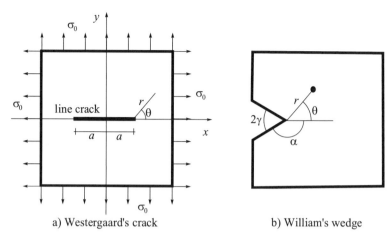

a) Westergaard's crack                     b) William's wedge

**Figure 2.6**  (a) Westergaard infinite plate with a crack subjected to uniform normal tractions, (b) Williams wedge problem (Mohammadi, 2008). (Reproduced by permission of John Wiley & Sons, Ltd.)

The complex function $\phi$ is determined so that the boundary conditions are satisfied along the crack and at infinity ($\sigma_0$):

$$\phi(z) = \frac{\sigma_0}{\sqrt{1 - \dfrac{a^2}{z^2}}} \tag{2.47}$$

Then, the final near crack tip solutions are obtained ($r \ll a$) (Meguid, 1989):

$$\sigma_{xx} = \sigma_0 \sqrt{\frac{a}{2r}} \cos\frac{\theta}{2}\left(1 - \sin\frac{\theta}{2}\sin\frac{3\theta}{2}\right) + \cdots \tag{2.48}$$

$$\sigma_{yy} = \sigma_0 \sqrt{\frac{a}{2r}} \cos\frac{\theta}{2}\left(1 + \sin\frac{\theta}{2}\sin\frac{3\theta}{2}\right) + \cdots \tag{2.49}$$

$$\sigma_{xy} = \sigma_0 \sqrt{\frac{a}{2r}} \sin\frac{\theta}{2}\cos\frac{\theta}{2}\cos\frac{3\theta}{2} + \cdots \tag{2.50}$$

Neglecting the higher order terms in Eqs. (2.48)–(2.50) is only acceptable for small values of $r$ around the crack tip.

### 2.3.5  Williams Solution of a Wedge Corner

Williams (1952) proposed a new set of solutions for a plate with angular corners (wedges), as depicted in Figure 2.6b. The method is capable of determining the stress and displacement fields near and far from a crack tip. It can also be extended to a crack at the interface of two dissimilar materials, and even to a general solution for a contact stick/slip crack between dissimilar materials (as will be briefly addressed in Chapter 3).

Williams' solution can be defined as,

$$\Phi(r,\theta) = r^{\lambda+1}F(\theta,\lambda) = r^{\lambda+1}e^{m(\lambda)\theta}$$

$$= r^{\lambda+1}[A\cos(\lambda-1)\theta + B\cos(\lambda+1)\theta + C\sin(\lambda-1)\theta + D\sin(\lambda+1)\theta] \quad (2.51)$$

where $m(\lambda) = \pm i(\lambda \pm 1)$.

Imposing the traction-free boundary conditions at the corner edges, results in the following characteristic equation (Saouma, 2000)

$$\sin 2\lambda_n\alpha + \lambda_n\sin 2\alpha = 0 \quad (2.52)$$

or for the simplified case of a sharp crack ($\alpha = \pi$)

$$\sin 2\pi\lambda_n = 0, \quad \lambda_n = \frac{n}{2}, \quad n = 1, 3, 4, \ldots \quad (2.53)$$

The final polar components of the stress field for the crack problem are obtained as,

$$\sigma_{rr} = \frac{1}{\sqrt{r}}\sum_n\left\{b_n\left[\cos\frac{\theta}{2}\left(1+\sin^2\frac{\theta}{2}\right)\right] + a_n\left[-\frac{5}{4}\sin\frac{\theta}{2} + \frac{3}{4}\sin\frac{3\theta}{2}\right]\right\} \quad (2.54)$$

$$\sigma_{\theta\theta} = \frac{1}{\sqrt{r}}\sum_n\left\{b_n\left[\cos\frac{\theta}{2}\left(1-\sin^2\frac{\theta}{2}\right)\right] + a_n\left[-\frac{3}{4}\sin\frac{\theta}{2} - \frac{3}{4}\sin\frac{3\theta}{2}\right]\right\} \quad (2.55)$$

$$\sigma_{r\theta} = \frac{1}{\sqrt{r}}\sum_n\left\{b_n\left[\sin\frac{\theta}{2}\cos^2\frac{\theta}{2}\right] + a_n\left[\frac{1}{4}\cos\frac{\theta}{2} + \frac{3}{4}\cos\frac{3\theta}{2}\right]\right\} \quad (2.56)$$

where $b_n$ and $a_n$ are

$$a_n = \frac{B_n}{A_n} = -\frac{\cos(\lambda_n-1)\pi}{\cos(\lambda_n+1)\pi}, \quad \lambda_n = \frac{n}{2}, \quad n = 1, 3, 4, \ldots \quad (2.57)$$

$$b_n = \frac{C_n}{D_n} = -\frac{\sin(\lambda_n-1)\pi}{\sin(\lambda_n+1)\pi}, \quad \lambda_n = \frac{n}{2}, \quad n = 1, 3, 4, \ldots \quad (2.58)$$

Equations (2.54)–(2.56) include both opening and shearing modes of crack response, associated with $b_n$ and $a_n$ terms, respectively. The first term of the series ($\lambda_1 = 1/2$) represents the Westergaard approximate solution for the opening mode near the crack tip. Similar radial and angular functions are obtained in both approaches. Several more terms, however, have to be computed in order to compute accurate solutions for points far from the crack tip.

## 2.4 Stress Intensity Factor, K

### 2.4.1 Definition of the Stress Intensity Factor

Irwin (1957) introduced the concept of stress intensity factor $K$(SIF), as a measure of the strength of the singularity. He illustrated that all elastic stress fields around a crack tip are distributed similarly, and $K \propto \sigma\sqrt{\pi r}$ controls the local stress intensity.

Recalling Eqs. (2.48)–(2.50), the elastic stress state around a crack can now be represented by:

$$\sigma_{xx} = \sigma_0 \sqrt{\frac{a}{2r}} \cos\frac{\theta}{2} \left(1 - \sin\frac{\theta}{2} \sin\frac{3\theta}{2}\right) + \cdots = \sigma_0 \sqrt{\frac{1}{r}} f(\theta) + \cdots \qquad (2.59)$$

or in the more general form of:

$$\sigma_{ij} = r^{-\frac{1}{2}} \left\{ K_I f_{ij}^I(\theta) + K_{II} f_{ij}^{II}(\theta) + K_{III} f_{ij}^{III}(\theta) \right\} + \text{higher order terms} \qquad (2.60)$$

where $\sigma_{ij}$ are the near crack tip stresses, and $K_I$, $K_{II}$, $K_{III}$ are the stress intensity factors associated with three independent modes of movement of crack surfaces (Figure 2.2),

$$K_I = \lim_{\substack{r \to 0 \\ \theta = 0}} \sigma_{yy} \sqrt{2\pi r} \qquad (2.61)$$

$$K_{II} = \lim_{\substack{r \to 0 \\ \theta = 0}} \sigma_{xy} \sqrt{2\pi r} \qquad (2.62)$$

$$K_{III} = \lim_{\substack{r \to 0 \\ \theta = 0}} \sigma_{yz} \sqrt{2\pi r} \qquad (2.63)$$

For example, the first mode stress intensity factor of the infinite tensile plate with a central crack, based on Eq. (2.61), can then be simplified to:

$$K_I = \lim_{\substack{r \to 0 \\ \theta = 0}} \sigma_{yy} \sqrt{2\pi r} = \lim_{\substack{r \to 0 \\ \theta = 0}} \sqrt{2\pi r} \sigma_0 \sqrt{\frac{a}{2r}} \cos\frac{\theta}{2} \left(1 - \sin\frac{\theta}{2} \sin\frac{3\theta}{2}\right) = \sigma_0 \sqrt{\pi a} \qquad (2.64)$$

Similar to the conventional theory of strength of materials where the existing stress is compared with an allowable material stress/strength, fracture mechanics states that unstable fracture occurs when the stress intensity factor $K$ reaches a critical value $K_c$, called the fracture toughness, which represents the potential ability of a material to withstand a given stress field at the tip of a crack and to resist progressive tensile crack extension. While for each pure fracture mode, the stress intensity factor $K_i$ should be compared with its relevant toughness $K_{ic}$, a mixed mode criterion is required to account for the combined effect of all individual modes.

Substituting Eqs. (2.61)–(2.63) into Eqs. (2.48)–(2.50) at the crack tip $\theta = 0$, allows the final stress and displacement fields to be described in terms of the stress intensity factors. They are categorized into three pure modes of fracture (Saouma, 2000).

For pure opening mode I, the displacement fields are given by:

$$u_x = \frac{K_I}{2\mu} \sqrt{\frac{r}{2\pi}} \cos\frac{\theta}{2} \left(\kappa - 1 + 2\sin^2\frac{\theta}{2}\right) \qquad (2.65)$$

$$u_y = \frac{K_I}{2\mu} \sqrt{\frac{r}{2\pi}} \sin\frac{\theta}{2} \left(\kappa + 1 - 2\cos^2\frac{\theta}{2}\right) \qquad (2.66)$$

$$u_z = \begin{cases} 0 & \text{plane strain} \\ -\dfrac{\nu z}{E}(\sigma_{xx} + \sigma_{yy}) & \text{plane stress} \end{cases} \qquad (2.67)$$

where parameter $\kappa$ is defined as,

$$\kappa = \begin{cases} \dfrac{3 - v}{1 + v} & \text{plane stress} \\[2mm] 3 - 4v & \text{plane strain} \end{cases} \tag{2.68}$$

and the stress field:

$$\sigma_{xx} = \frac{K_I}{\sqrt{2\pi r}} \cos \frac{\theta}{2} \left( 1 - \sin \frac{\theta}{2} \sin \frac{3\theta}{2} \right) \tag{2.69}$$

$$\sigma_{yy} = \frac{K_I}{\sqrt{2\pi r}} \cos \frac{\theta}{2} \left( 1 + \sin \frac{\theta}{2} \sin \frac{3\theta}{2} \right) \tag{2.70}$$

$$\sigma_{xy} = \frac{K_I}{\sqrt{2\pi r}} \sin \frac{\theta}{2} \cos \frac{\theta}{2} \cos \frac{3\theta}{2} \tag{2.71}$$

$$\sigma_{zz} = \begin{cases} v(\sigma_{xx} + \sigma_{yy}) & \text{plane strain} \\[2mm] 0 & \text{plane stress} \end{cases} \tag{2.72}$$

$$\sigma_{xz} = \sigma_{yz} = 0 \tag{2.73}$$

The pure mode II is governed by the following displacement fields:

$$u_x = \frac{K_{II}}{2\mu} \sqrt{\frac{r}{2\pi}} \sin \frac{\theta}{2} \left( \kappa + 1 + 2 \cos^2 \frac{\theta}{2} \right) \tag{2.74}$$

$$u_y = -\frac{K_{II}}{2\mu} \sqrt{\frac{r}{2\pi}} \cos \frac{\theta}{2} \left( \kappa - 1 - 2 \sin^2 \frac{\theta}{2} \right) \tag{2.75}$$

$$u_z = \begin{cases} 0 & \text{plane strain} \\[2mm] -\dfrac{vz}{E} (\sigma_{xx} + \sigma_{yy}) & \text{plane stress} \end{cases} \tag{2.76}$$

and the stress fields,

$$\sigma_{xx} = -\frac{K_{II}}{\sqrt{2\pi r}} \sin \frac{\theta}{2} \left( 2 + \cos \frac{\theta}{2} \cos \frac{3\theta}{2} \right) \tag{2.77}$$

$$\sigma_{yy} = \frac{K_{II}}{\sqrt{2\pi r}} \sin \frac{\theta}{2} \cos \frac{\theta}{2} \cos \frac{3\theta}{2} \tag{2.78}$$

$$\sigma_{xy} = \frac{K_{II}}{\sqrt{2\pi r}} \cos \frac{\theta}{2} \left( 1 - \sin \frac{\theta}{2} \sin \frac{3\theta}{2} \right) \tag{2.79}$$

$$\sigma_{zz} = \begin{cases} v(\sigma_{xx} + \sigma_{yy}) & \text{plane strain} \\[2mm] 0 & \text{plane stress} \end{cases} \tag{2.80}$$

$$\sigma_{xz} = \sigma_{yz} = 0 \tag{2.81}$$

The tearing mode III, has only one non-zero displacement and two non-zero stress components:

$$u_x = u_y = 0 \tag{2.82}$$

$$u_z = \frac{K_{\text{III}}}{\mu} \sqrt{\frac{r}{2\pi}} \sin \frac{\theta}{2} \tag{2.83}$$

$$\sigma_{xx} = \sigma_{yy} = \sigma_{zz} = \sigma_{xy} = 0 \tag{2.84}$$

$$\sigma_{xz} = -\frac{K_{\text{III}}}{\sqrt{2\pi r}} \sin \frac{\theta}{2} \tag{2.85}$$

$$\sigma_{yz} = \frac{K_{\text{III}}}{\sqrt{2\pi r}} \cos \frac{\theta}{2} \tag{2.86}$$

Finally, Eqs. (2.69)–(2.81) can be expressed in polar coordinates $\sigma_{rr}$, $\sigma_{\theta\theta}$ and $\sigma_{r\theta}$. In the opening mode I:

$$\sigma_{rr} = \frac{K_{\text{I}}}{\sqrt{2\pi r}} \cos \frac{\theta}{2} \left(1 + \sin^2 \frac{\theta}{2}\right) = \frac{K_{\text{I}}}{\sqrt{2\pi r}} \left(\frac{5}{4} \cos \frac{\theta}{2} - \frac{1}{4} \cos \frac{3\theta}{2}\right) \tag{2.87}$$

$$\sigma_{\theta\theta} = \frac{K_{\text{I}}}{\sqrt{2\pi r}} \cos^3 \frac{\theta}{2} = \frac{K_{\text{I}}}{\sqrt{2\pi r}} \left(\frac{3}{4} \cos \frac{\theta}{2} + \frac{1}{4} \cos \frac{3\theta}{2}\right) \tag{2.88}$$

$$\sigma_{r\theta} = \frac{K_{\text{I}}}{\sqrt{2\pi r}} \sin \frac{\theta}{2} \cos^2 \frac{\theta}{2} = \frac{K_{\text{I}}}{\sqrt{2\pi r}} \left(\frac{1}{4} \sin \frac{\theta}{2} + \frac{1}{4} \sin \frac{3\theta}{2}\right) \tag{2.89}$$

and for the shear mode II:

$$\sigma_{rr} = \frac{K_{\text{II}}}{\sqrt{2\pi r}} \sin \frac{\theta}{2} \left(1 - 3\sin^2 \frac{\theta}{2}\right) = \frac{K_{\text{II}}}{\sqrt{2\pi r}} \left(-\frac{5}{4} \sin \frac{\theta}{2} + \frac{3}{4} \sin \frac{3\theta}{2}\right) \tag{2.90}$$

$$\sigma_{\theta\theta} = \frac{-3K_{\text{II}}}{2\sqrt{2\pi r}} \sin \theta \cos \frac{\theta}{2} = \frac{K_{\text{II}}}{\sqrt{2\pi r}} \left(-\frac{3}{4} \sin \frac{\theta}{2} - \frac{3}{4} \sin \frac{3\theta}{2}\right) \tag{2.91}$$

$$\sigma_{r\theta} = \frac{K_{\text{II}}}{\sqrt{2\pi r}} \cos \frac{\theta}{2} \left(1 - 3\sin^2 \frac{\theta}{2}\right) = \frac{K_{\text{II}}}{\sqrt{2\pi r}} \left(\frac{1}{4} \cos \frac{\theta}{2} + \frac{3}{4} \cos \frac{3\theta}{2}\right) \tag{2.92}$$

## 2.4.2   Examples of Stress Intensity Factors for LEFM

In this section some of the basic problems of fracture mechanics with available analytical SIF solutions are illustrated. In all problems, the thickness of the plate is assumed to be constant $t$.

a. Finite tensile plate problem with a central crack (Figure 2.7a):

$$K_{\text{I}} = \left[1 - 0.1\left(\frac{a}{b}\right)^2 + 0.96\left(\frac{a}{b}\right)^4\right] \sqrt{\sec \frac{\pi a}{b}} \, \sigma_0 \sqrt{\pi a} \tag{2.93}$$

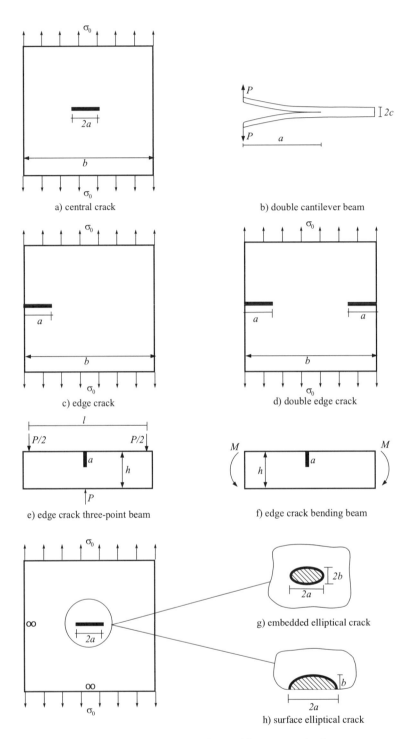

**Figure 2.7**   Classical problems of fracture mechanics.

b. Plane stress (thin-section) double cantilever beam (Figure 2.7b):

$$K_{\mathrm{I}} = 2\sqrt{3}\,\frac{Pa}{c^{1.5}} \tag{2.94}$$

c. Finite tensile plate problem with an edge crack (Figure 2.7c):

$$K_{\mathrm{I}} = \left[ 1.12 - 0.23\left(\frac{a}{b}\right) + 10.56\left(\frac{a}{b}\right)^2 - 21.74\left(\frac{a}{b}\right)^3 + 30.42\left(\frac{a}{b}\right)^4 \right] \sigma_0\sqrt{\pi a} \tag{2.95}$$

d. Finite tensile plate problem with double edge cracks (Figure 2.7d):

$$K_{\mathrm{I}} = \left[ 1.12 + 0.43\left(\frac{a}{b}\right) - 4.79\left(\frac{a}{b}\right)^2 + 15.46\left(\frac{a}{b}\right)^3 \right] \sigma_0\sqrt{\pi a} \tag{2.96}$$

e. Three-point beam with an edge crack (Figure 2.7e):

$$K_{\mathrm{I}} = \frac{1.5\dfrac{l}{h}\sqrt{\dfrac{a}{h}}}{\left(1+2\dfrac{a}{h}\right)\left(1-\dfrac{a}{h}\right)^{1.5}} \left[ 1.99 - \frac{a}{h}\left(1-\frac{a}{h}\right)\left\{ 2.15 - 3.93\left(\frac{a}{h}\right) + 2.7\left(\frac{a}{h}\right)^2 \right\} \right] \frac{P}{th^{0.5}} \tag{2.97}$$

f. Edge cracked beam in bending (Figure 2.7f):

$$K_{\mathrm{I}} = \frac{6M}{th^{1.5}} \sqrt{\frac{2\tan\left(\dfrac{\pi a}{2h}\right)}{\cos\left(\dfrac{\pi a}{2h}\right)}} \left[ 0.923 + 0.199\left\{1 - \sin\left(\frac{\pi a}{2h}\right)\right\}^4 \right] \tag{2.98}$$

g. Embedded elliptical flaw under normal tension (Figure 2.7g):

$$K_{\mathrm{I}}(\theta) = \left(1 + 1.464\left(\frac{b}{a}\right)^{1.65}\right)^{-\frac{1}{2}} \left(\sin^2\theta + \frac{b^2}{a^2}\cos^2\theta\right)^{\frac{1}{4}} \sigma_0\sqrt{\pi b} \tag{2.99}$$

h. Elliptical surface flaw under normal tension (Figure 2.7h):

$$K_{\mathrm{I}}(\theta) = \lambda_s\left(1 + 1.464\left(\frac{b}{a}\right)^{1.65}\right)^{-\frac{1}{2}} \left(\sin^2\theta + \frac{b^2}{a^2}\cos^2\theta\right)^{\frac{1}{4}} \sigma_0\sqrt{\pi b} \tag{2.100}$$

$$\lambda_s = \left(1.13 - 0.09\left(\frac{b}{a}\right)\right)(1 + 0.1(1 - \sin\theta)^2) \tag{2.101}$$

A comprehensive list of different problems with their analytic or approximate stress intensity factors can be found in handbooks and textbooks on fracture mechanics.

## 2.4.3 Griffith Energy Theories

While exploring the theoretical strength of solids by performing a series of experiments on glass rods of various diameters, Griffith observed that the tensile strength of glass decreased with increasing diameter. He realized that something different from a simple inherent material property had caused the size dependency of the tensile strength.

While studying the solution of the elliptical hole problem, Inglis initiated the idea that the theoretical strength of a solid has to be reduced due to the presence of internal flaws. In other words, he assumed that the theoretical strength of a material must be compared with the concentrated stress field, which is much higher than the average stress or the flawless-based stress field.

Instead of using a stress-based criterion, Griffith derived a thermodynamic criterion for fracture by considering the total change in energy of a cracked body in terms of the crack length increase. His model described the failure of a solid material in terms of satisfying a critical energy criterion rather than a maximum stress-based failure control.

Consider a crack in a deformable continuum subjected to arbitrary loading. The first law of thermodynamics for an adiabatic quasi-static system states that the change in total energy is proportional to the amount of performed work:

$$\frac{d}{dt}(U_s + U_\Gamma) = \frac{d}{dt}(W) \tag{2.102}$$

where $U_s$ is the total internal strain energy, $U_\Gamma$ is the surface energy and $W$ is the external work.

Equation (2.102) can be rewritten in terms of virtual crack extension,

$$\frac{\partial W}{\partial a} = \frac{\partial U_s}{\partial a} + \frac{\partial U_\Gamma}{\partial a} \tag{2.103}$$

This equation represents the energy balance during crack growth, and indicates that the work rate supplied to the continuum by the applied external load is equal to the surface energy dissipated during crack propagation, $U_\Gamma$, plus the rate of strain energy, $U_s$, decomposed into elastic $U_s^e$ and plastic $U_s^p$ parts

$$U_s = U_s^e + U_s^p \tag{2.104}$$

Equation (2.103) can be expressed in terms of the potential energy, $\Pi$:

$$\Pi = U_s^e - W \tag{2.105}$$

$$-\frac{\partial \Pi}{\partial a} = -\frac{\partial U_s^e}{\partial a} + \frac{\partial W}{\partial a} = -\frac{\partial U_s^e}{\partial a} + \frac{\partial U_s}{\partial a} + \frac{\partial U_\Gamma}{\partial a} = \frac{\partial U_s^p}{\partial a} + \frac{\partial U_\Gamma}{\partial a} \tag{2.106}$$

Therefore, the energy available for crack growth is compared with the resistance of the material that must be overcome for crack growth. It also indicates that the decrease rate of potential energy during crack growth is equal to the rate of energy dissipated in plastic deformation and crack growth.

For a perfectly brittle material, $U_s^p$ vanishes and Eq. (2.106) reduces to:

$$-\frac{\partial \Pi}{\partial a} = \frac{\partial W}{\partial a} - \frac{\partial U_s^e}{\partial a} = \frac{\partial U_\Gamma}{\partial a} = 2\gamma_s \tag{2.107}$$

where $\gamma_s$ is the surface energy and the factor 2 represents the existence of two material surfaces upon fracture. The value of $\gamma_s$ is experimentally measured for different materials. Water has a value of $\gamma_s = 0.077$, while for most metals, $\gamma_s = 1$ (Saouma, 2000).

Equation (2.107) can also be rewritten as:

$$\frac{\partial}{\partial a}(\Pi + U_\Gamma) = 0 \tag{2.108}$$

Therefore, a crack's growth can be considered unstable or stable when the energy at equilibrium is a maximum or minimum, respectively. Therefore, a sufficient condition for crack stability is obtained from the second derivative of $(\Pi + U_\Gamma)$ (Gtoudos, 1993):

$$\frac{\partial^2 (\Pi + U_\Gamma)}{\partial a^2} \begin{cases} < 0 & \text{unstable fracture} \\ > 0 & \text{stable fracture} \\ = 0 & \text{neutral equilibrium} \end{cases} \tag{2.109}$$

Equation (2.107) is defined as the Griffith crack growth energy, $G$

$$G = -\frac{\partial \Pi}{\partial a} = 2\gamma_s \tag{2.110}$$

Thus a criterion for crack propagation can be expressed as an inequality between the energy release rate per unit crack extension and the surface energy:

$$\frac{\mathrm{d}\Pi}{\mathrm{d}a} \geq 2\gamma_s \tag{2.111}$$

Griffith has shown that for a plane stress infinite plate with a central crack of length $2a$ and unit thickness subjected to unilateral tensile loading, the strain energy required to introduce the crack is equal in magnitude to the work required to close the crack by the stresses acting on it. Using the solutions for displacements of the free surface of one of the crack faces,

$$\Pi = 4\int_0^a \frac{1}{2}\sigma u_y(x)\mathrm{d}x = \frac{\pi\sigma^2 a^2}{2E'} \tag{2.112}$$

where the effective Young's modulus, $E'$, is defined as:

$$E' = \begin{cases} E & \text{plane stress} \\ \dfrac{E}{1-v^2} & \text{plane strain} \end{cases} \tag{2.113}$$

Therefore, going back to Eq. (2.110) gives,

$$G = -\frac{\mathrm{d}\Pi}{\mathrm{d}a} = \frac{\pi a\sigma^2}{E'} = 2\gamma_s \tag{2.114}$$

The critical stress for cracking that satisfies Eq. (2.114) is denoted by $\sigma_{cr}$,

$$\sigma_{cr} = \sqrt{\frac{2E'\gamma_s}{\pi a}} \tag{2.115}$$

The critical stress intensity factor $K_c$ for the infinite tensile plate with a central crack can then be defined as:

$$K_c = \sigma_{cr}\sqrt{\pi a} \tag{2.116}$$

and an unstable crack extension occurs if $K = K_c$.

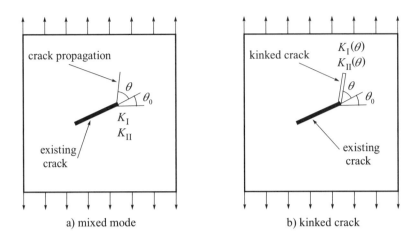

a) mixed mode                                    b) kinked crack

**Figure 2.8**   Mixed mode crack propagation.

## 2.4.4   Mixed Mode Crack Propagation

The problems discussed so far were mainly in fracture mode I. Practical engineering problems, however, rarely fall into this category. They usually include inclined or curvilinear crack propagations and are subjected to multiaxial loadings (Figure 2.8), creating non-zero $K_I$ and $K_{II}$ stress intensity factors. The effect of the tearing mode is neglected in this section.

In an infinite uniaxial tensile plate with an inclined crack (Figure 2.8a), the mode I and II stress intensity factors can be written as,

$$K_I = \sigma_0 \cos^2 \theta_0 \sqrt{\pi a} \tag{2.117}$$

$$K_{II} = \sigma_0 \sin \theta_0 \cos \theta_0 \sqrt{\pi a} \tag{2.118}$$

Clearly, these simple rules cannot be used in general mixed mode problems.

For a crack to extend in a linear elastic flawed structure, two criteria can be considered: First, a global approach for comparing the energy release rate, $G$, determined from the crack global transfer of energy, with a material property, $G_c$ (critical energy release rate). The second approach is based on a rather local criterion by comparing the stress intensity factor, $K$, determined from the near crack-tip stress field, with a material property $K_c$ (critical stress intensity factor or fracture toughness).

Generalization of the original SIF-based collinear crack propagation criterion ($K_I > K_{Ic}$) to include mixed mode effects can be defined in terms of $K_I$ and $K_{II}$ stress intensity factors, and $K_{Ic}$ and $K_{IIc}$ fracture toughness factors:

$$f(K_I, K_{Ic}, K_{II}, K_{IIc}) = 0 \tag{2.119}$$

Equation (2.126) is further simplified because usually only the mode I fracture toughness $K_{Ic}$ is experimentally measured:

$$f(K_I, K_{Ic}, K_{II}) = 0 \tag{2.120}$$

In the following sections (2.4.8.1–2.4.8.4), some of the most widely used mixed mode criteria are briefly reviewed.

### 2.4.4.1   Maximum Circumferential/Hoop Tensile Stress

Erdogan and Sih (1963) developed the first theory of mixed mode stress intensity factor based on the solution of stress state near a crack tip. They assumed that the crack propagates at its tip in a radial direction within a plane perpendicular to the direction of maximum tension when the maximum circumferential tensile stress, $(\sigma_\theta)_{max}$, reaches a critical material constant.

The mixed mode criterion for a crack angle $\theta$ can then be defined as (Figure 2.8a):

$$K_{\theta\theta} = \sigma_{\theta\theta}\sqrt{2\pi r} = K_I \cos^3\frac{\theta}{2} - \frac{3}{2}K_{II}\cos\frac{\theta}{2}\sin\theta = K_{Ic} \tag{2.121}$$

### 2.4.4.2   Minimum Strain Energy Density Criterion

The minimum strain energy density criterion is based on this idea that a crack propagates along the minimum resistance path. It determines the crack propagation from the crack tip in a direction $\theta$, along which the strain energy density at a critical distance is a minimum. Occurrence of crack propagation is controlled by checking such a minimum resistance until it reaches a critical value.

The final form of the criterion according to Sih (1973, 1974) can be defined as (Chang, Xu and Mutoh, 2006):

$$W_s = \frac{1}{16\pi\mu r}\left[a_{11}K_I^2 + 2a_{12}\left(K_I K_{II}\right) + a_{22}K_{II}^2\right] \tag{2.122}$$

where

$$\begin{cases} a_{11} = (1+\cos\theta)(\kappa-\cos\theta) \\ a_{12} = \sin\theta\left[2\cos\theta - (\kappa-1)\right] \\ a_{22} = (\kappa+1)(1-\cos\theta) + (1+\cos\theta)(3\cos\theta-1) \end{cases} \tag{2.123}$$

and the propagation occurs subject to $(W_s(\theta) = W_c)$

$$\begin{cases} \dfrac{\partial^2 W_s(\theta)}{\partial\theta^2} > 0 & \text{instability condition} \\[2mm] \dfrac{\partial W_s(\theta)}{\partial\theta} = 0 & \text{propagation angle} \end{cases} \tag{2.124}$$

### 2.4.4.3   Maximum Energy Release Rate

An alternative model, the maximum energy release rate, is based on the original work of Erdogan and Sih (1963) and Hussain, Pu and Underwood (1974) who solved for the stress intensity factors $K_I(\theta)$ and $K_{II}(\theta)$ of a major existing crack with an infinitesimal tip kink at an angle $\theta$ (Figure 2.8b) in terms of the stress intensity factors of the original crack $K_I$ and $K_{II}$:

$$K_I(\theta) = g(\theta)\left(K_I\cos\theta + \frac{3}{2}K_{II}\sin\theta\right) \tag{2.125}$$

$$K_{II}(\theta) = g(\theta)\left(K_{II}\cos\theta - \frac{1}{2}K_I\sin\theta\right) \tag{2.126}$$

with

$$g(\theta) = \left(\frac{4}{3 + \cos^2 \theta}\right)\left(\frac{1 - \theta/\pi}{1 + \theta/\pi}\right)^{\frac{\theta}{2\pi}} \tag{2.127}$$

Adopting Irwin's generalized expression for the energy release rate, evaluation of $G(\theta)$ for the kinked crack becomes

$$\begin{aligned}
G(\theta) &= \frac{1}{E'}\left(K_{\mathrm{I}}^2(\theta) + K_{\mathrm{II}}^2(\theta)\right) \\
&= \frac{1}{4E'}g^2(\theta)\left[(1 + 3\cos^2\theta)K_{\mathrm{I}}^2 + 8\sin\theta\cos\theta K_{\mathrm{I}}K_{\mathrm{II}} + (9 - 5\cos^2\theta)K_{\mathrm{II}}^2\right]
\end{aligned} \tag{2.128}$$

and the propagation occurs subject to $(G(\theta) \geq G_c)$

$$\begin{cases}
\dfrac{\partial^2 G(\theta)}{\partial\theta^2} < 0 & \text{instability condition} \\[3mm]
\dfrac{\partial G(\theta)}{\partial\theta} = 0 & \text{propagation angle}
\end{cases} \tag{2.129}$$

### 2.4.4.4  Non-Local Stress Fracture Criterion

A non-local stress fracture criterion based on the damage model of an elastic solid, containing growing microcracks, was proposed by Romanowicz and Seweryn (2008). According to this criterion, crack initiation and propagation would occur when the mean value of the function $R(\sigma_n, \tau_n)$ of decohesive normal and shear stresses over a segment $d$, the length of the damaged zone, reaches its critical value,

$$\max_\theta \bar{R}(\sigma_n, \tau_n) = \max_\theta \left[\frac{1}{d}\int_0^d R(\sigma_n, \tau_n)\mathrm{d}x\right] = 1 \tag{2.130}$$

$R(\sigma_n, \tau_n)$ is the local stress function obtained by the microcrack damage model, which states that the propagation of microcracks takes place when the stress energy release rate equals its resistance to microcrack growth (Hallai, 2008).

### 2.4.4.5  Other Empirical Models

The following typical relation defines a number of mixed mode stress intensity criteria that are basically developed from experimental observations rather than theoretical bases. Many of them are valid only for specific problems in concrete, rock and composites, where the models have been experimentally obtained or calibrated. They are usually of the general form of:

$$\alpha_1\left(\frac{K_{\mathrm{I}}}{K_{\mathrm{Ic}}}\right)^{\beta_1} + \alpha_2\left(\frac{K_{\mathrm{I}}K_{\mathrm{II}}}{K_{\mathrm{Ic}}K_{\mathrm{IIc}}}\right)^{\beta_2} + \alpha_3\left(\frac{K_{\mathrm{II}}}{K_{\mathrm{IIc}}}\right)^{\beta_3} + \alpha_4\left(\frac{K_{\mathrm{II}}}{K_{\mathrm{Ic}}}\right)^{\beta_4} = 1 \tag{2.131}$$

Table 2.1 presents the coefficients of a number of adopted empirical formulae.

**Table 2.1** Constant parameters of empirical mixed mode criteria.

| Model | $\alpha_1$ | $\alpha_2$ | $\alpha_3$ | $\alpha_4$ | $\beta_1$ | $\beta_2$ | $\beta_3$ | $\beta_4$ |
|---|---|---|---|---|---|---|---|---|
| Linear | 1 | — | 1 | — | 1 | — | 1 | — |
| Elliptical | 1 | — | 1 | — | 2 | — | 2 | — |
| Quadratic | 1 | $c \neq 2$ | 1 | — | 2 | 1 | 2 | — |
| Jernkvist | 1 | — | 1 | — | 1 | — | 2–3.4 | — |
| Awaji/Sato | 1 | — | 1 | — | 1.6 | — | 1.6 | — |
| Advani/Lee | 1 | — | — | 0.25 | 1 | — | — | 2 |
| Palanisawamy/ Knauss | 1 | — | — | 1.5 | 1 | — | — | 2 |

## 2.5 Classical Solution Procedures for *K* and *G*

For most practical problems no analytical solution is available, and numerical techniques such as the finite element method, boundary element method and, more recently, the meshless methods should be used. In this section, a review of available solutions related to the finite element method is provided.

### 2.5.1 Displacement Extrapolation/Correlation Method

The stress field at a crack tip is singular and conventional finite elements cannot represent it no matter how fine they are. Nevertheless, it was recognized in earlier simulations of LEFM that a very fine mesh is required at the crack tip if it is to be used to approximate the stress field for evaluation of the stress intensity factor (Chan, Tuba and Wilson, 1970).

Stress intensity factors can be determined at different radial distances from the crack tip by equating the numerically obtained displacements with their analytical expression in terms of the SIF. For plane problems in the *xy* plane (Figure 2.9):

$$K_{\mathrm{I}} = \mu \sqrt{\frac{2\pi}{r}} \frac{u_y^b - u_y^a}{\kappa + 1} \tag{2.132}$$

$$K_{\mathrm{II}} = \mu \sqrt{\frac{2\pi}{r}} \frac{u_x^b - u_x^a}{\kappa + 1} \tag{2.133}$$

Note that $u^a$ represents the total crack tip displacement (obtained from the numerical analysis) and is different from the local asymptotic displacement at the crack tip.

A simple extrapolation technique, as depicted in Figure 2.9, can then be used to approximately evaluate the value of SIF at the crack tip. The same procedure for stresses can also be used, although it is likely to yield less accurate predictions.

### 2.5.2 Mode I Energy Release Rate

A direct method for evaluation of the mode I stress intensity factor and the energy release rate is based on the direct definition of *G*,

$$G = \frac{K_{\mathrm{I}}^2}{E'} = -\frac{\partial \Pi}{\partial a} \cong \frac{\Delta \Pi}{\Delta a} \tag{2.134}$$

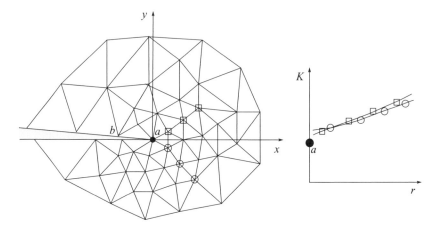

**Figure 2.9** Extrapolation of stress intensity factor (Mohammadi, 2008). (Reproduced by permission of John Wiley & Sons, Ltd.)

and finding the difference in the total strain energy, $\Delta \Pi$, of the system for initial and extended crack lengths $a$ and $a + \Delta a$, respectively. Therefore, the algorithm requires two completely separate analyses, which is a computationally expensive task. The strain energy can be determined from either $\Pi = \mathbf{u}^\mathrm{T} \mathbf{K} \mathbf{u}$ in terms of the nodal displacement $\mathbf{u}$ and the global stiffness matrix $\mathbf{K}$, or $\Pi = \mathbf{u}^\mathrm{T} P$ in terms of the external load $P$ and displacements $\mathbf{u}$.

### 2.5.3 Mode I Stiffness Derivative/Virtual Crack Model

A major problem with the previous method is the requirement for two complete analyses for the evaluation of $G$. It should be noted, however, that the stiffness matrix of the second analysis (associated with $a + \Delta a$) is only slightly altered from the first one (associated with $a$). A remedy, therefore, is to use relaxation methods to reduce the computational costs for the second analysis, as independently proposed by Parks (1974) and Hellen and Blackburn (1975), who called the method stiffness derivative and virtual crack model, respectively.

Assuming that the loading is unaltered during the crack extension, Equation (2.114) leads to

$$G = -\frac{\partial \Pi}{\partial a} = -\frac{\partial}{\partial a} \left( \frac{1}{2} \mathbf{u}^\mathrm{T} \mathbf{K} \mathbf{u} - \mathbf{u}^\mathrm{T} \mathbf{P} \right) = -\frac{1}{2} \mathbf{u}^\mathrm{T} \frac{\partial \mathbf{K}}{\partial a} \mathbf{u} \tag{2.135}$$

Therefore, the derivative of the stiffness matrix is required for evaluation of $G$. Instead of cumbersome numerical differentiations, this is usually computed by perturbing the elements around the crack tip (Figure 2.10a) and evaluating the modified part of the stiffness matrix associated with the nodes of elements along the crack extension.

### 2.5.4 Two Virtual Crack Extensions for Mixed Mode Cases

A natural extension to the virtual crack model to include the effects of mixed mode fracture is to use the two virtual crack extensions model. A rather simple case with a closed form

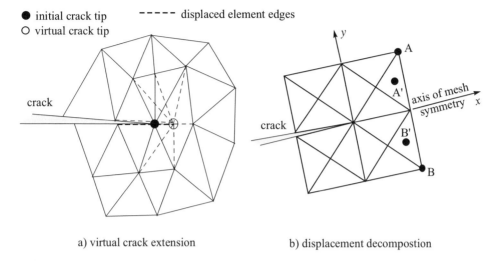

a) virtual crack extension                  b) displacement decompostion

**Figure 2.10**    Virtual crack extension model and the displacement decomposition technique.

solution was proposed by Hellen and Blackburn (1975) based on the following expressions of
the energy release rates for virtual crack extensions at $\theta = 0$ and $\theta = \pi/2$:

$$G_1 = \frac{K_I^2 + K_{II}^2}{E'} + \frac{K_{III}^2}{2\mu} \tag{2.136}$$

$$G_2 = \frac{-2K_I K_{II}}{E'} \tag{2.137}$$

The procedure can be started by computing the total strain energy $U_s$ for an initial crack
length $a$. Then the crack length is extended to $a + \Delta a$ along $\theta = 0$ and $\theta = \pi/2$ to determine
$G_1$ and $G_2$, respectively. The stress intensity factors are computed from solving simultaneous
Eqs. (2.136) and (2.137):

$$K_I, K_{II} = \frac{1}{2}\left(s \pm \sqrt{s^2 + \frac{8G_2}{\kappa'}}\right) \tag{2.138}$$

where

$$s = 2\sqrt{\frac{G_1 - G_2}{\kappa'}} \tag{2.139}$$

$$\kappa' = \frac{(1+v)(1+\kappa)}{E} \tag{2.140}$$

## 2.5.5    Single Virtual Crack Extension Based
on Displacement Decomposition

Previous techniques were computationally expensive and inefficient because they required at
least one complete finite element analysis, followed by two separate analyses or two virtual
crack extensions. The present approach, proposed by Ishikawa (1980) and Sha (1984), requires

only one analysis (or one virtual crack extension) but it is limited to symmetrical local elements around the crack tip.

With reference to Figure 2.10b, the nodal displacements for a symmetrical local mesh around the crack tip can be decomposed into two local symmetric $U^{\mathrm{I}}$ and anti-symmetric $U^{\mathrm{II}}$ components about the crack plane:

$$U = U^{\mathrm{I}} + U^{\mathrm{II}} \tag{2.141}$$

with

$$U^{\mathrm{I}} = \left\{ \begin{array}{c} u_x^{\mathrm{I}} \\ u_y^{\mathrm{I}} \end{array} \right\} = \frac{1}{2} \left\{ \begin{array}{c} u_x^A + u_x^B \\ u_y^A - u_y^B \end{array} \right\} \tag{2.142}$$

$$U^{\mathrm{II}} = \left\{ \begin{array}{c} u_x^{\mathrm{II}} \\ u_y^{\mathrm{II}} \end{array} \right\} = \frac{1}{2} \left\{ \begin{array}{c} u_x^A - u_x^B \\ u_y^A + u_y^B \end{array} \right\} \tag{2.143}$$

where $(x, y)$ are the local crack-tip coordinate system. Then, the local mode I $(U^{\mathrm{I}})$ and mode II $(U^{\mathrm{II}})$ components can be used directly to determine $K_{\mathrm{I}}$ and $K_{\mathrm{II}}$, respectively. Similar rules can be used to decompose the stress field into mode I and mode II stress components for evaluation of the energy release rate $G$ and the $J$ integral (Nikishkov and Atluri, 1987).

It should be noted that this method can be used with any two symmetric points with respect to the crack direction, even on any arbitrarily generated mesh (such as A′ and B′ in Figure 2.10b).

## 2.6    Quarter Point Singular Elements

One of the easiest and most powerful techniques for modelling a stress singularity is by quarter point singular elements. Development of these singular elements had an enormous impact on the wide range application of fracture mechanics by the finite element method while increasing the accuracy of various numerical analyses of LEFM. It was first proposed by Barsoum (1974, 1975, 1976a, 1976b, 1977, 1981) and independently by Henshell and Shaw (1975), as they demonstrated that the $r^{-1/2}$ singularity characteristic of LEFM can be obtained by two-dimensional 8-node isoparametric elements when the midside nodes near the crack tip are placed at the quarter points. A direct consequence was the additional capability of an existing continuum-based finite element code for modelling a stress singularity just by shifting the midside nodes of elements adjacent to the crack tip to their quarter point positions.

Hibbit (1977) and Ying (1982) studied the singularity of rectangular and triangular quarter elements and concluded that the singularity of a rectangular quarter point element is along the edges and diagonal only, whereas in a triangular element it occurs in all radial directions emanating from the crack tip. In general quadrilateral singular elements, however, such a conclusion for edge-only direction of singularity was put in doubt by a number of other researchers. Later, a complete study, including finite element implementation, related numerical issues and comparison of various singular crack-tip elements were comprehensively presented by Owen and Fawkes (1983).

In order to demonstrate the way a singular element performs, a simple one-dimensional element with three nodes is considered, as depicted in Figure 2.11. The physical location of

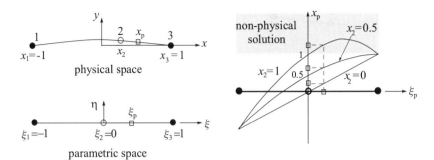

**Figure 2.11**  1D finite element with three nodes (Mohammadi, 2008). (Reproduced by permission of John Wiley & Sons, Ltd.)

the middle node $x_2$ may be anywhere in between the other nodes: $-1 < x_2 < 1$, while its location in the mapping parametric space is always at $\xi_2 = 0$. Assuming a second-order basis function, leads to the following position vector $x$ (Macneal, 1994),

$$x = \xi + (1 - \xi^2)x_2 \tag{2.144}$$

Equation (2.144) illustrates that, for values of $x_2 > 1/2$, the value of $x$ can lie outside the range of $[-1, +1]$. Physically, the middle node crosses the corner nodes and the element spills over its boundaries.

It can also be observed that placing the middle node at the quarter point ($x_2 = \pm 1/2$) allows the element to simulate a stress singularity at the corner. The strain can be determined from:

$$\varepsilon_x = u_{,x} = u_{,\xi}\xi_{,x} \tag{2.145}$$

which becomes infinity at the corner point $\xi = 1$, provided that the middle point is at the quarter point $x_2 = 1/2$:

$$x_{,\xi} = 0, \quad \xi_{,x} = \infty, \tag{2.146}$$

The same conclusion can be made for two- and three-dimensional elements. The stress tensor in an isoparametric finite element is given by:

$$\sigma = \mathbf{DB}\bar{u}_i \tag{2.147}$$

where the components of the $\mathbf{B}$ matrix are evaluated from the components of the Jacobian $\mathbf{J}$. In order to simulate a singular stress field, Eq. (2.147) requires $\mathbf{B}$ to be singular, as the other two components are constants. Consequently, the determinant of $\mathbf{J}$ must vanish at the crack tip. This is possible only if either one of the diagonal terms becomes zero.

$$\mathbf{J} = \begin{bmatrix} \dfrac{\partial x}{\partial \xi} & \dfrac{\partial y}{\partial \xi} \\[2ex] \dfrac{\partial x}{\partial \eta} & \dfrac{\partial y}{\partial \eta} \end{bmatrix}, \quad \mathbf{J}^{-1} = \dfrac{1}{\det \mathbf{J}} \begin{bmatrix} \dfrac{\partial y}{\partial \eta} & -\dfrac{\partial y}{\partial \xi} \\[2ex] -\dfrac{\partial x}{\partial \eta} & \dfrac{\partial x}{\partial \xi} \end{bmatrix} \tag{2.148}$$

Now referring to Eq. (2.148), it is found that this is achieved only if the midside node of the element is located at $x_2 = 1/2$ $(l/4)$ instead of $x_2 = 0$ $(l/2)$. As a result, the stresses and strains at the nearby corner node will become singular (Figure 2.11).

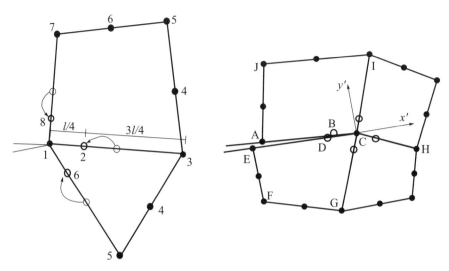

a) quadrilateral and triangular singular elements        b) singular elements around a crack tip

**Figure 2.12**    (a) Quarter point (singular) quadrilateral and triangular finite elements and (b) four quarter point elements around a crack tip (Mohammadi, 2008). (Reproduced by permission of John Wiley & Sons, Ltd.)

It can be proved that the quarter point element approximates the displacement field along $\eta = -1$ in the general form of:

$$u = A_1 + A_2 x + A_3 \left(\frac{x}{l}\right)^{\frac{1}{2}} \tag{2.149}$$

which is an indication of the order of singularity of $r^{-1/2}$ for the strain (derivative of displacement) and so the stress field.

It is reiterated that there are conflicting arguments that general quarter point singular quadrilateral elements provide radial singular strain and stress fields along the corresponding edges only, whereas triangular quarter elements create singularity in all radial directions (inside the element) emanating from the crack tip.

Singular finite elements allow an efficient and fast approach for evaluation of the stress intensity factors (Shih, de Lorenzi and German, 1976). Referring to Figure 2.12b, the basic idea is to equate the local displacement field in the quarter point singular element with the theoretical solution. For a general mixed mode problem,

$$K_I = \frac{1}{2}\frac{2\mu}{\kappa + 1}\sqrt{\frac{2\pi}{l}}[-3u_{y'C} + 4(u_{y'B} - u_{y'D}) - (u_{y'A} - u_{y'E})] \tag{2.150}$$

$$K_{II} = \frac{1}{2}\frac{2\mu}{\kappa + 1}\sqrt{\frac{2\pi}{l}}[-3u_{x'C} + 4(u_{x'B} - u_{x'D}) - (u_{x'A} - u_{x'E})] \tag{2.151}$$

where $u_{x'}$ and $u_{y'}$ are the local displacements of the nodes along the crack in the singular element, with $x'$ aligned with the crack axis, as depicted in Figure 2.12b.

The extension of the method to the case of more than one singular element, and the possible discrepancy of the results obtained from different singular elements, remains unresolved. Nevertheless, the method has remained popular because of the advantage of being exceptionally simple and fast.

## 2.7  *J* Integral

In a pioneering work, Eshelby (1956, 1974) defined a number of contour integrals that were path independent by virtue of the theorem of energy conservation. This was achieved while he was studying dislocations in elastic domains, and he did not realize its importance or applications in fracture mechanics. It was up to Rice and Rosengren (1968) to notice the importance of the *J* integral as a criterion for crack growth in fracture mechanics (Anderson, 1995). While the original introduction of the *J* integral was limited to problems with no unloading, no internal stress/strains and no crack face tractions, the new developments have now been well extended to cohesive crack and dynamic problems. Path independence of *J* also allows the evaluation of linear and nonlinear elastic energy release rates and elastoplastic work far from the crack tip.

First, the problem is considered without the presence of the body force and crack tractions ($\mathbf{f}^b = \mathbf{f}^c = 0$). The two-dimensional form of one of these integrals can be written as:

$$J = \oint_\Gamma \left( w_s dy - \mathbf{t}\frac{\partial \mathbf{u}}{\partial x} d\Gamma \right) \tag{2.152}$$

or in component form,

$$J = \oint_\Gamma \left( w_s \delta_{1j} - \sigma_{ij}\frac{\partial u_i}{\partial x} \right) n_j d\Gamma \tag{2.153}$$

where

$$w_s = \int_0^{\varepsilon_{ij}} \sigma_{ij} d\varepsilon_{ij} = \frac{1}{2}\sigma_{ij}\varepsilon_{ij} \tag{2.154}$$

is the strain energy density, $\Gamma$ is a closed counter-clockwise contour, $d\Gamma$ is the differential element of the arc along the path $\Gamma$, $\mathbf{t} = \sigma\mathbf{n}$ is the traction vector on a plane defined by the outward normal $\mathbf{n}$, and $\mathbf{u}$ is the displacement vector (Figure 2.13a).

It can be concluded that the absolute values of the *J* integral evaluated over arbitrary paths $\Gamma_1$ and $\Gamma_3$ remain identical; an indication of the path independency of *J*. As a result, if a contour begins from one crack surface and ends at the other face, it can be used to determine the *J* integral.

Selection of the size and shape of the appropriate contour curve for a specific problem requires complementary numerical studies. Practically, they should also be related to the geometry of the crack and the finite element model, as will be further discussed in the following sections.

It can be shown that when *J* is applied along a contour around a crack tip, it represents the change in potential energy for a virtual crack extension d*a* (for details see Mohammadi, 2008)

$$J = -\frac{d\Pi}{da} = G \tag{2.155}$$

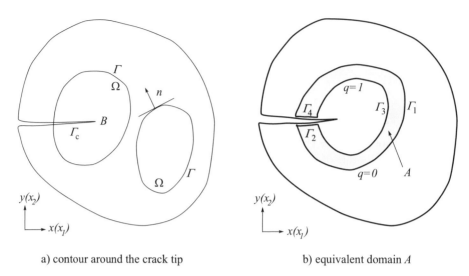

a) contour around the crack tip                              b) equivalent domain $A$

**Figure 2.13** Definition of the $J$ integral around a crack, and the equivalent domain $A$ (Mohammadi, 2008). (Reproduced by permission of John Wiley & Sons, Ltd.)

which is equivalent to the definition of the fracture energy release for linear elastic materials. Eq. (2.155) can also be used to define the energy release rate for a unit crack extension in nonlinear elastic materials.

### 2.7.1   Generalization of J

The original definition of $J$ can be regarded as the first component of a more general path-independent vector with the following components:

$$J_1 = \int_\Gamma \left\{ w_s dy - \mathbf{t} \frac{\partial \mathbf{u}}{\partial x} d\Gamma \right\} \tag{2.156}$$

$$J_2 = \int_\Gamma \left\{ w_s dx - \mathbf{t} \frac{\partial \mathbf{u}}{\partial y} d\Gamma \right\} \tag{2.157}$$

The generalized $J$ also satisfies the following relation for crack extensions parallel and perpendicular to the crack (Hellen and Blackburn, 1975)

$$J = J_1 - iJ_2 = \frac{1}{E'} \left( K_I^2 + K_{II}^2 + 2iK_I K_{II} \right) \tag{2.158}$$

### 2.7.2   Effect of Crack Surface Traction

Karlsson and Backlund (1978) extended the concept of the $J$ integral to account for the effect of crack surface tractions by simply extending the definition of the contour path to include the crack surfaces:

$$J = \int_\Gamma \left\{ w_s dy - \mathbf{t} \frac{\partial \mathbf{u}}{\partial x} d\Gamma \right\} - \int_{\Gamma_c} \mathbf{f}^c \frac{\partial \mathbf{u}}{\partial x} d\Gamma \qquad (2.159)$$

where $\Gamma_c$ is the portion of the crack surfaces between the points in between the two ends of $\Gamma$ and $\mathbf{f}^c$ is the crack surface traction vector.

### 2.7.3 Effect of Body Force

In the case of present body force, $\mathbf{f}^b$, the equilibrium equation can be written as:

$$\sigma_{ij,j} + \mathbf{f}_i^b = 0 \qquad (2.160)$$

and a modified $J$ integral has to be defined (Atluri, 1982):

$$J = \int_\Gamma \left\{ w_s dy - \mathbf{t} \frac{\partial \mathbf{u}}{\partial x} d\Gamma \right\} - \int_\Omega \mathbf{f}^b \frac{\partial \mathbf{u}}{\partial x} d\Omega \qquad (2.161)$$

### 2.7.4 Equivalent Domain Integral (EDI) Method

Li, Shih and Needleman (1985) proposed the equivalent domain integral method as an alternative approach to the original contour $J$ integral. According to Figure 2.13b, the $J$ integral can be defined as (Li, Shih and Needleman, 1985; Babuska and Miller, 1984):

$$J = \int_A \left[ \sigma_{ij} \frac{\partial u_i}{\partial x_1} - w_s \delta_{1j} \right] \frac{\partial q}{\partial x_j} dA \qquad (2.162)$$

where $q$ is an arbitrary smoothing function which is equal to unity on $\Gamma_3$ and zero on the boundary of the $J$ domain $\Gamma_1$. In fact, the contour integral has been replaced by an equivalent area integral, which is more suited to finite element solutions. The function $q$ may be interpreted as the kinematically admissible virtual extension field due to a unit extension of the crack (Li, Shih and Needleman, 1985; Moran and Shih, 1987; Combescure et al., 2008).

In FEM, the inner contour $\Gamma_3$ is often taken as the crack tip, and so $A$ naturally corresponds to the area inside $\Gamma_1$. The boundary $\Gamma_1$ should also coincide with element boundaries to facilitate numerical calculations. Alternatively, a fine evaluation is obtained with a $q$ field whose spatial derivatives are zero close to the crack tip, avoiding poorly evaluated stress calculations in the vicinity of the crack tip (Combescure et al., 2008).

### 2.7.5 Interaction Integral Method

In the interaction integral method, auxiliary fields are introduced and superimposed onto the actual fields satisfying the boundary value problem (Sih, Paris and Irwin, 1965). Stresses and strains for the auxiliary state should be chosen so as to satisfy both the equilibrium equation and the traction-free boundary condition on the crack surface in the $A$ area. These auxiliary fields are suitably selected in order to find a relationship between the mixed mode stress intensity factors and the interaction integrals. The contour $J$ integral for the sum of the two states can be defined as

$$J = J^{\text{act}} + J^{\text{aux}} + M \tag{2.163}$$

where $J^{\text{act}}$ and $J^{\text{aux}}$ are associated with the actual and auxiliary states, respectively, and $M$ is the interaction integral:

$$J^{\text{act}} = \int_A \left[ \sigma_{ij} \frac{\partial u_i}{\partial x_1} - w_s \delta_{1j} \right] \frac{\partial q}{\partial x_j} \, dA \tag{2.164}$$

$$J^{\text{aux}} = \int_A \left[ \sigma_{ij}^{\text{aux}} \frac{\partial u_i^{\text{aux}}}{\partial x_1} - w^{\text{aux}} \delta_{1j} \right] \frac{\partial q}{\partial x_j} \, dA \tag{2.165}$$

$$M = \int_A \left[ \sigma_{ij} \frac{\partial u_i^{\text{aux}}}{\partial x_1} + \sigma_{ij}^{\text{aux}} \frac{\partial u_i}{\partial x_1} - w^M \delta_{1j} \right] \frac{\partial q}{\partial x_j} \, dA \tag{2.166}$$

with the actual, auxiliary and interaction works defined as:

$$w_s = \frac{1}{2} \sigma_{ij} \varepsilon_{ij} \tag{2.167}$$

$$w^{\text{aux}} = \frac{1}{2} \sigma_{ij}^{\text{aux}} \varepsilon_{ij}^{\text{aux}} \tag{2.168}$$

$$w^M = \frac{1}{2} \left( \sigma_{ij} \varepsilon_{ij}^{\text{aux}} + \sigma_{ij}^{\text{aux}} \varepsilon_{ij} \right) \tag{2.169}$$

One of the choices for the auxiliary state is the displacement and stress fields in the vicinity of the crack tip. From the relation of the $J$ integral and mode I and II stress intensity factors,

$$J = \frac{1}{E'} \left( K_I^2 + K_{II}^2 \right) \tag{2.170}$$

the following relationship is obtained:

$$M = \frac{2}{E'} \left( K_I K_I^{\text{aux}} + K_{II} K_{II}^{\text{aux}} \right) \tag{2.171}$$

Therefore, the mode I and II stress intensity factors can be obtained from:

$$K = \frac{E'}{2} M \tag{2.172}$$

by choosing $K_I^{\text{aux}} = 1$, $K_{II}^{\text{aux}} = 0$ for mode I and $K_I^{\text{aux}} = 0$, $K_{II}^{\text{aux}} = 1$ for mode II.

It should be noted that evaluation of the interaction integral requires careful attention as the main fields are usually obtained from the finite element solution in a global or local (but crack-independent) coordinate system, while the auxiliary fields are defined in the local crack-tip polar coordinate system. Therefore, necessary transformations are required to use a unified coordinate system.

## 2.8  Elastoplastic Fracture Mechanics (EPFM)

Under the assumptions of linear elastic fracture mechanics, the stress at the crack tip is theoretically infinite. Consequently, it may usually lead to conservative and expensive solutions as it does not account for plastification at the crack tip. From a physical point of view, however, no material can withstand infinite stress, and a small plastic/fractured zone will be formed around the crack tip. As a result, an extension to ductile fracture or elastoplastic fracture mechanics (EFPM) is required.

In this section, first the problem of the size of the plastic zone is addressed using various levels of approximations, then the concepts of the crack opening displacement (COD) are discussed. Theoretically, the models can be extended to more sophisticated plasticity models for simulation of nonlinear material behaviour around the crack tip.

### 2.8.1  Plastic Zone

#### 2.8.1.1  First-Order Uniaxial Stress Criterion

Here, only the uniaxial stress state normal to the crack axis is considered. The simplest estimate of the size of the plastic zone is obtained by equating the uniaxial stress $\sigma_{yy}$ to the yield stress $\sigma_{yld}$. Recalling Eq. (2.49) (for plane stress problems),

$$\sigma_{yy} = \frac{K_I}{\sqrt{2\pi r}} \cos \frac{\theta}{2} \left( 1 + \sin \frac{\theta}{2} \sin \frac{3\theta}{2} \right) \tag{2.173}$$

the size of the plastic zone, $r = r_p$, can be obtained at $\theta = 0$ from

$$\sigma_{yy} = \frac{K_I}{\sqrt{2\pi r_p}} = \sigma_{yld} \tag{2.174}$$

$$r_p = \frac{1}{2\pi} \frac{K_I^2}{\sigma_{yld}^2} \tag{2.175}$$

or in terms of the stress intensity factor for a central crack in an infinite plate, $K_I = \sigma_0 \sqrt{\pi a}$

$$r_p = \frac{a}{2} \left( \frac{\sigma_0}{\sigma_{yld}} \right)^2 \tag{2.176}$$

Similarly, the size of the plastic zone for mode II for both plane stress and plane strain problems can be obtained from:

$$r_p = \frac{3}{2\pi} \frac{K_{II}^2}{\sigma_{yld}^2} \tag{2.177}$$

The main problem with this approximation is that it simply ignores all stresses exceeding $\sigma_{yld}$ (Figure 2.14a). As a result, equilibrium is no longer satisfied.

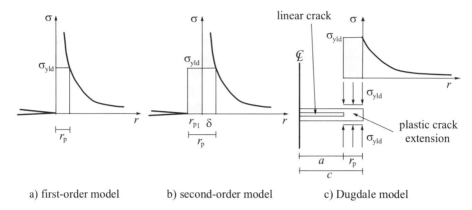

a) first-order model           b) second-order model           c) Dugdale model

**Figure 2.14** First- and second-order approximations of the plastic zone, and the Dugdale model (Mohammadi, 2008). (Reproduced by permission of John Wiley & Sons, Ltd.)

### 2.8.1.2  Second-Order Uniaxial Stress Criterion

In an alternative approach to avoid violation of equilibrium equations, Irwin (1961) developed a second-order approximation for the plastic zone based on the stress redistribution occurring at the crack tip. In this model, the area under the stress curve which was eliminated in the first order approach, is redistributed to satisfy equilibrium requirements (Figure 2.14b),

$$\int_0^\delta \frac{K_I}{\sqrt{2\pi r}} dr = r_p \sigma_{yld} = (\delta + r_{p_1}) \sigma_{yld} \tag{2.178}$$

The final solution is then obtained as:

$$\delta = r_{p_1} = \frac{1}{2\pi} \frac{K_I^2}{\sigma_{yld}^2} \tag{2.179}$$

or for a central crack in an infinite plate

$$\delta = r_{p_1} = \frac{a}{2} \left( \frac{\sigma_0}{\sigma_{yld}} \right)^2 \tag{2.180}$$

which means that

$$r_p = 2r_{p_1} \tag{2.181}$$

implying that the second-order size is twice the first-order size. Thus, the first-order solution $r_{p_1}$ (Eq. (2.175)) may still be used with an effective crack length of $\bar{a} = a + r_{p_1}$ extending to the centre of the plastic zone. Thus,

$$K_I = \sigma_0 \sqrt{\pi (a + r_{p_1})} \tag{2.182}$$

### 2.8.1.3 Dugdale Criterion

Another alternative to satisfy equilibrium equations was proposed by Dugdale (1960) based on the assumption of replacing the actual physical crack length ($2a$) by a total effective crack length $2c$, where $c = a + r_p$ (Figure 2.14c). Dugdale assumed that a closing constant stress $\sigma_{\text{yld}}$ is applied over the length $r_p$, causing a negative $K_{\text{yld}}$. As a result, the combined model requires the overall stress intensity factor to vanish: $K_{\text{combined}} = K + K_{\text{yld}} = 0$, determining the length $c$ or $r_p$.

Using the Westergaard's approach, the solution for $c$ is obtained from:

$$\frac{a}{c} = \cos\left(\frac{\pi}{2}\frac{\sigma_0}{\sigma_{\text{yld}}}\right) \tag{2.183}$$

Then, applying the Taylor's expansion and neglecting higher order terms results in (for plane stress):

$$r_p = \frac{\pi^2}{8}\left(\frac{\sigma_0}{\sigma_{\text{yld}}}\right)^2 a = \frac{\pi}{8}\left(\frac{K_I}{\sigma_{\text{yld}}}\right)^2 \tag{2.184}$$

and for plane strains $\sigma_{\text{yld}}$ is replaced by $\sqrt{3}\sigma_{\text{yld}}$.

### 2.8.1.4 First-Order Multiaxial Yield Criterion

In this section, the idea of a plastic zone is further extended to the first-order multiaxial conditions to include points other than $\theta = 0$. The idea is to assume that yielding would occur when the effective stress $\sigma_{\text{eff}}$ computed from any specified yield criterion reaches $\sigma_{\text{yld}}$.

In order to provide sample explicit solutions, the von Mises yield criterion in terms of principal stresses $\sigma_1$, $\sigma_2$ and $\sigma_3$ is considered:

$$\sigma_{\text{eff}} = \frac{1}{\sqrt{2}}[(\sigma_1 - \sigma_2)^2 + (\sigma_2 - \sigma_3)^2 + (\sigma_3 - \sigma_1)^2] \tag{2.185}$$

Relating the stress components to $K_1$ through Eqs. (2.69)–(2.71), the size of the plastic zone $r_p(\theta)$ can be evaluated for plane strain problems as:

$$r_p(\theta) = \frac{1}{4\pi}\frac{K_I}{\sigma_{\text{yld}}^2}\left[\frac{3}{2}\sin^2\theta + (1 - 2v)^2(1 + \cos\theta)\right] \tag{2.186}$$

and a much larger size for plane stress conditions:

$$r_p(\theta) = \frac{1}{4\pi}\frac{K_I}{\sigma_{\text{yld}}^2}\left[1 + \frac{3}{2}\sin^2\theta + \cos\theta\right] \tag{2.187}$$

## 2.8.2 Crack-Tip Opening Displacements (CTOD)

In a totally different approach, a local criterion based on the crack-tip opening displacement (CTOD) has been proposed to account for elastoplastic behaviour around the crack tip (Cottrell, 1961; Wells, 1963). In LEFM and brittle fracture, sharp cracks are considered and the CTOD

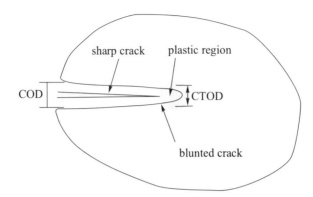

**Figure 2.15**  Estimation of the CTOD (Mohammadi, 2008). (Reproduced by permission of John Wiley & Sons, Ltd.)

is always zero. In contrast, CTOD is not negligible in ductile fracture and elastoplastic fracture mechanics due to relatively large deformation and blunting of the crack (Figure 2.15).

Similar to the method of determining the size of the plastic zone, two first- and second-order CTOD approaches are available.

### 2.8.2.1  First-Order CTOD

The first-order solution for CTOD is based on Irwin's solution for the mode I vertical displacement of a point next to the crack tip, as depicted in Figure 2.15.

$$u_y = \frac{K_I}{2\mu} \left[ \frac{r}{2\pi} \right]^{\frac{1}{2}} \sin \frac{\theta}{2} \left[ \kappa + 1 - 2\cos^2 \frac{\theta}{2} \right] \tag{2.188}$$

Substituting $\theta = \pm\pi$ results in evaluation of crack-tip opening (COD) at a distance $r$. Then, for $r = r_p$, where $r_p$ is the Irwin plastic zone, the final crack tip opening (CTOD) is estimated as:

$$\text{CTOD} = \frac{4}{\pi} \frac{K_I^2}{E\sigma_{\text{yld}}} \tag{2.189}$$

Recalling the definition of the energy release rate as $G = K^2/E'$, the following approximation can be assumed:

$$\text{CTOD} = \frac{G}{\sigma_{\text{yld}}} \tag{2.190}$$

### 2.8.2.2  Second-Order CTOD

The second-order CTOD formulation is based on the application of the second-order Dugdale model along the crack for determination of COD (Kanninen, 1984):

$$u_y(x) = \frac{2}{\pi} \frac{a\sigma_{\text{yld}}}{E} \left\{ \log \left| \frac{e + \sqrt{c^2 - x^2}}{e - \sqrt{c^2 - x^2}} \right| + \frac{x}{a} \log \left| \frac{xe + a\sqrt{c^2 - x^2}}{xe - a\sqrt{c^2 - x^2}} \right| \right\} \quad 0 \le x \le c \tag{2.191}$$

with $e^2 = c^2 - a^2$. Using Eqs. (2.191) and (2.183) for $x = a$ results in:

$$\text{CTOD} = 2u_y(x = a) = \frac{8}{\pi} \frac{a\sigma_{\text{yld}}}{E} \log\left[\sec\frac{\pi}{2}\frac{\sigma}{\sigma_{\text{yld}}}\right] = \frac{K^2}{E\sigma_{\text{yld}}}\left[1 + \frac{\pi^2}{24}\frac{\sigma^2}{\sigma_{\text{yld}}^2} + \cdots\right] \quad (2.192)$$

### 2.8.3   J Integral for EPFM

Although the $J$ integral was originally developed for linear elastic problems, it can be used for nonlinear monotonic problems, where no unloading or reverse loading is allowed. For details see Anderson (1995).

# 3

# Extended Finite Element Method

## 3.1  Introduction

The first enriched finite element model was proposed by Benzley (1974) who introduced the idea of enriching the near tip field in the finite element method using the asymptotic solution for static fracture problems. This work was continued by Atluri, Kobayashi and Nakagaki (1975a, 1975b) and Gifford and Hilton (1978) for stationary cracks. It took, however, until 1999 for a first practical enriched finite element method to be developed for general crack propagation problems in the form of the extended finite element method.

This chapter, which is an extended edition of a similar chapter in Mohammadi (2008), is devoted to a full discussion on various aspects of the extended finite element method. The goal has been set to describe in detail all theoretical and computational bases of the approach, including advantages and potential drawbacks. The discussion is limited to two-dimensional XFEM, although many of the basic formulations can be similarly extended to three-dimensional problems.

The introduction section begins with a broad review of general developments of the XFEM methodology, and will then be followed by specific examination of various composite applications, whose details will be comprehensively discussed in the next five chapters.

The main part of the chapter starts with the theoretical discussion on the fundamental bases of the partition of unity and the existing extensions to more advanced frameworks, primarily based on fracture mechanics applications. The concept of enrichment will then be explained in detail, based on the concept of partition of unity and the extended finite element method. It covers the basic techniques and definitions of enrichment functions, comprising crack-tip enrichments to reproduce asymptotic crack-tip fields, jump functions to approximate field discontinuity across a crack, and weak discontinuity functions to simulate a material discontinuity interface within a finite element. The XFEM approximation is then implemented to discretize the weak formulation of the boundary value problem and various standard and enriched components of the discretized stiffness matrix and the force vector are discussed. The section continues with addressing numerical implications regarding the Gauss quadrature rules, partitioning methods, and so on.

*XFEM Fracture Analysis of Composites*, First Edition. Soheil Mohammadi.
© 2012 John Wiley & Sons, Ltd. Published 2012 by John Wiley & Sons, Ltd.

Three comprehensive sections are devoted to discussing various enrichment functions. First, different options of the strong discontinuity enrichments are described, which include conventional definitions, tangential discontinuity, intersecting cracks, and so on. In the next section, the problem of weak discontinuity is analysed and appropriate enrichments are defined. Then, a very comprehensive list of crack-tip enrichments in a wide range of applications, from isotropic functions to orthotropic and inhomogeneous media, contact, viscoelasticity, piezoelectric applications, and dislocations within the static, dynamic and large deformation regimes, are presented and explained.

A brief section on tracking evolving boundaries, such as crack propagation paths, is necessary for most XFEM simulations. Among the existing algorithms, the Level set method, as a powerful alternative to classical approaches for representing crack paths and crack intersections, is examined.

The chapter concludes with the simulation and discussion of several engineering examples to illustrate the validity, robustness and efficiency of XFEM in dealing with various discontinuity and singularity problems. Specific problems related to cracking in composites will be dealt with in subsequent chapters.

## 3.2   Historic Development of XFEM

Despite the innovative idea of enriching the near tip field in the finite element method by Benzley (1974), Atluri, Kobayashi and Nakagaki (1975a, 1975b) and Gifford and Hilton (1978), the fundamentals of partition of unity finite element method and the extended finite element method were developed twenty years later by Melenk and Babuska (1996) and Belytschko and Black (1999). In less than a decade, XFEM has now become the main numerical approach for solution of all kinds of discontinuity problems.

### 3.2.1   A Review of XFEM Development

A brief review of the development of the basic idea of the extended finite element method is presented in this section. The issues related to composites will be dealt with in more detail in the next five chapters.

Melenk and Babuska (1996) discussed the mathematical bases of the partition of unity finite element method (PUFEM) and illustrated that PUFEM can be used to employ the structure of the differential equation under consideration to construct effective and robust methods. The same idea of PUFEM, but in a local nature, has become the theoretical basis of the local partition of unity finite element method, to be called later the extended finite element method.

The first effort at developing the extended finite element methodology was reported by Belytschko and Black (1999), presenting a minimal remeshing finite element method for crack growth. They added discontinuous enrichment functions to the finite element approximation to account for the presence of the crack. A local remeshing scheme was only required for severely curved arbitrary cracks to improve the solution.

The terminology of the extended finite element method (XFEM) was proposed by Moës, Dolbow and Belytschko (1999), as they improved the methodology by allowing mesh-independent representation of the entire crack based on the construction of the enriched approximation.

Another major step forward was achieved by Dolbow (1999) and Dolbow, Moës and Belytschko (2000a, 2000b, 2000c) for XFEM simulation of two-dimensional elasticity and Mindlin–Reissner plates using both a jump function and the asymptotic near tip fields. They also presented a technique to model arbitrary discontinuities in the finite element framework by locally enriching a displacement based approximation through a partition of unity method.

Further development was performed by Daux *et al.* (2000) by simulation of arbitrary branched and intersecting cracks with multiple branches, multiple holes and cracks emanating from holes, Sukumar and Prevost (2003), Huang, Sukumar and Prevost (2003) and Legay, Wang and Belytschko (2005) who extended XFEM to simulation of strong and weak discontinuities, Chessa, Wang and Belytschko (2003) by discussing the construction of blending elements, Lee *et al.* (2004) for modelling of stationary and growing cracks by superposition of the s-version FEM with XFEM, and Stazi *et al.* (2003) formulating higher order elements for curved cracks. Also, Liu, Xiao and Karihaloo (2004) proposed an improved XFEM by including higher order terms of the crack-tip asymptotic field in the enrichment approximation and applying a penalty function method to ensure the reproduction of the actual asymptotic fields. In addition, they adopted a direct approach for evaluation of mixed mode stress intensity factors (SIFs) without extra post processing.

The method was originally extended to three-dimensional crack modelling by Sukumar *et al.* (2000). Later, Areias and Belytschko (2005a) and Areias, Song and Belytschko (2006) developed new XFEM formulations for arbitrary crack propagation in shells with new definitions for rotational enrichments. Also, Bayesteh and Mohammadi (2011) studied the effect of various five crack tip enrichment terms for arbitrary cracking in general plates and pressurised shells.

Generalization of XFEM for arbitrary crack propagation problems required implementation of tracking techniques for moving boundaries. The level set method (LSM) has since become the primary approach for determining the crack path and the location of crack tips. Stolarska *et al.* (2001) introduced coupling the level set method with XFEM to model crack growth and Belytschko *et al.* (2001) proposed a technique for approximating arbitrary discontinuities in the main field and its derivatives in terms of a signed distance function, so level sets could be used to update the position of the discontinuities. A similar approach was adopted by Sukumar *et al.* (2001) to model holes and inclusions by level sets within an XFEM famwork. An extension to non-planar three-dimensional crack growth was developed by Moës, Gravouil and Belytschko (2002) and Gravouil, Moës and Belytschko (2002), based on the Hamilton–Jacobi equation to update the level sets with a velocity extension approach to preserve the old crack surface. Budyn *et al.* (2004) presented a combined XFEM/level set method for modelling homogeneous and inhomogeneous linear elastic media and Zi *et al.* (2004) discussed the junction of two cracks and analysed the growth and mixture of cracks in a quasi-brittle cell containing multiple cracks.

The level set method has not been the sole tracking technique within XFEM. The fast marching method (FMM) was coupled with the extended finite element method to develop a numerical technique for planar three-dimensional fatigue crack growth simulations (Sukumar, Chopp and Moran, 2003). Chopp and Sukumar (2003) extended the method to represent the entire multiple coplanar crack geometry by a single signed distance (level set) function. The method allowed merging of distinct cracks by FMM with no necessity of collision detection or mesh reconstruction procedures. A different approach for tackling the same set of problems was proposed by Ventura, Budyn and Belytschko (2003).

Following the initial success of the method, XFEM has been extensively applied in various engineering applications. The problem of cohesive cracks was studied by Moës and Belytschko (2002a), Zi and Belytschko (2003), Mergheim, Kuh and Steinmann (2005), Forghani (2005), de Borst, Remmers and Needleman (2004), de Borst *et al.* (2004a, 2004b) and Mahdavi and Mohammadi (2012) among others. Also, Bellec and Dolbow (2003) and Mariani and Perego (2003) adopted XFEM for simulation of cohesive crack propagation by assuming a cubic displacement discontinuity that allowed reproduction of the typical cusp-like shape of the process zone at the tip of a cohesive crack. Areias and Belytschko (2005a, 2005b) coupled XFEM with a viscosity-regularized continuum damage constitutive model to develop a regularized 'crack-band' version of XFEM. Recently, Mahdavi and Mohammadi (2012) have adopted a cohesive hydraulic crack model to simulate dynamic crack propagation in concrete dams.

Application of XFEM for crack analysis in composites was started by Dolbow and Nadeau (2002), Dolbow and Gosz (2002), Remmers, Wells and de Borst (2003), Sukumar *et al.* (2004), Nagashima, Omoto and Tani (2003) and Nagashima and Suemasu (2004), and then followed by Asadpoure, Mohammadi and Vafai (2006, 2007) and Asadpoure and Mohammadi (2007) by introducing new orthotropic enrichment functions, a beginning for future development of orthotropic enrichment functions in various applications, as presented in Section 3.2.2.

The localization problem has been studied by several authors including Jirásek and Zimmermann (2001a, 2001b), Jirásek (2002), Sukumar *et al.* (2003), Patzak and Jirásek (2003), Dumstorff and Meschke (2003), Samaniego and Belytschko (2005), Areias and Belytschko (2006), Song, Areias and Belytschko (2006), Ferrie *et al.* (2006), Ventura, Moran and Belytschko (2005), Larsson and Fagerström (2005) and Stolarska and Chopp (2003). Recently, Daneshyar and Mohammadi (2011, 2012a, 2012b) have developed a strong tangential discontinuity model for XFEM simulation of shear band initiation and propagation.

Modelling contact by XFEM was first introduced by Dolbow, Moës and Belytschko (2000c, 2001) and revisited by Belytschko, Daniel and Ventura (2002a). Khoei and Nikbakht (2006, 2007) applied the available formulation to modelling frictional contact problems. In addition, Ebrahimi, Mohammadi and Kani (2012) have recently proposed a numerical approach for determining the order of singularity of a sliding contact corner within the framework of partition of unity finite element method.

Introduction of plastic enrichment terms into XFEM was first reported by Elguedj, Gravouil and Combescure (2006) based on the Hutchinson–Rice–Rosengren (HRR) fields to represent the singularities in EPFM. Khoei, Anahid and Shahim (2008), Khoei *et al.* (2008) and Khoei, Biabanaki and Anahid (2008) have simulated several plasticity problems by XFEM. Further development into the viscoelastic regime was reported by Zhang, Rong and Li (2010) and TianTang and QingWen (2011) and followed by Hajikarimi, Mohammadi and Aflaki (2012) for simulation of crack healing in pavement applications.

Many researchers have tackled the XFEM within the large deformation regime, among them Dolbow and Devan (2004), Legrain, Moës and Verron (2005), Mergheim, Kuh and Steinmann (2006), Fagerström and Larsson (2006), Anahid and Khoei (2008), Khoei, Anahid and Shahim (2008), Khoei, Biabanaki and Anahid (2008), Khoei *et al.* (2008) and Rashetnia, Mohammadi and Mahmoudzadeh Kani (2012), presenting various aspects of geometrically nonlinear XFEM analyses.

Damage mechanics has recently attracted much attention within the XFEM community (Duarte, 2011; Jin *et al.*, 2011; Moes *et al.*, 2011). Combescure *et al.* (2011) discussed different problems of transition from brittle damage to ductile fracture with cohesive XFEM, and Moes (2011) proposed a thick level set approach for modelling damage growth and its transition to fracture. Also, Broumand and Khoei (2011) combined the damage plasticity and XFEM to simulate ductile fracture problems and Hatefi and Mohammadi (2012) adopted XFEM for damage analysis of orthotropic materials.

XFEM has been extensively adopted for dynamic fracture analysis. Dynamic XFEM was proposed by Belytschko *et al.* (2003), Belytschko and Chen (2004) and Zi *et al.* (2005) based on the singular enrichment finite element method for elastodynamic crack propagation. Also, Réthoré, Gravouil and Combescure (2005a) proposed a generalized XFEM to model dynamic fracture and time-dependent problems. Later, Menouillard *et al.* (2006) presented an explicit XFEM by introduction of a lumped mass matrix for enriched elements. In addition, Motamedi and Mohammadi (2010a, 2010b, 2012) developed new orthotropic enrichment functions for moving cracks, and Parchei, Mohammadi and Zafarani (2012) used the dynamic XFEM combined with a contact approach to simulate a near field solution for various induced fault rupture problems.

The concept of spatial discontinuity enrichment has also been extended to the time domain by introduction of a locally enriched space–time extended finite element method by Chessa and Belytschko (2004, 2006) or by solving hyperbolic problems with discontinuities. The coupling was implemented through a weak enforcement of the continuity of the flux between the space–time and semi-discrete domains in a manner similar to discontinuous Galerkin methods. They successfully applied the TXFEM to the Rankine–Hugoniot jump conditions for linear first order wave and nonlinear Burgers' equations. Furthermore, Réthoré, Gravouil and Combescure (2005b) proposed a combined space–time extended finite element method, based on the idea of the time extended finite element method, allowing a suitable form of the time stepping formulae to study stability and energy conservation. The proposed TXFEM was successfully adopted by Kabiri (2009) and Rezaei (2010) for simulating various dynamic fracture problems.

A number of XFEM investigations have been directed towards the study of accuracy, stability and convergence. Laborde *et al.* (2005) and Chahine, Laborde and Renard (2006) studied the convergence of a variety of XFEM models on cracked domains. Peters and Hack (2005) discussed the ways that a singular stiffness matrix may be avoided by deleting some of the enhanced degrees of freedom, while Béchet *et al.* (2005) proposed a geometrical enrichment instead of the usual topological one in which a given domain size would be enriched even if the elements did not touch the crack front. Ventura (2006) showed how the standard Gauss quadrature rule can be accurately used in the elements containing the discontinuity without splitting the elements into subdomains or introducing any additional approximation.

Xiao and Karihaloo (2006) discussed improving the accuracy of XFEM crack-tip fields using higher order quadrature and statically admissible stress recovery procedures. They proposed a statically admissible stress recovery (SAR) scheme to be constructed by basis functions and moving least squares (MLS) to fit the stresses at sampling points (e.g. quadrature points) obtained by XFEM.

In addition to the first text book on the subject (Mohammadi, 2008), there are also a number of available review papers and comprehensive references such as Moës and Belytschko

(2002b), Karihaloo and Xiao (2003), Bordas and Legay (2005), Rabczuk and Wall (2006), Chahine (2009) and Pommier *et al.* (2011), which can be referred to for further details.

### 3.2.2 A Review of XFEM Composite Analysis

The first applications of XFEM for composites were reported by Dolbow and Nadeau (2002), Dolbow and Gosz (2002) and Remmers, Wells and de Borst (2003), while Nagashima and Suemasu (2004, 2006) simulated thin layered composites and concluded that the isotropic enrichment functions cannot represent the asymptotic solution for a crack in an orthotropic material. Then, Sukumar *et al.* (2004), developed partition of unity based enrichment functions for bimaterial interface cracks between two isotropic media, as a special type of multilayer composites. Later, Hettich and Ramm (2006) simulated the delamination crack as a jump in the displacement field without using any crack tip enrichment.

Also, a number of XFEM simulations have focused on thermo-mechanical analysis of orthotropic FGMs (Dag, Yildirim and Sarikaya, 2007), 3D isotropic FGMs (Ayhan, 2009; Zhang *et al.*, 2011; Moghaddam, Ghajar and Alfano, 2011) and frequency analysis of cracked isotropic FGMs (Natarajan *et al.*, 2011). The most recent work is reported by Bayesteh and Mohammadi (2012) on general crack propagations in orthotropic FGM composites. They have discussed the effect of crack-tip enrichments and various types of interaction integral for determining the mixed mode stress intensity factors.

Development of orthotropic crack-tip enrichment functions was reported in a series of papers by Asadpoure, Mohammadi and Vafai (2006, 2007), Asadpoure and Mohammadi (2007) and Mohammadi and Asadpoure (2006). They developed three types of enrichment functions for various classes of composites. Ebrahimi, Mohammadi and Assadpoure (2008) used the same enrichments to analyse several composite crack propagation problems. Similar enrichment functions were implemented in meshless EFG method by Ghorashi, Mohammadi and Sabbagh-Yazdi (2011) to analyse crack propagation in composites.

Later, Motamedi (2008) and Motamedi and Mohammadi (2010a, 2010b) studied the dynamic crack stability and propagation in composites based on static orthotropic enrichment functions, while Motamedi and Mohammadi (2012) developed new orthotropic dynamic enrichment functions for moving cracks in orthotropic composites.

In addition, Esna Ashari (2009) and Esna Ashari and Mohammadi (2009, 2010b, 2011a, 2011b, 2012) further extended the method for orthotropic bimaterial interfaces. They also adopted the same functions to analyse delamination in FRP-strengthened structures and to predict the crack arrest mechanism and the load bearing capacity in terms of the debonded length.

## 3.3 Enriched Approximations

Enriched approximations have been the subject of several computational studies in the past two decades. Most of them are discussed within the framework of partition of unity.

### 3.3.1 Partition of Unity

Since the introduction of the concept of partition of unity, it has been frequently used in various disciplines as a means of enhancing an original numerical approximation. A partition of unity

(PU) is defined as a set of $m$ functions $f_k(\mathbf{x})$ within a domain $\Omega_{pu}$ such that (Melenk and Babuska, 1996),

$$\sum_{k=1}^{m} f_k(\mathbf{x}) = 1 \tag{3.1}$$

It can easily be shown that by selection of any arbitrary function $\psi(\mathbf{x})$, partition of unity functions $f_k(\mathbf{x})$ hold the following reproducing property,

$$\sum_{k=1}^{m} f_k(\mathbf{x})\psi(\mathbf{x}) = \psi(\mathbf{x}) \tag{3.2}$$

This is equivalent to the definition of completeness of a specific order $l$, where the approximating PU functions $f_k(\mathbf{x})$ can represent exactly the $l$-order polynomial $\psi(\mathbf{x}) = p^{(l)}(\mathbf{x})$. Then, zero completeness is achieved if Eq. (3.2) holds for a constant $p(\mathbf{x})$.

Isoparametric finite element shape functions, $N_j$, are the most frequently used PU functions,

$$\sum_{j=1}^{n} N_j(\mathbf{x}) = 1 \tag{3.3}$$

where $n$ is the number of nodes for each finite element.

The concept of partition of unity, as will be discussed in the next section, provides a mathematical framework for the development of enriched solutions such as the partition of unity finite element method and the extended finite element method.

### 3.3.2   Intrinsic and Extrinsic Enrichments

Theoretically, enrichment can be regarded as the principal of increasing the order of completeness or the type of reproduction that can be achieved. The choice of the enrichment functions depends on the *a priori* solution of the problem. For instance, in a crack analysis this is equivalent to an increase in accuracy of the approximation if analytical near crack-tip solutions are somehow included in the enrichment terms. Computationally, it aims at higher accuracy of the approximation by including the information extracted from the analytical solution into the basic numerical approximation.

Let us begin with the classical approximation of a field variable $\mathbf{u}$ within a finite element method:

$$\mathbf{u} = \sum_{j=1}^{n} \mathbf{N}_j \bar{\mathbf{u}}_j \tag{3.4}$$

or in a more appropriate form in terms of the $m$ basis functions $\mathbf{p}$,

$$\mathbf{u} = \mathbf{p}^T \mathbf{a} = \sum_{k=1}^{m} p_k \mathbf{a}_k \tag{3.5}$$

where unknowns $\mathbf{a}_k$ are determined from the approximation at nodal points. The basis function $\mathbf{p}$ may be defined for different orders of completeness:

$$
\begin{cases}
1D : \begin{cases}
\mathbf{p}^T = \{1, x\} & \text{1st order} \\
\mathbf{p}^T = \{1, x, x^2\} & \text{2nd order} \\
\mathbf{p}^T = \{1, x, x^2, x^3\} & \text{3rd order}
\end{cases} \\[2em]
2D : \begin{cases}
\mathbf{p}^T = \{1, x, y\} & \text{1st order} \\
\mathbf{p}^T = \{1, x, y, x^2, xy, y^2\} & \text{2nd order} \\
\mathbf{p}^T = \{1, x, y, x^2, xy, y^2, x^3, x^2y, xy^2, y^3\} & \text{3rd order}
\end{cases}
\end{cases}
\tag{3.6}
$$

The basic idea of the enrichment is to transform Eqs. (3.4) or (3.5) into a more appropriate form to enhance the accuracy of numerical approximation. The enhancement may be attributed to the degree of consistency of the approximation, or to the capability of approximation to reproduce a given complex field of interest.

There are basically two classes of enrichments: enriching the basis function $\mathbf{p}$ (intrinsic enrichment) and adding enrichment terms to the whole approximation (extrinsic enrichment). The following sections discuss both approaches.

To further elaborate the concept of enrichment, the classical crack problem of Section 2.4 is considered, and the enrichment concept is explained for this specific case. The asymptotic near tip displacement field can be written as:

$$
u_x = \frac{1}{\mu}\sqrt{\frac{r}{2\pi}}\left(K_I \cos\frac{\theta}{2}(\kappa - \cos\theta) + K_{II}\sin\frac{\theta}{2}(\kappa + \cos\theta + 2)\right)
\tag{3.7}
$$

$$
u_y = \frac{1}{\mu}\sqrt{\frac{r}{2\pi}}\left(K_I \sin\frac{\theta}{2}(\kappa - \cos\theta) - K_{II}\cos\frac{\theta}{2}(\kappa + \cos\theta - 2)\right)
\tag{3.8}
$$

where $r$ and $\theta$ are defined in Figure 3.1, and $K_I$ and $K_{II}$ are the mode I and II stress intensity factors, respectively.

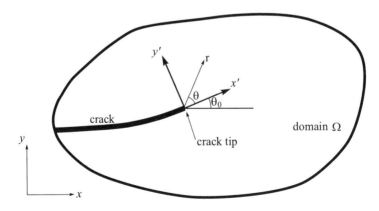

**Figure 3.1**  Global, local and polar coordinates at the crack tip.

### 3.3.2.1 Intrinsic Enrichment

In this approach, new terms are added to the basis function $\mathbf{p}$ of approximation (3.5) to satisfy a certain condition of reproducing a complex field (Fries and Belytschko, 2006). For instance, for a first-order 2D linear basis function $\mathbf{p}^{\text{lin}} = \{1, x, y\}$, new enrichment terms $\mathbf{p}^{\text{enr}} = \{f_1, f_2, f_3, f_4\}$ can be included:

$$\mathbf{p} = \{\mathbf{p}^{\text{lin}}, \mathbf{p}^{\text{enr}}\} = \{1, x, y, f_1, f_2, f_3, f_4\} \tag{3.9}$$

It can be shown that the asymptotic near crack-tip displacement field (Eqs. (3.7) and (3.8)) can be expressed by the following basis function $\mathbf{p}^{\text{enr}}(\mathbf{x})$, defined in the polar coordinate system,

$$\mathbf{p}^{\text{enr}}(\mathbf{x}) = \{f_1, f_2, f_3, f_4\} = \left[ \sqrt{r} \sin \frac{\theta}{2}, \sqrt{r} \cos \frac{\theta}{2}, \sqrt{r} \sin \theta \sin \frac{\theta}{2}, \sqrt{r} \sin \theta \cos \frac{\theta}{2} \right] \tag{3.10}$$

The basis function for the total solution must include the constant and linear terms:

$$\mathbf{p}^{\text{T}}(\mathbf{x}) = \left[ 1, x, y, \sqrt{r} \sin \frac{\theta}{2}, \sqrt{r} \cos \frac{\theta}{2}, \sqrt{r} \sin \theta \sin \frac{\theta}{2}, \sqrt{r} \sin \theta \cos \frac{\theta}{2} \right] \tag{3.11}$$

which is a familiar basis function previously used for fracture analysis by the meshless element free Galerkin (EFG) method (Belytschko, Lu and Gu, 1994),

$$\mathbf{u}^h(\mathbf{x}) = \mathbf{p}^{\text{T}}(\mathbf{x})\mathbf{a}(\mathbf{x}) \tag{3.12}$$

where $\mathbf{a}(\mathbf{x})$ is the vector of unknown coefficients, constant or variable, obtained from the weighted least squares (WLS) or moving least squares (MLS) techniques, respectively, for minimizing the overall error of approximation (Onate et al., 1995).

### 3.3.2.2 Extrinsic Enrichment

Another form of enrichment is based on the so-called extrinsic enrichment, which adds a set of extra terms $\mathbf{p}^{\text{enr}}\mathbf{a}^{\text{enr}}$ to the standard term $\mathbf{pa}$ to increase the order of completeness. It can then be defined in the following explicit form

$$\mathbf{u}^h(\mathbf{x}) = \sum_{j=1}^{n} N_j(\mathbf{x})\mathbf{u}_j + \sum_{k=1}^{m} p_k(\mathbf{x})\mathbf{a}_k \tag{3.13}$$

where $\mathbf{a}_k$ are additional unknowns or degrees of freedom associated with the enriched solution. In a general partition of unity enrichment, Eq. (3.13) is rewritten as,

$$\mathbf{u}^h(\mathbf{x}) = \sum_{j=1}^{n} N_j(\mathbf{x})\mathbf{u}_j + \sum_{k=1}^{m} f_k^{\text{pu}}(\mathbf{x})p(\mathbf{x})\mathbf{a}_k \tag{3.14}$$

where $f_k^{\text{pu}}(\mathbf{x})$ are the set of the partition of unity functions.

### 3.3.3  Partition of Unity Finite Element Method

The partition of unity finite element method (PUFEM) (Melenk and Babuska, 1996) adopts the same concept of extrinsic enrichment (3.14) by the use of classical finite element shape functions as the partition of unity functions: $f^{pu}(\mathbf{x}) = N(\mathbf{x})$. For a general point $\mathbf{x}$ within a finite element,

$$\mathbf{u}^h(\mathbf{x}) = \sum_{j=1}^{n} N_j(\mathbf{x}) \left( \mathbf{u}_j + \sum_{k=1}^{m} p_k(\mathbf{x}) \mathbf{a}_{jk} \right) \tag{3.15}$$

Approximation (3.15) is clearly a partition of unity, which guarantees a compatible solution. Examining the approximate solution (3.15) for a typical enriched node $\mathbf{x}_i$ leads to:

$$\mathbf{u}^h(\mathbf{x}_i) = \sum_{j=1}^{n} N_j(\mathbf{x}_i) \left( \mathbf{u}_j + \sum_{k=1}^{m} p_k(\mathbf{x}_i) \mathbf{a}_{jk} \right) \tag{3.16}$$

where the first part vanishes, except for $N_i(\mathbf{x}_i)\mathbf{u}_i = \mathbf{u}_i$. Therefore,

$$\mathbf{u}^h(\mathbf{x}_i) = \mathbf{u}_i + \sum_{k=1}^{m} p_k(\mathbf{x}_i) \mathbf{a}_{ik} \tag{3.17}$$

which is not a feasible interpolation. To satisfy interpolation at nodal points, Eq. (3.15) is transformed to:

$$\mathbf{u}^h(\mathbf{x}) = \sum_{j=1}^{n} N_j(\mathbf{x}) \left( \mathbf{u}_j + \sum_{k=1}^{m} \left( p_k(\mathbf{x}) - p_k(\mathbf{x}_j) \right) \mathbf{a}_j \right) \tag{3.18}$$

which ensures $\mathbf{u}^h(\mathbf{x}_i) = \mathbf{u}_i$.

### 3.3.4  MLS Enrichment

The meshless moving least-square approximation (MLS) can also be enriched in an extrinsic form, as proposed by Duarte and Oden (1995) in the form of the meshless Hp-clouds,

$$\mathbf{u}^h(\mathbf{x}) = \sum_{j=1}^{\bar{n}} \Phi_j(\mathbf{x}) \left( \mathbf{u}_j + \sum_{k=1}^{m} p_k(\mathbf{x}) \mathbf{a}_{jk} \right) \tag{3.19}$$

where $\Phi_j(\mathbf{x})$ are the MLS shape functions evaluated over a moving support domain procedure, $\mathbf{a}_{jk}$ are additional degrees of freedom introduced to enrich the domain of interest, and $\bar{n}$ is the number of nodes within each MLS support domain.

### 3.3.5  Generalized Finite Element Method

In the generalized finite element method (GFEM), different shape functions are used for the classical and enriched parts of the approximation. Beginning with Eq. (3.15),

$$\mathbf{u}^h(\mathbf{x}) = \sum_{j=1}^{n} N_j(\mathbf{x})\mathbf{u}_j + \sum_{j=1}^{n} N_j(\mathbf{x})\left(\sum_{k=1}^{m} p_k(\mathbf{x})\mathbf{a}_{jk}\right) \tag{3.20}$$

the generalized form can then be written as:

$$\mathbf{u}^h(\mathbf{x}) = \sum_{j=1}^{n} N_j(\mathbf{x})\mathbf{u}_j + \sum_{j=1}^{n} \bar{N}_j(\mathbf{x})\left(\sum_{k=1}^{m} p_k(\mathbf{x})\mathbf{a}_{jk}\right) \tag{3.21}$$

where $\bar{N}_j(\mathbf{x})$ are the new set of shape functions associated with the enrichment part of the approximation.

Despite the fact that approximation (3.21) does not satisfy the partition of unity condition, it provides some flexibility for the solution by appropriate selection of $\bar{N}$ to enforce specific conditions or constraints.

### 3.3.6  Extended Finite Element Method

In contrast to PUFEM and GFEM, where the enrichments are usually employed on a global level and over the entire domain, the extended finite element method adopts the same procedure on a local level.

Assumption of the approximation (3.20) generates a compatible solution even if a local partition of unity is adopted. This is a considerable computational advantage as it is equivalent to enriching only nodes close to a crack tip; a basis for the extended finite element solution. The extended finite element method will be comprehensively discussed in Section 3.4.

### 3.3.7  Generalized PU Enrichment

The original PU enrichment (3.15) can be further generalized if a number of different PU support domains $\Omega_{pu}^l$ and associated partition of unity functions $f_k^l(\mathbf{x})$ are used for the enrichment:

$$\mathbf{u}^h(\mathbf{x}) = \sum_{j=1}^{n} N_j(\mathbf{x})\mathbf{u}_j + \sum_{l=1}^{np} \sum_{k=1}^{m} f_k^l(\mathbf{x})p^l(\mathbf{x})\mathbf{a}_k^l \tag{3.22}$$

where $\mathbf{a}_k^l$ are the additional unknowns associated with each set of $np$ domains of partition of unity.

## 3.4  XFEM Formulation

The basic concept of XFEM is to enrich the approximation space so that it becomes capable of reproducing certain features of the problem of interest, in particular discontinuities such as cracks or interfaces. Although it is a local version of the partition of unity finite element enrichment applied only in a certain local subdomain, it has strongly relied on the development of extrinsic enrichments for crack simulations by a number of meshless methods such as EFG and Hp-clouds. Naturally, the first XFEM approximations were also developed for simulation of strong discontinuities in fracture mechanics. This was later extended to include weak discontinuity and interface problems. XFEM can be assumed to be a classical FEM capable of handling arbitrary strong and weak discontinuities.

In the extended finite element method, first, the usual finite element mesh is produced. Then, by considering the location of discontinuities, a few degrees of freedom are added to the classical finite element model in selected nodes near to the discontinuities to provide a higher level of accuracy.

### 3.4.1  Basic XFEM Approximation

Consider **x,** a point in a finite element model. Also assume there is a discontinuity or singularity in the arbitrary domain discretized into some $n$ node finite elements. In the extended finite element method, the following approximation is utilized to calculate the displacement for the point **x** locating within the domain (Belytschko and Black, 1999)

$$\mathbf{u}^h(\mathbf{x}) = \mathbf{u}^{\text{FE}} + \mathbf{u}^{\text{enr}} = \sum_{j=1}^{n} N_j(\mathbf{x})\mathbf{u}_j + \sum_{k=1}^{m} N_k(\mathbf{x})\psi(\mathbf{x})\mathbf{a}_k \qquad (3.23)$$

where $\mathbf{u}_j$ is the vector of regular degrees of nodal freedom in the finite element method, $\mathbf{a}_k$ is the set of degrees of freedom added to the standard finite element model and $\psi(\mathbf{x})$ is the set of enrichment functions defined for the set of nodes included in the influence (support) domain of the discontinuity or singularity. For instance, the influence domain associated with node A in Figure 3.2 consists of the elements containing that node, whereas for an interior node B (usually in higher order elements) it is the element surrounding the node.

The first term on the right-hand side of Eq. (3.23) is the classical finite element approximation to determine the displacement field, while the second term is the enrichment approximation which takes into account the existence of any discontinuities. The second term utilizes additional degrees of freedom to facilitate modelling of the existence of any discontinuous field, such as a crack, without modelling it explicitly in the finite element mesh (see Figure 3.2).

The enrichment function $\psi(\mathbf{x})$ can be chosen by applying appropriate analytical solutions according to the type of discontinuity or singularity. Some of the main objectives for using various types of enrichment functions within an XFEM procedure can be expressed as:

1. Reproducing the singular field around a crack tip.
2. Continuity in displacement between adjacent finite elements.
3. Independent strain fields in two different sides of a crack surface or a material interface.
4. Other features according to the specific discontinuity problem.

For $np$ multiple discontinuities within a finite element, the approximation (3.23) can be further extended to:

$$\mathbf{u}^h(\mathbf{x}) = \mathbf{u}^{\text{FE}} + \mathbf{u}^{\text{enr}} = \sum_{j=1}^{n} N_j(\mathbf{x})\mathbf{u}_j + \sum_{l=1}^{np}\sum_{k=1}^{m} N_k(\mathbf{x})\psi^l(\mathbf{x})\mathbf{a}_k^l \qquad (3.24)$$

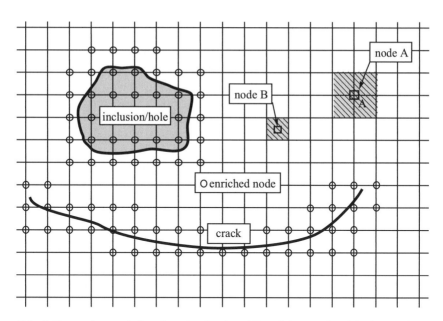

**Figure 3.2**  Influence (support) domains of nodes A and B and the set of enriched nodes for a typical crack and inclusion.

## 3.4.2   Signed Distance Function

The distance $d$ from a point $\mathbf{x}$ to an interface $\Gamma$ is defined as,

$$d = \|\mathbf{x} - \mathbf{x}_\Gamma\| \tag{3.25}$$

where $\mathbf{x}_\Gamma$ is the normal projection of $\mathbf{x}$ on $\Gamma$ (Figure 3.3). The signed distance function $\xi(\mathbf{x})$ can then be defined as,

$$\xi(\mathbf{x}) = \underbrace{\min \|\mathbf{x} - \mathbf{x}_\Gamma\|}_{\mathbf{x}_\Gamma \in \Gamma} \, sign\,(\mathbf{n} \cdot (\mathbf{x} - \mathbf{x}_\Gamma)) \tag{3.26}$$

where $\mathbf{n}$ is the unit normal vector.

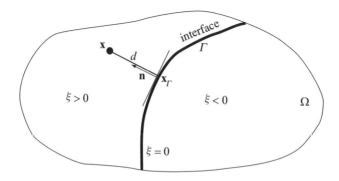

**Figure 3.3**  Definition of the signed distance function.

### 3.4.3    Modelling the Crack

In order to model crack surfaces and crack tips in the extended finite element method, the approximate displacement function $\mathbf{u}^h$ can be expressed in terms of the classical $\mathbf{u}$, crack-split $\mathbf{u}^{\mathrm{H}}$, crack-tip $\mathbf{u}^{\mathrm{tip}}$ and material interface $\mathbf{u}^{\mathrm{mat}}$ components as

$$\mathbf{u}^h(\mathbf{x}) = \mathbf{u}(\mathbf{x}) + \mathbf{u}^{\mathrm{H}}(\mathbf{x}) + \mathbf{u}^{\mathrm{tip}}(\mathbf{x}) + \mathbf{u}^{\mathrm{mat}}(\mathbf{x}) \tag{3.27}$$

or more explicitly

$$\mathbf{u}^h(\mathbf{x}) = \left[ \sum_{j=1}^{n} N_j(\mathbf{x})\mathbf{u}_j \right] + \left[ \sum_{h=1}^{mh} N_h(\mathbf{x})H(\mathbf{x})\mathbf{a}_h \right]$$
$$+ \left[ \sum_{k=1}^{mt} N_k(\mathbf{x}) \left( \sum_{l=1}^{mf} F_l(\mathbf{x})\mathbf{b}_k^l \right) \right] + \left[ \sum_{m=1}^{mm} N_m(\mathbf{x})\chi(\mathbf{x})\mathbf{c}_m \right] \tag{3.28}$$

and in a simple shifted form for $H$ and $F$ functions to guarantee the interpolation:

$$\mathbf{u}^h(\mathbf{x}) = \left[ \sum_{j=1}^{n} N_j(\mathbf{x})\mathbf{u}_j \right] + \left[ \sum_{h=1}^{mh} N_h(\mathbf{x}) \left( H(\mathbf{x}) - H(\mathbf{x}_h) \right) \mathbf{a}_h \right]$$
$$+ \left[ \sum_{k=1}^{mt} N_k(\mathbf{x}) \left( \sum_{l=1}^{mf} [F_l(\mathbf{x}) - F_l(\mathbf{x}_k)] \mathbf{b}_k^l \right) \right] + \left[ \sum_{m=1}^{mm} N_m(\mathbf{x})\chi_m(\mathbf{x})\mathbf{c}_m \right] \tag{3.29}$$

In Eqs. (3.28) and (3.29) $n$ is the number of nodes of each finite element with classical degrees of freedom $\mathbf{u}_j$ and shape functions $N_j(\mathbf{x})$, $mh$ is the number of nodes that have crack face (but not crack tip) in their support domain, $\mathbf{a}_h$ is the vector of additional degrees of nodal freedom for modelling crack faces (not crack tips) by the Heaviside function $H(\mathbf{x})$, $mt$ is the number of nodes associated with the crack tip in its influence domain, $\mathbf{b}_k^l$ is the vector of additional degrees of nodal freedom for modelling crack tips, $F_l(\mathbf{x})$ are crack-tip enrichment functions, $mm$ is the number of nodes that are directly affected by the weak discontinuity, $\mathbf{c}_m$ is the vector of additional degrees of nodal freedom for modelling weak discontinuity interfaces and $\chi_m(\mathbf{x})$ is the enrichment function used for modelling weak discontinuities.

The most conventional form of the Heaviside function $H$ is defined as,

$$H(\xi) = \mathrm{sign}\,(\xi) = \begin{cases} 1 & \forall \xi > 0 \\ -1 & \forall \xi < 0 \end{cases} \tag{3.30}$$

and for the crack-tip enrichments $F_\alpha$ in the local crack-tip polar coordinate system,

$$F_\alpha(r, \theta) = \left\{ \sqrt{r} \sin \frac{\theta}{2}, \sqrt{r} \cos \frac{\theta}{2}, \sqrt{r} \sin \theta \sin \frac{\theta}{2}, \sqrt{r} \sin \theta \cos \frac{\theta}{2} \right\} \tag{3.31}$$

Finally, the weak discontinuity function $\chi_m(\mathbf{x})$ is defined as,

$$\chi_m(\mathbf{x}) = |\xi(\mathbf{x})| - |\xi(\mathbf{x}_m)| \tag{3.32}$$

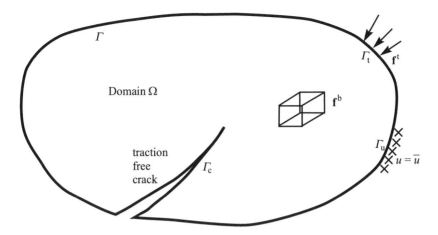

**Figure 3.4** A body in a state of elastostatic equilibrium.

Details of various options for the Heaviside strong discontinuity, material weak discontinuity and crack-tip functions are comprehensively discussed in Sections 3.5, 3.6 and 3.7, respectively.

### 3.4.4 Governing Equation

Consider a body in the state of equilibrium with the boundary conditions in the form of traction and displacement conditions, as depicted in Figure 3.4.

The strong form of the equilibrium equation can be written as:

$$\nabla \cdot \boldsymbol{\sigma} + \mathbf{f}^{\mathrm{b}} = 0 \quad \text{in } \Omega \tag{3.33}$$

with the following boundary conditions:

$$\boldsymbol{\sigma} \cdot \mathbf{n} = \mathbf{f}^{\mathrm{t}} \quad \text{on } \Gamma_{\mathrm{t}}: \text{external traction} \tag{3.34}$$

$$\mathbf{u} = \bar{\mathbf{u}} \quad \text{on } \Gamma_{\mathrm{u}}: \text{prescribed displacement} \tag{3.35}$$

$$\boldsymbol{\sigma} \cdot \mathbf{n} = 0 \quad \text{on } \Gamma_{\mathrm{c}}: \text{traction free crack} \tag{3.36}$$

where $\Gamma_{\mathrm{t}}$, $\Gamma_{\mathrm{u}}$ and $\Gamma_{\mathrm{c}}$ are traction, displacement and crack boundaries, respectively, $\boldsymbol{\sigma}$ is the stress tensor and $\mathbf{f}^{\mathrm{b}}$ and $\mathbf{f}^{\mathrm{t}}$ are the body force and external traction vectors, respectively.

The variational formulation of the boundary value problem can be defined as:

$$W^{\mathrm{int}} = W^{\mathrm{ext}} \tag{3.37}$$

or

$$\int_{\Omega} \boldsymbol{\sigma} \cdot \delta\boldsymbol{\varepsilon} \, \mathrm{d}\Omega = \int_{\Omega} \mathbf{f}^{\mathrm{b}} \cdot \delta\mathbf{u} \, \mathrm{d}\Omega + \int_{\Gamma_t} \mathbf{f}^{\mathrm{t}} \cdot \delta\mathbf{u} \, \mathrm{d}\Gamma \tag{3.38}$$

### 3.4.5   XFEM Discretization

Discretization of Eq. (3.38) using the XFEM procedure (3.28) results in a discrete system of linear equilibrium equations:

$$\mathbf{K}\mathbf{u}^h = \mathbf{f} \tag{3.39}$$

where $\mathbf{K}$ is the stiffness matrix, $\mathbf{u}^h$ is the vector of degrees of nodal freedom (for both classical and enriched ones) and $\mathbf{f}$ is the vector of external force. The global matrix and vectors are calculated by assembling the matrix and vectors of each element. $\mathbf{K}$ and $\mathbf{f}$ for each element $e$ are defined as

$$\mathbf{K}_{ij}^e = \begin{bmatrix} \mathbf{K}_{ij}^{uu} & \mathbf{K}_{ij}^{ua} & \mathbf{K}_{ij}^{ub} & \mathbf{K}_{ij}^{uc} \\ \mathbf{K}_{ij}^{au} & \mathbf{K}_{ij}^{aa} & \mathbf{K}_{ij}^{ab} & \mathbf{K}_{ij}^{ac} \\ \mathbf{K}_{ij}^{bu} & \mathbf{K}_{ij}^{ba} & \mathbf{K}_{ij}^{bb} & \mathbf{K}_{ij}^{bc} \\ \mathbf{K}_{ij}^{cu} & \mathbf{K}_{ij}^{ca} & \mathbf{K}_{ij}^{cb} & \mathbf{K}_{ij}^{cc} \end{bmatrix} \tag{3.40}$$

$$\mathbf{f}_i^e = \left\{ \mathbf{f}_i^u \quad \mathbf{f}_i^a \quad \mathbf{f}_i^{b1} \quad \mathbf{f}_i^{b2} \quad \mathbf{f}_i^{b3} \quad \mathbf{f}_i^{b4} \quad \mathbf{f}_i^c \right\}^T \tag{3.41}$$

and $\mathbf{u}^h$ is the vector of nodal parameters:

$$\mathbf{u}^h = \{ \mathbf{u} \quad \mathbf{a} \quad \mathbf{b}_1 \quad \mathbf{b}_2 \quad \mathbf{b}_3 \quad \mathbf{b}_4 \quad \mathbf{c} \}^T \tag{3.42}$$

with

$$\mathbf{K}_{ij}^{rs} = \int_{\Omega^e} (\mathbf{B}_i^r)^T \mathbf{D} \mathbf{B}_j^s \, d\Omega \quad (r, s = \mathbf{u}, \mathbf{a}, \mathbf{b}, \mathbf{c}) \tag{3.43}$$

$$\mathbf{K}_{ij}^{ac} = \mathbf{K}_{ij}^{ca} = 0 \tag{3.44}$$

and

$$\mathbf{f}_i^u = \int_{\Gamma_t} N_i \mathbf{f}^t \, d\Gamma + \int_{\Omega^e} N_i \mathbf{f}^b \, d\Omega \tag{3.45}$$

$$\mathbf{f}_i^a = \int_{\Gamma_t} N_i H \mathbf{f}^t \, d\Gamma + \int_{\Omega^e} N_i H \mathbf{f}^b \, d\Omega \tag{3.46}$$

$$\mathbf{f}_i^{b\alpha} = \int_{\Gamma_t} N_i F_\alpha \mathbf{f}^t \, d\Gamma + \int_{\Omega^e} N_i F_\alpha \mathbf{f}^b \, d\Omega \quad (\alpha = 1, 2, 3 \text{ and } 4) \tag{3.47}$$

$$\mathbf{f}_i^c = \int_{\Gamma_t} N_i \chi \mathbf{f}^t \, d\Gamma + \int_{\Omega^e} N_i \chi \mathbf{f}^b \, d\Omega \tag{3.48}$$

Evaluation of (3.48) becomes important when a $\chi$-enriched element (weak discontinuity or material interface) is subjected to a traction boundary condition.

In Eq. (3.43), $\mathbf{B}$ is the matrix of shape function derivatives,

$$\mathbf{B}_i^u = \begin{bmatrix} N_{i,x} & 0 \\ 0 & N_{i,y} \\ N_{i,y} & N_{i,x} \end{bmatrix} \tag{3.49}$$

$$\mathbf{B}_i^a = \begin{bmatrix} (N_i H(\xi))_{,x} & 0 \\ 0 & (N_i H(\xi))_{,y} \\ (N_i H(\xi))_{,y} & (N_i H(\xi))_{,x} \end{bmatrix} \tag{3.50}$$

$$\mathbf{B}_i^b = \begin{bmatrix} \mathbf{B}_i^{b1} & \mathbf{B}_i^{b2} & \mathbf{B}_i^{b3} & \mathbf{B}_i^{b4} \end{bmatrix} \tag{3.51}$$

$$\mathbf{B}_i^\alpha = \begin{bmatrix} (N_i F_\alpha)_{,x} & 0 \\ 0 & (N_i F_\alpha)_{,y} \\ (N_i F_\alpha)_{,y} & (N_i F_\alpha)_{,x} \end{bmatrix} (\alpha = 1,2,3 \text{ and } 4) \tag{3.52}$$

$$\mathbf{B}_i^c = \begin{bmatrix} (N_i \chi)_{,x} & 0 \\ 0 & (N_i \chi)_{,y} \\ (N_i \chi)_{,y} & (N_i \chi)_{,x} \end{bmatrix} \tag{3.53}$$

To include the effects of interpolation, the following shifting amendments are required:

$$\mathbf{B}_i^a = \begin{bmatrix} (N_i [H(\xi) - H(\xi_i)])_{,x} & 0 \\ 0 & (N_i [H(\xi) - H(\xi_i)])_{,y} \\ (N_i [H(\xi) - H(\xi_i)])_{,y} & (N_i [H(\xi) - H(\xi_i)])_{,x} \end{bmatrix} \tag{3.54}$$

$$\mathbf{B}_i^\alpha = \begin{bmatrix} [N_i (F_\alpha - F_{\alpha i})]_{,x} & 0 \\ 0 & [N_i (F_\alpha - F_{\alpha i})]_{,y} \\ [N_i (F_\alpha - F_{\alpha i})]_{,y} & [N_i (F_\alpha - F_{\alpha i})]_{,x} \end{bmatrix} (\alpha = 1,2,3 \text{ and } 4) \tag{3.55}$$

and no shifting procedure is required for derivatives of weak enrichment functions. In addition, the following vectors of nodal forces are modified to:

$$\mathbf{f}_i^a = \int_{\Gamma_t} N_i [H(\xi) - H(\xi_i)] \mathbf{f}^t \, d\Gamma + \int_{\Omega^e} N_i [H(\xi) - H(\xi_i)] \mathbf{f}^b \, d\Omega \tag{3.56}$$

$$\mathbf{f}_i^{b\alpha} = \int_{\Gamma_t} N_i (F_\alpha - F_{\alpha i}) \mathbf{f}^t \, d\Gamma + \int_{\Omega^e} N_i (F_\alpha - F_{\alpha i}) \mathbf{f}^b \, d\Omega \quad (\alpha = 1,2,3 \text{ and } 4) \tag{3.57}$$

## 3.4.6 Evaluation of Derivatives of Enrichment Functions

Computation of the matrix of shape function derivatives $\mathbf{B}$ depends on the definition of the enrichment function. The following types are considered:

### 3.4.6.1 Derivatives of Strong Discontinuity Enrichment $\psi(x) = H(\xi)$

Derivative of the Heaviside function is the Dirac delta function:

$$H_{,i}(\xi) = \delta(\xi) \tag{3.58}$$

which vanishes except at the position of the crack interface:

$$H_{,i}(\xi) = \begin{cases} 1 & \text{at crack} \\ 0 & \text{otherwise} \end{cases} \tag{3.59}$$

As a result, Eq. (3.50) can be rewritten as:

$$\mathbf{B}_i^a = \begin{bmatrix} N_{i,x}H(\xi) & 0 \\ 0 & N_{i,y}H(\xi) \\ N_{i,y}H(\xi) & N_{i,x}H(\xi) \end{bmatrix} \tag{3.60}$$

To include the effects of interpolation, $H(\xi)$ should be replaced by $H(\xi) - H(\xi_i)$.

### 3.4.6.2 Derivatives of Weak Discontinuity Enrichment $\psi(\mathbf{x}) = \chi(\mathbf{x}) = |\xi(\mathbf{x})|$

The derivative of the weak discontinuity enrichment function can be obtained from

$$\psi_{,i}(\mathbf{x}) = sign(\xi)\xi_{,i}(\mathbf{x}) \tag{3.61}$$

Derivatives of $\xi(\mathbf{x})$ are calculated from the derivatives of the shape functions if a standard finite element interpolation is adopted to define the function in terms of its nodal values:

$$\xi(\mathbf{x}) = \sum_{j=1}^{4} N_j(\mathbf{x})\xi_j \tag{3.62}$$

$$\xi_{,i}(\mathbf{x}) = \sum_{j=1}^{4} N_{j,i}(\mathbf{x})\xi_j \tag{3.63}$$

### 3.4.6.3 Derivatives of Crack-Tip Enrichment $\psi = F_\alpha(r, \theta)$

The near tip enrichment functions have already been defined in terms of the local crack tip polar coordinate system $(r, \theta)$, (Figure 3.1)

$$F_\alpha(r, \theta) = \left\{ \sqrt{r}\sin\frac{\theta}{2}, \sqrt{r}\cos\frac{\theta}{2}, \sqrt{r}\sin\theta\sin\frac{\theta}{2}, \sqrt{r}\sin\theta\cos\frac{\theta}{2} \right\} \tag{3.64}$$

Derivatives of $F_\alpha(r, \theta)$ with respect to the crack tip polar coordinates $(r, \theta)$ become

$$F_{1,r} = \frac{1}{2\sqrt{r}}\sin\frac{\theta}{2}, \qquad F_{1,\theta} = \frac{\sqrt{r}}{2}\cos\frac{\theta}{2} \tag{3.65}$$

$$F_{2,r} = \frac{1}{2\sqrt{r}}\cos\frac{\theta}{2}, \qquad F_{2,\theta} = -\frac{\sqrt{r}}{2}\sin\frac{\theta}{2} \tag{3.66}$$

$$F_{3,r} = \frac{1}{2\sqrt{r}}\sin\frac{\theta}{2}\sin\theta, \qquad F_{3,\theta} = \sqrt{r}\left(\frac{1}{2}\cos\frac{\theta}{2}\sin\theta + \sin\frac{\theta}{2}\cos\theta\right) \tag{3.67}$$

$$F_{4,r} = \frac{1}{2\sqrt{r}} \cos \frac{\theta}{2} \sin \theta, \qquad F_{4,\theta} = \sqrt{r}\left(-\frac{1}{2} \sin \frac{\theta}{2} \sin \theta + \cos \frac{\theta}{2} \cos \theta\right) \tag{3.68}$$

and the derivatives of $F_\alpha(r, \theta)$ with respect to the local crack coordinate system $(x', y')$ can then be defined as (see Figure 3.1):

$$F_{1,x'} = -\frac{1}{2\sqrt{r}} \sin \frac{\theta}{2}, \qquad F_{1,y'} = \frac{1}{2\sqrt{r}} \cos \frac{\theta}{2} \tag{3.69}$$

$$F_{2,x'} = \frac{1}{2\sqrt{r}} \cos \frac{\theta}{2}, \qquad F_{2,y'} = \frac{1}{2\sqrt{r}} \sin \frac{\theta}{2} \tag{3.70}$$

$$F_{3,x'} = -\frac{1}{2\sqrt{r}} \sin \frac{3\theta}{2} \sin \theta, \qquad F_{3,y'} = \frac{1}{2\sqrt{r}}\left(\sin \frac{\theta}{2} + \sin \frac{3\theta}{2} \cos \theta\right) \tag{3.71}$$

$$F_{4,x'} = -\frac{1}{2\sqrt{r}} \cos \frac{3\theta}{2} \sin \theta, \qquad F_{4,y'} = \frac{1}{2\sqrt{r}}\left(\cos \frac{\theta}{2} + \cos \frac{3\theta}{2} \cos \theta\right) \tag{3.72}$$

Finally, the derivatives in the global coordinate system are obtained,

$$F_{\alpha,x} = F_{\alpha,x'} \cos(\theta_0) - F_{\alpha,y'} \sin(\theta_0) \tag{3.73}$$

$$F_{\alpha,y} = F_{\alpha,x'} \sin(\theta_0) + F_{\alpha,y'} \cos(\theta_0) \tag{3.74}$$

where $\theta_0$ is the angle of crack path with respect to the $x$ axis.

### 3.4.7  Selection of Nodes for Discontinuity Enrichment

There have been different approaches for the selection of nodes to be enriched. One method allows the discontinuity to be modelled across the crack over the points along the crack surface. The value of the modified (enriched) shape function remains zero at all nodes and edges that do not intersect with the crack. This is important in satisfying the inter-element continuity requirements. This method only affects the element containing a crack, and does not directly influence other elements, even if they share a common node with the enriched element.

Figure 3.5a illustrates a simple procedure for selection of nodes for enrichment. At each stage of the propagation, nodes on edges cut by the crack path are enriched. Even if the crack tip locates just on an edge, the corresponding nodes are not enriched. A potential source of instability is when a crack path passes along the finite element edges. This technique adds two enrichment degrees of freedom to an element per any enriched node. As a result, for a quadrilateral element on the path of a crack, sixteen degrees of freedom (DOF) are assigned: eight classical DOFs and eight enriched DOFs. This approach is frequently used in complicated problems, such as cracking in shells, where an asymptotic analytical solution around the crack tip may not exist or may be difficult to be derived.

In an alternative procedure, the enrichment is applied onto all nodal points of elements that contain part of the crack. Figure 3.5b illustrates the procedure of node selection for enrichment based on this formulation.

The most appropriate choice, however, is to use both Heaviside and crack-tip enrichments where required. Accordingly, the crack-tip enrichment functions in the crack-tip element also allow accurate representation of discontinuity in that element. In a finite element mesh, as

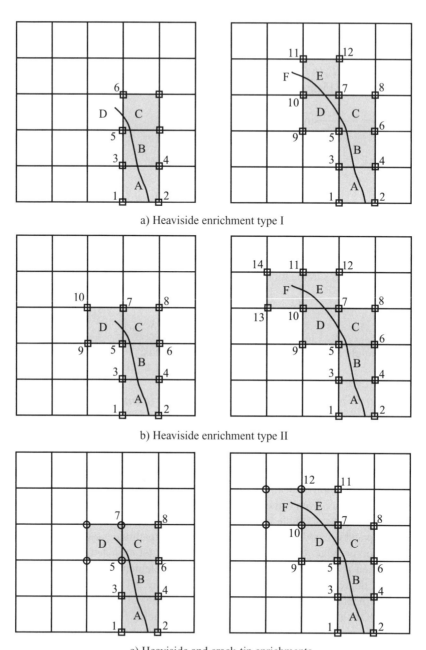

a) Heaviside enrichment type I

b) Heaviside enrichment type II

c) Heaviside and crack-tip enrichments

**Figure 3.5**   Node selection for Heaviside and crack-tip enrichment procedures at different stages of a crack propagation: (a) Heaviside enrichment type I, (b) Heaviside enrichment type II, (c) Heaviside and crack-tip enrichments. Nodes marked by circles are enriched by crack-tip functions and those marked by squares are enriched by the Heaviside function.

depicted in Figure 3.5c, the sets of nodes that must be enriched with Heaviside or crack-tip functions are marked by squares and circles, respectively. The crack does not affect other nodes and their associated classical finite element degrees of freedom.

## 3.4.8   Numerical Integration

The Gauss quadrature rule is widely used in finite element analysis for numerical evaluation of various integrals over a specified domain of interest such as a finite element. For polynomial integrands, the Gauss quadrature is proved to be exact. However, for non-polynomial integrands, it may result in substantial accuracy reduction.

Introduction of singularity within a finite element transforms the displacement and stress fields into highly nonlinear fields. This is further complicated as the crack path turns to be substantially curved. On the other hand, existence of discontinuity within a finite element may similarly complicate the accurate integration procedure, even by conventional higher order quadrature rules. As a result, an efficient approach is required to define the necessary points needed for the integration within an enriched element. The approach has to be consistent with the geometry of the crack as well as the order of the enrichment functions.

### 3.4.8.1   Sub-Quad Technique

The first method, proposed by Dolbow (1999), is to subdivide the element into sub-quads (Figure 3.6a). Each sub-quad is then integrated by a conventional Gauss integration rule. A different number of sub-quads can be used for elements that are cut by a crack or include a crack tip.

### 3.4.8.2   Sub-Triangle Approach

The second approach, also proposed by Dolbow (1999), is to subdivide the element at both sides of the crack into sub-triangles whose edges are adapted to crack faces (Figure 3.6b).

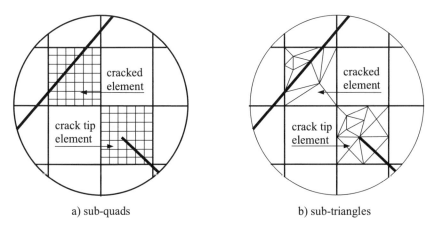

a) sub-quads                                                    b) sub-triangles

**Figure 3.6**   Two methods for partitioning the cracked element.

For a crack-tip element, more sub-triangles are required in front of the crack tip because of the existence of a highly nonlinear and singular stress field. This method has been the most adopted integration approach in XFEM fracture analysis.

### 3.4.8.3  Polar Distribution

In recent years, a new idea has been developed to better fit the polar nature of stress distribution around the crack tip. Accordingly, the distribution of Gauss points in a crack-tip element is designed in a polar format, with closer points in the vicinity of the crack tip becoming further apart towards the edges of the element. Such an approach has also been successfully used in crack analysis by the extended isogeometric analysis (XIGA) (Ghorashi, Valizadeh and Mohammadi, 2012).

### 3.4.8.4  Limitations of Enrichments

It is well-known that the XFEM approximation may become ill-conditioned, or even fail, if the enrichment is adopted for a case where the crack passes through or very close to a node. To avoid such a numerical difficulty, Dolbow (1999) proposed a simple geometric criterion to determine the candidacy of a node for enrichment. Figure 3.7 illustrates two different cases of the relative position of a crack within a cracked element.

In both methods, if values of $A^-/(A^+ + A^-)$ or $A^+/(A^+ + A^-)$, where $A^+$ and $A^-$ are the area of the influence domain of a node above and below the crack, respectively, are smaller than an allowable tolerance value, the node must not be enriched. A tolerance of 0.01% was proposed by Dolbow (1999), although it should be set according to each specific problem. In an alternative approach, a node is enriched if each side of the crack in its influence domain includes at least one Gauss point. Figure 3.7b shows a mesh that contains a crack with the sub-quad partitioning. Although the crack cuts the element in Figure 3.7b, node J must not be

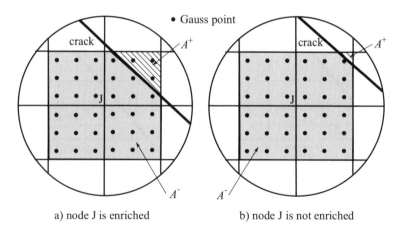

a) node J is enriched                    b) node J is not enriched

**Figure 3.7**  Criteria for node enrichment: (a) based on definitions of $A^+$ and $A^-$ in its support domain, (b) based on the existence of Gauss points within its support domain (Mohammadi, 2008). (Reproduced by permission of John Wiley & Sons, Ltd.)

enriched because there is no Gaussian point above the crack. On the contrary, node J in Figure 3.22a has to be enriched.

## 3.5   XFEM Strong Discontinuity Enrichments

In the extended finite element method, approximation of a discontinuous displacement field is based on the definition of specially designed shape functions by the use of enrichment functions. The method operates on additional independent virtual degrees of freedom for the definition of the crack geometry and approximation of the displacement field. It will then be combined with the classical finite element method to approximate the overall solution.

In order to discuss various effects of the modelling, a one-dimensional problem is considered which consists of four nodes and three linear 2-node finite elements with a strong discontinuity (crack) in an arbitrary location $x_c(\xi_c)$ within the middle element, as depicted in Figure 3.8. Similar one-dimensional examples can be found in almost all basic XFEM-related documents, references and textbooks.

Only nodes 2 and 3 are required to be enriched, whereas nodes 1 and 4 are not influenced by the crack.

There have been a number of possible choices for the enrichment function $\psi(\mathbf{x})$ in Eq. (3.23). The following sections explain the basic ideas and discuss the various effects of enrichment on this simple one-dimensional problem.

### 3.5.1   A Modified FE Shape Function

Earlier models used a simple modified shape function in the form of:

$$N_i^h = \begin{cases} N_i - 1 & \mathbf{x} \in \Omega_i \\ N_i & \mathbf{x} \notin \Omega_i \end{cases} \tag{3.75}$$

where $N_i$ is the conventional finite element shape function and $\Omega_i$ is the part of the element in between the crack and node $i$, as illustrated in Figure 3.9.

Figure 3.10 shows how this jump enrichment can affect the shape functions for the simple one-dimensional case of Figure 3.8.

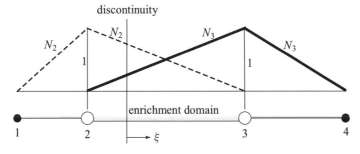

**Figure 3.8**   Simulation of a crack in a one-dimensional problem using the standard linear finite element shape functions (Mohammadi, 2008). (Reproduced by permission of John Wiley & Sons, Ltd.)

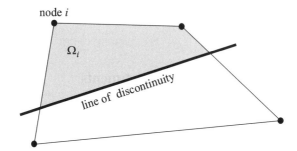

**Figure 3.9**   An element cut across by a crack.

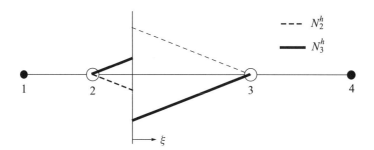

**Figure 3.10**   A simple description of discontinuity by a jump function (Mohammadi, 2008). (Reproduced by permission of John Wiley & Sons, Ltd.)

One problem with this type of jump function is that it provides similar strain fields (derivative of the displacement field) in both sides of the discontinuity (Figure 3.11). This is in contrast to the independent physical response of the segments, anticipated in a cracked element.

Another drawback is the lower number of degrees of freedom required by approximation (3.75) than other recently available techniques. This may directly affect the quality of the approximation field and the crack analysis.

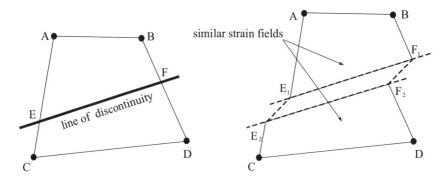

**Figure 3.11**   Deformation of a quadrilateral element with the simple rule (3.75).

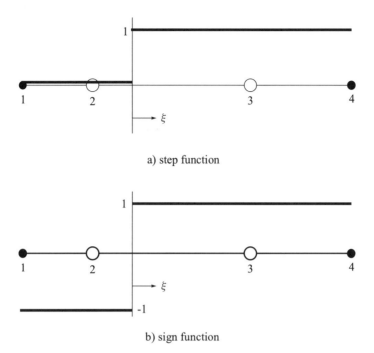

a) step function

b) sign function

**Figure 3.12**   Different types of Heaviside function $H(\xi)$ (Mohammadi, 2008). (Reproduced by permission of John Wiley & Sons, Ltd.)

## 3.5.2   The Heaviside Function

Different definitions have been adopted for the Heaviside function over the years. The first type of Heaviside function $H(\xi)$ can be defined as a step function,

$$H(\xi) = \begin{cases} 1 & \forall \xi > 0 \\ 0 & \forall \xi < 0 \end{cases} \tag{3.76}$$

A simple one-dimensional representation of this step function is depicted in Figure 3.12a. Approximation (3.23) then reads,

$$\mathbf{u}^h(\mathbf{x}) = \sum_{j=1}^{n} N_j(\mathbf{x})\mathbf{u}_j + \sum_{k=1}^{m} N_k(\mathbf{x})H(\xi)\mathbf{a}_k \tag{3.77}$$

Figure 3.13a illustrates the way the step function simulates the discontinuity.

To examine whether the approximation (3.77) is an interpolation, the value of the field variable $\mathbf{u}(\mathbf{x})$ on an enriched node $i$ can be obtained as:

$$\mathbf{u}^h(\mathbf{x}_i) = \mathbf{u}_i + H(\xi_i)\mathbf{a}_i \tag{3.78}$$

which means that approximation (3.77) is not an interpolation and the nodal parameter $\mathbf{u}_i$ is not the real displacement value on the enriched node $i$. A simple remedy to this shortcoming

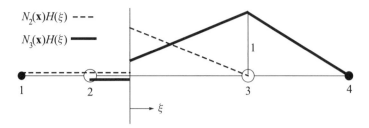

a) effect of the step function on shape functions

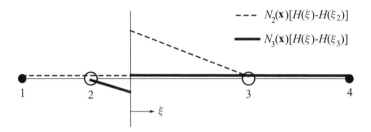

b) effect of shifting on shape functions

**Figure 3.13**  Enriched shape functions for nodes 2 and 3 and application of the shifting Heaviside function (Mohammadi, 2008). (Reproduced by permission of John Wiley & Sons, Ltd.)

is to shift the step function around the node of interest:

$$\mathbf{u}^h(\mathbf{x}) = \sum_{j=1}^{n} N_j(\mathbf{x})\mathbf{u}_j + \sum_{k=1}^{m} N_k(\mathbf{x}) \left( H(\xi) - H(\xi_k) \right) \mathbf{a}_k \tag{3.79}$$

Consequently, interpolation can be automatically guaranteed. Figure 3.13b illustrates the effect of the modified approximation on the one-dimensional crack problem. The overall jump in the displacement field can be obtained from:

$$\left\langle \mathbf{u}^h(\mathbf{x}) \right\rangle = \mathbf{u}^h(\mathbf{x}^+) - \mathbf{u}^h(\mathbf{x}^-) = \cdots = \sum_{k=1}^{m} N_k(\mathbf{x})\mathbf{a}_k \tag{3.80}$$

Figure 3.13b illustrates that in the shifted case, the enriched shape functions vanish at both nodes 2 and 3. As a result, the direct effect of enrichment is limited only to the middle element which contains a discontinuity.

Application of the aforementioned jump function on a quadrilateral element may lead to a discontinuous field, as depicted in Figure 3.14. The deformation field includes all potential displacement fields independently for both sides of the crack. The strain fields also remain independent for both sides of the crack, compared with the previous case illustrated in Figure 3.11.

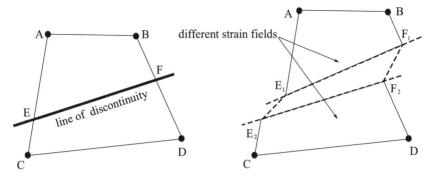

**Figure 3.14**   Deformation of a quadrilateral element with a jump function.

In order to avoid numerical instabilities, the following smoothed Heaviside functions can also be used for a small value of $\beta$ less than the element size (Bordas and Legay, 2005):

$$
H(\xi) = \begin{cases} 0 & \xi < -\beta \\ \dfrac{1}{2} + \dfrac{\xi}{2\beta} + \dfrac{1}{2\pi} \sin \dfrac{\pi \xi}{\beta} & -\beta < \xi < \beta \\ 1 & \xi > \beta \end{cases} \tag{3.81}
$$

or

$$
H(\xi) = \begin{cases} 0 & \xi < -\beta \\ \dfrac{1}{2} + \dfrac{1}{8}\left(9\dfrac{\xi}{\beta} - 5\dfrac{\xi^3}{\beta^3}\right) & -\beta < \xi < \beta \\ 1 & \xi > \beta \end{cases} \tag{3.82}
$$

The derivative of the Heaviside function is the Dirac delta function $\delta(\xi)$

$$
H_{,i}(\xi) = \delta(\xi) \tag{3.83}
$$

which can be approximated by the following smoothed functions

$$
\delta(\xi) = \begin{cases} \dfrac{1}{2\beta} + \dfrac{1}{2\beta} \cos \dfrac{\pi \xi}{\beta} & -\beta < \xi < \beta \\ 0 & \text{otherwise} \end{cases} \tag{3.84}
$$

or

$$
\delta(\xi) = \begin{cases} \dfrac{1}{8}\left(\dfrac{9}{\beta} - 15\dfrac{\xi^2}{\beta^3}\right) & -\beta < \xi < \beta \\ 0 & \text{otherwise} \end{cases} \tag{3.85}
$$

An alternative form has also been proposed by Chessa and Belytschko (2003).

### 3.5.3   The Sign Function

An alternative Heaviside enrichment function is the sign function

$$H(\xi) = \text{sign}(\xi) = \begin{cases} 1 & \forall \xi > 0 \\ -1 & \forall \xi < 0 \end{cases}$$   (3.86)

A simple one-dimensional representation of the sign function is depicted in Figure 3.12b. Figure 3.15a illustrates the way the sign function simulates the discontinuity.

Again, approximation (3.77) is no longer an interpolation and the value of the field variable $u(\mathbf{x})$ on an enriched node $i$ is not equal to the nodal value $\mathbf{u}_i$

$$\mathbf{u}^h(\mathbf{x}_i) = \mathbf{u}_i + H(\xi_i)\mathbf{a}_i \neq \mathbf{u}_i$$   (3.87)

A simple shifting procedure guarantees the interpolation:

$$\mathbf{u}^h(\mathbf{x}) = \sum_{j=1}^{n} N_j(\mathbf{x})\mathbf{u}_j + \sum_{k=1}^{m} N_k(\mathbf{x})\,(H(\xi) - H(\xi_k))\,\mathbf{a}_k$$   (3.88)

Figure 3.15b illustrates the effect of the modified approximation on the one-dimensional crack problem. Again, the enriched shape functions in the shifted form vanish at both nodes 2

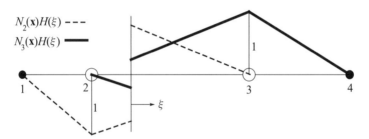

a) effect of the sign function on shape functions

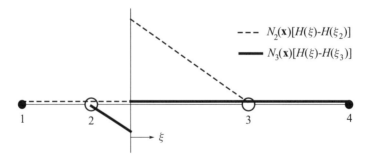

b) effect of shifting on shape functions

**Figure 3.15**   Enriched shape functions for nodes 2 and 3 and application of the shifting Heaviside function (Mohammadi, 2008). (Reproduced by permission of John Wiley & Sons, Ltd.)

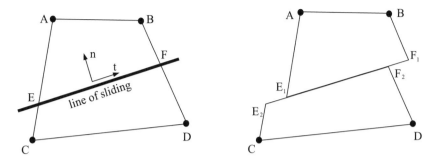

**Figure 3.16**   Strong tangential discontinuity.

and 3. As a result, the direct effect of enrichment is limited only to the middle element which contains a discontinuity.

The overall jump in the displacement field can be obtained from:

$$\langle \mathbf{u}^h(\mathbf{x}) \rangle = \mathbf{u}^h(\mathbf{x}^+) - \mathbf{u}^h(\mathbf{x}^-) = \cdots = 2 \sum_{k=1}^{m} N_k(\mathbf{x})\mathbf{a}_k \tag{3.89}$$

It is important to note that the choice of the jump in the enrichment function does not affect the overall solution. Similar forms to Eqs. (3.81)–(3.82) can be derived for smoothed equivalent forms of the sign function.

### 3.5.4   Strong Tangential Discontinuity

In a strong tangential discontinuity, a jump exists in the tangential direction whereas the displacement remains continuous in the normal direction. The conventional approach for modelling a strong discontinuity by the Heaviside function is now modified to simulate the tangential discontinuity (see Figure 3.16). This is achieved by defining only one extra enrichment degree of freedom (per node) in the tangential direction (instead of standard two DOFs) (Daneshyar and Mohammadi, 2012a).

The XFEM approximation can then be written as,

$$\mathbf{u}^h(\mathbf{x}) = \sum_{j=1}^{n} N_j(\mathbf{x})\mathbf{u}_j + \sum_{h=1}^{m} N_h(\mathbf{x}) \left( H(\xi) - H(\xi_h) \right) a_h \mathbf{t}_h \tag{3.90}$$

where $\mathbf{t}_h$ is the tangential direction unit vector. The effect of approximation (3.90) can be best viewed in Figure 3.16. All nodes of the finite elements that include a tangential discontinuity should be enriched by the tangential Heaviside enrichment functions.

### 3.5.5   Crack Intersection

The basic equation for XFEM enrichment requires further modification if two or more cracks intersect within a finite element, as illustrated in Figure 3.17.

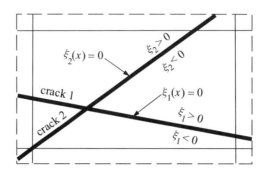

**Figure 3.17** Intersecting cracks within a finite element (Mohammadi, 2008).

The XFEM approximation of the displacement for an element cut by two intersecting cracks may be written as (Daux *et al.*, 2000):

$$\mathbf{u}^h(\mathbf{x}) = \sum_{j=1}^{n} N_j(\mathbf{x})\mathbf{u}_j + \sum_{h=1}^{m} N_h(\mathbf{x})H(\xi_1)\mathbf{a}_h^1 + \sum_{h=1}^{m} N_h(\mathbf{x})H(\xi_2)\mathbf{a}_h^2$$

$$+ \sum_{h=1}^{m} N_h(\mathbf{x})H(\xi_1)H(\xi_2)\mathbf{a}_h^3 \tag{3.91}$$

Some researchers have proposed a more efficient approach that avoids the cross terms by introduction of modified signed distance functions (Zi *et al.*, 2004).

## 3.6   XFEM Weak Discontinuity Enrichments

The XFEM approximation (3.77) can now be used for weak discontinuity problems by replacing the Heaviside function $H(\xi)$ with an appropriate enrichment function $\chi_k(\mathbf{x})$ (Bordas and Legay, 2005)

$$\mathbf{u}^h(\mathbf{x}) = \sum_{j=1}^{n} N_j(\mathbf{x})\mathbf{u}_j + \sum_{k=1}^{m} N_k(\mathbf{x})\chi_k(\mathbf{x})\mathbf{a}_k \tag{3.92}$$

where $\chi_k(\mathbf{x})$ is the weak discontinuous enrichment function defined in terms of the signed distance function $\xi(\mathbf{x})$:

$$\chi_k(\mathbf{x}) = |\xi(\mathbf{x})| - |\xi(\mathbf{x}_k)| \tag{3.93}$$

The same one-dimensional problem of Section 3.6.3 is considered (Figure 3.8). The only difference is the assumption of a weak discontinuity in an arbitrary location $\mathbf{x}_c(\xi_c)$ within the middle element.

Figure 3.18a illustrates these signed distance functions for the simple problem of Figure 3.8.

Figure 3.18b depicts how the original shape functions are transformed as an effect of enrichment by the weak discontinuous enrichment functions. According to this figure, a kink in the displacement field is introduced. As a result, a jump in its derivative, that is, a discontinuity in the gradient of the function is anticipated.

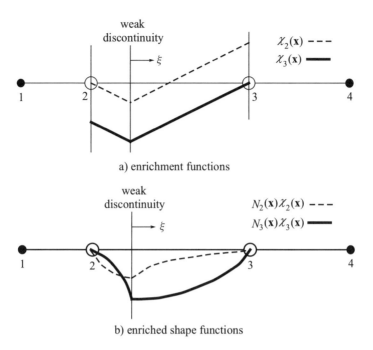

a) enrichment functions

b) enriched shape functions

**Figure 3.18** Weak discontinuous enrichment functions and final enriched shape functions (Moham-madi, 2008). (Reproduced by permission of John Wiley & Sons, Ltd.)

The normal jump in the gradient field, seen clearly in Figure 3.18b at the position of weak discontinuity, can be computed as:

$$\langle \nabla \mathbf{u}^h(\mathbf{x}) \rangle = 2 \sum_{k=1}^{m} N_k(\mathbf{x}) \mathbf{a}_k \tag{3.94}$$

The weak discontinuity enriched shape functions vanish at both nodes 2 and 3. As a result, the interpolation is guaranteed without any shifting. In addition, the direct effect of enrichment on displacements is limited only to the middle element which contains a weak discontinuity. However, an indirect influence exists due to different values of derivatives of enriched shape functions. A number of studies have directed towards new definitions of weak discontinuity enrichments to remove such indirect effects by ensuring vanishing derivatives of the enriched shape functions at nodal points (Esna Ashari and Mohammadi, 2011a).

## 3.7 XFEM Crack-Tip Enrichments

In this section, a brief review of major available tip enrichments is presented.

### 3.7.1 Isotropic Enrichment

Enrichment functions (3.10) are the basic and most used crack-tip enrichment functions,

$$\{F_l(r,\theta)\}_{l=1}^{4} = \left\{ \sqrt{r}\cos\frac{\theta}{2}, \sqrt{r}\sin\frac{\theta}{2}, \sqrt{r}\sin\theta\cos\frac{\theta}{2}, \sqrt{r}\sin\theta\sin\frac{\theta}{2} \right\} \tag{3.95}$$

It is important to note that the term $\sqrt{r}\sin\theta/2$ is discontinuous along the crack. As a result, once the enrichments (3.95) are adopted for the crack-tip element, the discontinuous field can be automatically represented and there is no need for inclusion of the Heaviside part of the approximation.

Despite being obtained from an asymptotic crack-tip analysis in a single isotropic homogeneous medium under static conditions for a traction-free crack, it has been frequently used for enhancement of the analyses in a wide variety of problems, including orthotropic materials, non-homogeneous FGMs, hydraulic fracture, dynamic problems, bimaterials, elastoplastic solutions, electromechanical problems, and so on.

### 3.7.2  Orthotropic Enrichment Functions

General crack-tip enrichment functions presented in Asadpoure and Mohammadi (2007) in the local crack tip polar coordinate system $(r, \theta)$ are defined as,

$$\{F_l(r, \theta)\}_{l=1}^4 = \left\{ \sqrt{r}\cos\frac{\theta_1}{2}\sqrt{g_1(\theta)}, \ \sqrt{r}\cos\frac{\theta_2}{2}\sqrt{g_2(\theta)}, \ \sqrt{r}\sin\frac{\theta_1}{2}\sqrt{g_1(\theta)}, \ \sqrt{r}\sin\frac{\theta_2}{2}\sqrt{g_2(\theta)} \right\}$$

(3.96)

where functions $g_k(\theta)$ and $\theta_k$, in their general form for all orthotropic composites, are defined as:

$$g_k(\theta) = \sqrt{(\cos\theta + s_{kx}\sin\theta)^2 + (s_{ky}\sin\theta)^2}$$

(3.97)

$$\theta_k = \arctan\left(\frac{s_{ky}\sin\theta}{\cos\theta + s_{kx}\sin\theta}\right)$$

(3.98)

where $s_{kx}$ and $s_{ky}$ have been defined in Eq. (4.113).

### 3.7.3  Bimaterial Enrichments

Sukumar et al. (2004) developed the following enrichment functions from the analytical solution near an interface crack tip in isotropic bimaterials,

$$\begin{aligned}
\{F_l(r, \theta)\}_{l=1}^{12} = \Big\{ & \sqrt{r}\cos(\varepsilon\ln r)e^{-\varepsilon\theta}\sin\frac{\theta}{2}, \ \sqrt{r}\cos(\varepsilon\ln r)e^{-\varepsilon\theta}\cos\frac{\theta}{2}, \\
& \sqrt{r}\cos(\varepsilon\ln r)e^{\varepsilon\theta}\sin\frac{\theta}{2}, \ \sqrt{r}\cos(\varepsilon\ln r)e^{\varepsilon\theta}\cos\frac{\theta}{2}, \\
& \sqrt{r}\sin(\varepsilon\ln r)e^{-\varepsilon\theta}\sin\frac{\theta}{2}, \ \sqrt{r}\sin(\varepsilon\ln r)e^{-\varepsilon\theta}\cos\frac{\theta}{2}, \\
& \sqrt{r}\sin(\varepsilon\ln r)e^{\varepsilon\theta}\sin\frac{\theta}{2}, \ \sqrt{r}\sin(\varepsilon\ln r)e^{\varepsilon\theta}\cos\frac{\theta}{2}, \\
& \sqrt{r}\cos(\varepsilon\ln r)e^{\varepsilon\theta}\sin\frac{\theta}{2}\sin\theta, \ \sqrt{r}\cos(\varepsilon\ln r)e^{\varepsilon\theta}\cos\frac{\theta}{2}\sin\theta, \\
& \sqrt{r}\sin(\varepsilon\ln r)e^{\varepsilon\theta}\sin\frac{\theta}{2}\sin\theta, \ \sqrt{r}\sin(\varepsilon\ln r)e^{\varepsilon\theta}\cos\frac{\theta}{2}\sin\theta \Big\}
\end{aligned}$$

(3.99)

$\varepsilon$ is the index of oscillation defined as

$$\varepsilon = \frac{1}{2\pi} \ln \left( \frac{1-\beta}{1+\beta} \right) \tag{3.100}$$

where $\beta$ is the second Dundurs parameter for isotropic materials (Dundurs, 1969)

$$\beta = \frac{\mu_1(\kappa_2 - 1) - \mu_2(\kappa_1 - 1)}{\mu_1(\kappa_2 + 1) + \mu_2(\kappa_1 + 1)} \tag{3.101}$$

### 3.7.4   Orthotropic Bimaterial Enrichments

The analytical solutions (7.17) and (7.18) can be used to derive the necessary enrichment functions for orthotropic bimaterial interface crack tip (Esna Ashari and Mohammadi, 2011a)

$$\{F_l(r,\theta)\}_{l=1}^8 = \left[ e^{-\varepsilon\theta_l} \cos\left( \varepsilon \ln(r_l) + \frac{\theta_l}{2} \right) \sqrt{r_l}, \; e^{-\varepsilon\theta_l} \sin\left( \varepsilon \ln(r_l) + \frac{\theta_l}{2} \right) \sqrt{r_l}, \right.$$

$$e^{\varepsilon\theta_l} \cos\left( \varepsilon \ln(r_l) - \frac{\theta_l}{2} \right) \sqrt{r_l}, \; e^{\varepsilon\theta_l} \sin\left( \varepsilon \ln(r_l) - \frac{\theta_l}{2} \right) \sqrt{r_l},$$

$$e^{-\varepsilon\theta_s} \cos\left( \varepsilon \ln(r_s) + \frac{\theta_s}{2} \right) \sqrt{r_s}, \; e^{-\varepsilon\theta_s} \sin\left( \varepsilon \ln(r_s) + \frac{\theta_s}{2} \right) \sqrt{r_s},$$

$$\left. e^{\varepsilon\theta_s} \cos\left( \varepsilon \ln(r_s) - \frac{\theta_s}{2} \right) \sqrt{r_s}, \; e^{\varepsilon\theta_s} \sin\left( \varepsilon \ln(r_s) - \frac{\theta_s}{2} \right) \sqrt{r_s} \right] \tag{3.102}$$

with

$$r_j = r\sqrt{\cos^2\theta + Z_j^2 \sin^2\theta}, \quad j = l, s, \quad \begin{cases} Z_l = p \\ Z_s = q \end{cases} \tag{3.103}$$

$$\theta_j = \tan^{-1}\left( Z_j \tan\theta \right), \quad j = l, s, \quad \begin{cases} Z_l = p \\ Z_s = q \end{cases} \tag{3.104}$$

where $\varepsilon$ is the index of oscillation defined in (7.22) in terms of the second Dundurs parameter $\beta$ for orthotropic materials and all other parameters have been defined in Section 7.2.2.

### 3.7.5   Dynamic Enrichment

Belytschko and Chen (2004) used the analytical asymptotic solutions near a propagating crack in an isotropic medium to develop the dynamic enrichment functions. A simplified form of these functions can be written as,

$$\{F_l(r,\theta)\}_{l=1}^4 = \left\{ r_d \cos\frac{\theta_d}{2}, r_s \cos\frac{\theta_s}{2}, r_d \sin\frac{\theta_d}{2}, r_d \sin\frac{\theta_s}{2} \right\} \tag{3.105}$$

where

$$r_d^2 = x^2 + \beta_d^2 y^2 \tag{3.106}$$

$$r_s^2 = x^2 + \beta_s^2 y^2 \tag{3.107}$$

$$\tan \theta_d = \beta_d \tan \theta \tag{3.108}$$

$$\tan \theta_s = \beta_s \tan \theta \tag{3.109}$$

with

$$\beta_d^2 = 1 - \frac{v_c^2}{c_d^2} \tag{3.110}$$

$$\beta_s^2 = 1 - \frac{v_c^2}{c_s^2} \tag{3.111}$$

where $v_c$ is the crack tip velocity and $c_d$ and $c_s$ are the dilatational and shear wave speeds, respectively,

$$c_d = \sqrt{\frac{\lambda + 2\mu}{\rho}} \tag{3.112}$$

$$c_s = \sqrt{\frac{\mu}{\rho}} \tag{3.113}$$

and $\mu$ and $\lambda$ are the Lame coefficients and $\rho$ is the material density.

Alternatively, Kabiri (2009) proposed the following enrichment functions for dynamic problems based on a simple idea of adding the influence of the crack-tip velocity on classical isotropic enrichment functions,

$$\{F_l(r, \theta)\}_{l=1}^4 = \left\{ \sqrt{r} \cos \frac{\theta}{2}, \sqrt{r} \sin \frac{\theta}{2}, \sqrt{r} \sin \theta \sin \frac{\theta}{2}, \sqrt{r} \cos \theta \sin \frac{\theta}{2} \right\} \sqrt{\frac{\gamma_d^2 + \gamma_s^2}{2}} \tag{3.114}$$

with

$$\gamma_d^2 = 1 - \frac{v_c^2}{c_d^2} \sin^2 \theta \tag{3.115}$$

$$\gamma_s^2 = 1 - \frac{v_c^2}{c_s^2} \sin^2 \theta \tag{3.116}$$

### 3.7.6   Orthotropic Dynamic Enrichments for Moving Cracks

The following dynamic orthotropic enrichment functions can be derived for moving crack-tip problems (Motamedi and Mohammadi, 2012):

$$\{F_l(r, \theta)\}_{l=1}^4 = \left\{ \sqrt{r} \cos \frac{\theta_1}{2} \sqrt{g_1(\theta)}, \sqrt{r} \cos \frac{\theta_2}{2} \sqrt{g_2(\theta)}, \right.$$

$$\left. \sqrt{r} \sin \frac{\theta_1}{2} \sqrt{g_1(\theta)}, \sqrt{r} \sin \frac{\theta_2}{2} \sqrt{g_2(\theta)} \right\} \tag{3.117}$$

where $g_k(\theta)$ and $\theta_k$, $(k = 1, 2)$ are defined as

$$g_k(\theta) = \sqrt{(\cos\theta + m_{kx}\sin\theta)^2 + (m_{ky}\sin\theta)^2}, \quad k = 1, 2 \qquad (3.118)$$

$$\theta_k = arctg\left(\frac{m_{ky}\sin\theta}{\cos\theta + m_{kx}\sin\theta}\right), \quad k = 1, 2 \qquad (3.119)$$

and $(r, \theta)$ are considered from the crack tip position, as depicted in Figure 3.1. $m_{kx}$ and $m_{ky}$ are the real and imaginary parts of $m_k$ (roots of Eq. (5.36)):

$$m_{kx} = \text{Re}\,(m_k), \quad k = 1, 2 \qquad (3.120)$$

$$m_{ky} = \text{Im}\,(m_k), \quad k = 1, 2 \qquad (3.121)$$

### 3.7.7 Bending Plates

For crack problems in bending plates and shells, the displacement fields around the crack tip can be decomposed to in-plane and out of plane components. Accordingly, a different set of enrichment functions are assigned to each component. For in-plane deformation, the enrichment functions are:

$$\{F_l(r, \theta)\}_{l=1}^{4} = \left\{\sqrt{r}\cos\frac{\theta}{2}, \sqrt{r}\sin\frac{\theta}{2}, \sqrt{r}\sin\theta\cos\frac{\theta}{2}, \sqrt{r}\sin\theta\sin\frac{\theta}{2}\right\} \qquad (3.122)$$

The out of plane and rotational enrichments can be defined as (Bayesteh and Mohammadi, 2011),

$$\{F_l(r, \theta)\}_{l=1}^{4} = \left\{\sqrt{r}\sin\frac{\theta}{2}, r^{3/2}\sin\frac{\theta}{2}, r^{3/2}\cos\frac{\theta}{2}, r^{3/2}\sin\frac{3\theta}{2}, r^{3/2}\cos\frac{3\theta}{2}\right\} \qquad (3.123)$$

$$\{F_l(r, \theta)\}_{l=1}^{4} = \left\{\sqrt{r}\cos\frac{\theta}{2}, \sqrt{r}\sin\frac{\theta}{2}, \sqrt{r}\sin\theta\cos\frac{\theta}{2}, \sqrt{r}\sin\theta\sin\frac{\theta}{2}\right\} \qquad (3.124)$$

### 3.7.8 Crack-Tip Enrichments in Shells

Folias (1999) has illustrated that a correlation function, independent of angle or distance from the crack tip, exists between plates and shells:

$$\sigma_{\text{shell}} \approx \sigma_{\text{plate}}\left\{1 + \left[c_1 + c_2\left(\ln\frac{c}{\sqrt{Rh}}\right) + O\left(\frac{1}{R^2}\right)\right]\right\}, \quad c_1, c_2, c_3 = \text{constant} \qquad (3.125)$$

Similarly, Zehnder and Viz (2005) have stated that: "Near the crack tip, the stress distribution in shells is the same as that of plates and the shape of the shell or plate will come in only through the stress intensity factors". Accordingly, similar enrichment functions can be used for approximation of the displacement field around a crack tip (Bayesteh and Mohammadi, 2011).

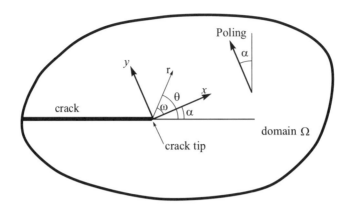

**Figure 3.19**   Material axes at the crack tip and its polarization in an infinite domain.

### 3.7.9   Electro-Mechanical Enrichment

Beginning with the constitutive equation in an electro-mechanical plane strain state (Figure 3.19), the mechanical strain tensor $\varepsilon_{ij}$ and the electric field $E_i$ can be defined in terms of the Cauchy stress tensor $\sigma_{ij}$ and the electric displacement vector $D_i$ (Sharma *et al.*, 2012),

$$
\begin{Bmatrix} \varepsilon_{xx} \\ \varepsilon_{yy} \\ \gamma_{xy} \\ E_x \\ E_y \end{Bmatrix} = \begin{bmatrix} c_{11} & c_{12} & 0 & 0 & b_{21} \\ c_{12} & c_{22} & 0 & 0 & b_{22} \\ 0 & 0 & c_{33} & b_{13} & 0 \\ 0 & 0 & -b_{13} & \delta_{11} & 0 \\ -b_{21} & -b_{22} & 0 & 0 & \delta_{22} \end{bmatrix} \begin{Bmatrix} \sigma_{xx} \\ \sigma_{yy} \\ \sigma_{xy} \\ D_x \\ D_y \end{Bmatrix} \tag{3.126}
$$

where $c_{ij}$, $b_{ij}$ and $\delta_{ij}$ are mechanical and electrical material constants.

The characteristic equation can be written in terms of $s = s_1 + is_2$ (Sharma *et al.*, 2012),

$$
c_{11}\delta_{11}s^6 + \left\{ c_{11}\delta_{22} + (2c_{12} + c_{33})\delta_{11} + b_{12}(b_{12} + 2b_{13}) + b_{13}^2 \right\} s^4
$$
$$
+ \left\{ c_{22}\delta_{11} + (2c_{12} + c_{33})\delta_{22} + 2b_{22}(b_{12} + b_{13}) \right\} s^2 + \left\{ c_{22}\delta_{22} + b_{22}^2 \right\} = 0 \tag{3.127}
$$

with six roots $s_m = s_{m1} + is_{m2}$. XFEM crack-tip enrichment functions for the mechanical displacement **u** and electric potential $\varphi$ can be written as (Bechet, Scherzer and Kuna, 2009; Sharma *et al.*, 2012),

$$
\{F_l(r, \theta, \mu_k)\}_{l=1}^6 = \left\{ \sqrt{r}g_1(\theta), \sqrt{r}g_2(\theta), \sqrt{r}g_3(\theta), \sqrt{r}g_4(\theta), \sqrt{r}g_5(\theta), \sqrt{r}g_6(\theta) \right\} \tag{3.128}
$$

where

$$
g_m(\theta) = \begin{cases} \rho_m(\omega, s_m) \cos\left( \dfrac{\psi(\omega, s_m)}{2} \right) & s_{m2} > 0 \\[4mm] \rho_m(\omega, s_m) \sin\left( \dfrac{\psi(\omega, s_m)}{2} \right) & s_{m2} < 0 \end{cases} \tag{3.129}
$$

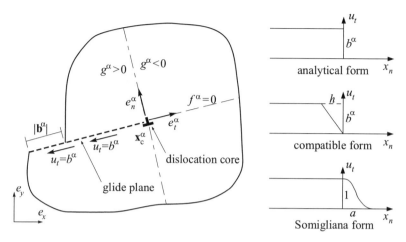

**Figure 3.20** Definition of functions $f(\mathbf{x})$ and $g(\mathbf{x})$ for an edge dislocation, and the incompatible, linearly regularized and tapered Somigliana profiles for a dislocation glide plane.

$$\psi(\omega, s_m) = \frac{\pi}{2} + \pi \operatorname{int}\left(\frac{\omega}{\pi}\right) - \tan^{-1}\left(\frac{\cos\left[\theta - \pi \operatorname{int}\left(\frac{\omega}{\pi}\right)\right] + s_{m1} \sin\left(\omega - \pi \operatorname{int}\left(\frac{\omega}{\pi}\right)\right)}{|s_{m2}| \sin\left(\omega - \pi \operatorname{int}\left(\frac{\omega}{\pi}\right)\right)}\right)$$

(3.130)

$$\rho_m(\omega, s_m) = \frac{1}{\sqrt{2}}\{|s_{m2}|^2 + s_{m1}\sin(2\omega) - [|s_m|^2 - 1]\cos(2\omega)\}^{\frac{1}{4}}$$

(3.131)

### 3.7.10 Dislocation Enrichment

Earlier simulations used a simple Heaviside function to represent the glide plane of edge dislocations. Such a model, in fact, somehow ignores the effect of dislocation core. Alternative functions are the compatible form (Figure 3.20),

$$\boldsymbol{\Psi}_H^\alpha = \mathbf{b}^\alpha \sum_{h=1}^{mh} N_h(\mathbf{x})[H(f^\alpha(\mathbf{x})) - H(f^\alpha(\mathbf{x}_I))]H(g^\alpha(\mathbf{x}))$$

(3.132)

and the tapered Somigliana dislocation proposed by Gracie, Ventura and Belytschko (2007) (Figure 3.20),

$$\boldsymbol{\Psi}_H^\alpha = \mathbf{b}^\alpha \sum_{h=1}^{mh} N_h(\mathbf{x})[H(f^\alpha(\mathbf{x})) - H(f^\alpha(\mathbf{x}_I))]\psi_H(g^\alpha(\mathbf{x}))$$

(3.133)

$$\psi_H(g(\mathbf{x})) = \begin{cases} 1 & g(\mathbf{x}) > 0 \\ \dfrac{(a^m - g(\mathbf{x})^m)^n}{a^{mn}} & -a < g(\mathbf{x}) < 0 \\ 0 & g(\mathbf{x}) < -a \end{cases}$$

(3.134)

where parameters $n$, $m$ and $a$ are material properties which have to be obtained from experimental results or atomistic simulations.

It is important to note that unlike conventional XFEM solutions, no additional degrees of freedom are introduced in the XFEM approximation because the problem is solved for known prescribed sliding $\mathbf{b}^\alpha = b_\alpha \mathbf{e}_t^\alpha$ along the sliding plane (Malekafzali, 2010; Mohammadi and Malekafzali, 2011).

## 3.7.11  Hydraulic Fracture Enrichment

Crack-tip enrichment functions for a hydraulic fracture problem can be defined as (Lecampion, 2009; Goodarzi, Mohammadi and Jafari, 2011),

$$\{F_l(r, \theta)\}_{l=1}^4 = \{r^\lambda \cos(\lambda\theta), r^\lambda \sin(\lambda\theta), r^\lambda \sin\theta \sin(\lambda\theta), r^\lambda \sin\theta \cos(\lambda\theta)\} \tag{3.135}$$

where $\lambda$ is the power of singularity, determined from $K_m$ which includes the effects of solid material modulus $E'$, defined in (2.113), fracture toughness $K_c$, fluid viscosity $\mu$ and crack propagation velocity $v_c$,

$$K_m = \frac{K_c}{\sqrt{12\mu E' v_c}} \tag{3.136}$$

For large values of $K_m$, the problem is toughness dominated, (Lecampion, 2009)

$$\lambda = \frac{1}{2} \tag{3.137}$$

and the fluid pressure $p$ crack can be represented by,

$$p^h(\mathbf{x}) = \sum_{j=1}^n N_j(\mathbf{x}) p_j + \sum_{k=1}^{mt} N_k(\mathbf{x}) \left( \sum_{l=1}^{mf} F_l(x) \mathbf{b}_k^l \right) \tag{3.138}$$

$$\{F_l(r, \theta)\}_{l=1}^4 = \{0, H(\mathbf{x} - \mathbf{x}_{tip}), 0, 0\} \tag{3.139}$$

In contrast, the hydraulic facture problem is viscosity controlled for low values of $K_m$ (Lecampion, 2009),

$$\lambda = \frac{2}{3} \tag{3.140}$$

and

$$\{F_l(r, \theta)\}_{l=1}^4 = \left\{0, (\mathbf{x} - \mathbf{x}_{tip})^{-\frac{1}{3}} H(\mathbf{x} - \mathbf{x}_{tip}), 0, 0\right\} \tag{3.141}$$

## 3.7.12  Plastic Enrichment

Elguedj, Gravouil and Combescure (2006) proposed the concept of plastic enrichment to include effects of crack-tip plastification based on the Ramberg–Osgood plasticity model. In principle, other plasticity models may also be adopted with some modifications. Beginning with the Hutchinson–Rice–Rosengren power law hardening material model (Hutchinson, 1968):

$$\frac{\varepsilon}{\varepsilon_{yld}} = \frac{\sigma}{\sigma_{yld}} + k_0 \left( \frac{\sigma}{\sigma_{yld}} \right)^n \tag{3.142}$$

Asymptotic crack-tip displacement, strain and stress fields can be defined as (Elguedj, Gravouil and Combescure, 2006):

$$u_i = k_0 \varepsilon_{yld} r \left( \frac{J}{k_0 \sigma_{yld} \varepsilon_{yld} I_n r} \right)^{\frac{n}{n+1}} \bar{u}_i(\theta, n) \tag{3.143}$$

$$\varepsilon_{ij} = k_0 \varepsilon_{yld} \left( \frac{J}{k_0 \sigma_{yld} \varepsilon_{yld} I_n r} \right)^{\frac{n}{n+1}} \bar{\varepsilon}_{ij}(\theta, n) \tag{3.144}$$

$$\sigma_{ij} = \sigma_{yld} \left( \frac{J}{k_0 \sigma_{yld} \varepsilon_{yld} I_n r} \right)^{\frac{1}{n+1}} \bar{\sigma}_{ij}(\theta, n) \tag{3.145}$$

where $J$ is the well-known contour integral and $I_n$ is a dimensionless constant that depends on $n$. The terms $\bar{\sigma}_{ij}$, $\bar{\varepsilon}_{ij}$ and $\bar{u}_{ij}$ are dimensionless angular functions. Applying a Fourier decomposition on $\bar{u}_{ij}$ for modes I and II allows the pure mode I and II displacement fields to be defined from the following basis function

$$r^{\frac{1}{n+1}} \left\{ \left( \cos \frac{k\theta}{2}, \sin \frac{k\theta}{2} \right); k \in [1, 3, 5, 7] \right\} \tag{3.146}$$

In practice, Elguedj, Gravouil and Combescure (2006) have used and compared the following options derived from Eq. (3.146):

$$r^{\frac{1}{n+1}} \left\{ \sin \frac{\theta}{2}, \cos \frac{\theta}{2}, \sin \frac{\theta}{2} \sin \theta, \cos \frac{\theta}{2} \sin \theta, \sin \frac{\theta}{2} \sin 2\theta, \cos \frac{\theta}{2} \sin 2\theta \right\} \tag{3.147}$$

$$r^{\frac{1}{n+1}} \left\{ \sin \frac{\theta}{2}, \cos \frac{\theta}{2}, \sin \frac{\theta}{2} \sin \theta, \cos \frac{\theta}{2} \sin \theta, \sin \frac{\theta}{2} \sin 3\theta, \cos \frac{\theta}{2} \sin 3\theta \right\} \tag{3.148}$$

$$r^{\frac{1}{n+1}} \left\{ \sin \frac{\theta}{2}, \cos \frac{\theta}{2}, \sin \frac{\theta}{2} \sin \theta, \cos \frac{\theta}{2} \sin \theta, \right.$$
$$\left. \sin \frac{\theta}{2} \sin 2\theta, \cos \frac{\theta}{2} \sin 2\theta, \sin \frac{\theta}{2} \sin 3\theta, \cos \frac{\theta}{2} \sin 3\theta \right\} \tag{3.149}$$

### 3.7.13   Viscoelastic Enrichment

The correspondence theory has been frequently used to co-relate the viscoelastic response to its equivalent linear elastic behaviour. Accordingly, Zhang, Rong and Li (2010) and TianTang and QingWen (2011) have illustrated that the displacement fields around a crack tip within a viscoelastic domain can be written in the general forms of,

$$
u_x = (1 + \nu)I(t)K_{\mathrm{I}}(t)\sqrt{\frac{r}{2\pi}}\left[\kappa - 1 + 2\sin^2\frac{\theta}{2}\right]\cos\frac{\theta}{2}
$$
$$
+ (1 + \nu)I(t)K_{\mathrm{II}}(t)\sqrt{\frac{r}{2\pi}}\left[\kappa + 1 + 2\cos^2\frac{\theta}{2}\right]\sin\frac{\theta}{2} \qquad (3.150)
$$

$$
u_x = (1 + \nu)I(t)K_{\mathrm{I}}(t)\sqrt{\frac{r}{2\pi}}\left[\kappa + 1 - 2\cos^2\frac{\theta}{2}\right]\sin\frac{\theta}{2}
$$
$$
- (1 + \nu)I(t)K_{\mathrm{II}}(t)\sqrt{\frac{r}{2\pi}}\left[\kappa - 1 - 2\sin^2\frac{\theta}{2}\right]\cos\frac{\theta}{2} \qquad (3.151)
$$

where $I(t)$ is the corresponding creep compliance, determined from the viscoelastic material properties (Zhang, 2006).

As a result, despite the time dependency of the response and the stress intensity factors, the isotropic enrichment functions (3.95) can be used for radial and angular enhancement of the crack-tip displacement approximation.

$$
\{F_l(r, \theta)\}_{l=1}^{4} = \left\{\sqrt{r}\cos\frac{\theta}{2}, \sqrt{r}\sin\frac{\theta}{2}, \sqrt{r}\sin\theta\cos\frac{\theta}{2}, \sqrt{r}\sin\theta\sin\frac{\theta}{2}\right\} \qquad (3.152)
$$

### 3.7.14   Contact Corner Enrichment

Ebrahimi (2012) and Ebrahimi, Mohammadi and Kani (2012) developed an iterative approach to solve for the dominant singularity power $\lambda$ at the corner of a contact wedge at various free/stick/slip contact boundary conditions (see Figure 3.21).

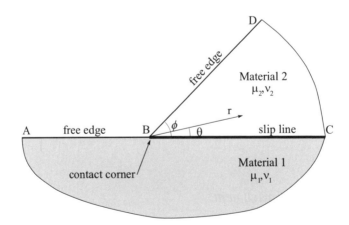

**Figure 3.21**   Configuration of a complete sliding contact.

The singularity power $\lambda$ for a complete sliding punch on a dissimilar plane surface can be obtained from the following characteristic equation (see Ebrahimi, 2012; Ebrahimi, Mohammadi and Kani, 2012)

$$8(1+\lambda)\sin\lambda\pi\left\{(1+\alpha)\cos\lambda\pi(\sin^2\lambda\phi-\lambda^2\sin^2\phi)+\frac{1}{2}(1-\alpha)\sin\lambda\pi(\sin 2\lambda\phi+\lambda\sin 2\phi)\right.$$
$$\left.+\mu\sin\lambda\pi\left[(1-\alpha)\lambda(1+\lambda)\sin^2\phi-2\beta(\sin^2\lambda\phi-\lambda^2\sin^2\phi)\right]\right\}=0$$

(3.153)

where $\phi$ is the wedge angle and $\alpha$ and $\beta$ are the Dundurs parameters for dissimilar bimaterials,

$$\alpha=\frac{(\mu_2/\mu_1)(\kappa_1+1)-(\kappa_2+1)}{(\mu_2/\mu_1)(\kappa_1+1)+(\kappa_2+1)}$$

(3.154)

$$\beta=\frac{(\mu_2/\mu_1)(\kappa_1-1)-(\kappa_2-1)}{(\mu_2/\mu_1)(\kappa_1+1)+(\kappa_2+1)}$$

(3.155)

Ebrahimi (2012) and Ebrahimi, Mohammadi and Kani (2012) developed a local Newton's iterative method to solve for the dominant $\lambda$. The contact corner enrichment functions are then defined as,

$$\{F_l(r,\theta)\}_{l=1}^4=\left\{r^\lambda\cos(\lambda\theta),r^\lambda\sin(\lambda\theta),r^\lambda\sin\theta\sin(\lambda\theta),r^\lambda\sin\theta\cos(\lambda\theta)\right\}$$

(3.156)

### 3.7.15   Modification for Large Deformation Problems

Bouchbinder, Livne and Finebergm (2009, 2010) have computed the complete weakly non-linear quasi-static solution for the displacement field for neo-Hookean materials,

$$u_x=\frac{K_I}{4\mu}\sqrt{\frac{r}{2\pi}}\left[\frac{7}{3}\cos\frac{\theta}{2}-\cos\frac{3\theta}{2}\right]$$
$$+\left(\frac{K_I}{4\mu\sqrt{2\pi}}\right)^2\left[-\frac{1}{15}\ln r-\frac{52}{45}\left(\ln r+\frac{3}{4}\sin^2\theta\right)-\frac{103}{48}\cos\theta+\frac{26}{15}\cos 2\theta-\frac{3}{16}\cos 3\theta\right]$$

(3.157)

$$u_y=\frac{K_I}{4\mu}\sqrt{\frac{r}{2\pi}}\left[\frac{13}{3}\sin\frac{\theta}{2}-\sin\frac{3\theta}{2}\right]$$
$$+\left(\frac{K_I}{4\mu\sqrt{2\pi}}\right)^2\left[\frac{\theta}{15}-\frac{52}{45}\left(\frac{\theta}{4}-\frac{3}{8}\sin 2\theta\right)-\frac{61}{48}\sin\theta+\frac{26}{15}\sin 2\theta-\frac{3}{16}\sin 3\theta\right]$$

(3.158)

As a result, in addition to the term $\sqrt{r}$, a new term $\ln r$ can be included in the enrichment function.

In a total Lagrangian formulation for large deformation analysis, the basic XFEM stiffness formulation (3.43) is modified to (Rashetnia, 2012):

$$(\mathbf{K}_{ab})_{ij}=\sum_{I,J,K,L=1}^3\int_{\Omega^e}\mathbf{F}_{iI}N_{a,J}\mathbf{D}_{IJKL}N_{b,K}\mathbf{F}_{jL}\,d\Omega,\quad(r,s=\mathbf{u},\mathbf{a},\mathbf{b})$$

(3.159)

or in a more appropriate form of

$$\mathbf{K}_{ij}^{rs} = \int_{\Omega^e} (\mathbf{B}_{0a}^r)^{\mathrm{T}} \mathbf{D} \mathbf{B}_{0b}^s \, d\Omega \qquad (3.160)$$

where $\mathbf{B}_0$ is defined in terms of the derivatives of the shape functions with respect to the initial configuration $\mathbf{X} = (X, Y)$ and the components of the deformation gradient $\mathbf{F}$ (Rashetnia, 2012),

$$\mathbf{B}_{0i}^{\mathbf{u}} = \begin{bmatrix} N_{i,X}\mathsf{F}_{11} & N_{i,X}\mathsf{F}_{21} \\ N_{i,Y}\mathsf{F}_{11} & N_{i,Y}\mathsf{F}_{22} \\ N_{i,X}\mathsf{F}_{12} + N_{i,Y}\mathsf{F}_{11} & N_{i,X}\mathsf{F}_{22} + N_{i,Y}\mathsf{F}_{21} \end{bmatrix} \qquad (3.161)$$

$$\mathbf{B}_i^{\mathbf{a}} = \begin{bmatrix} N_{i,X}\mathsf{F}_{11}H & N_{i,X}\mathsf{F}_{21}H \\ N_{i,Y}\mathsf{F}_{11}H & N_{i,Y}\mathsf{F}_{22}H \\ \left[N_{i,X}\mathsf{F}_{12} + N_{i,Y}\mathsf{F}_{11}\right]H & \left[N_{i,X}\mathsf{F}_{22} + N_{i,Y}\mathsf{F}_{21}\right]H \end{bmatrix} \qquad (3.162)$$

$$\mathbf{B}_i^{\mathbf{b}} = \begin{bmatrix} \mathbf{B}_i^{\mathbf{b}1} & \mathbf{B}_i^{\mathbf{b}2} & \mathbf{B}_i^{\mathbf{b}3} & \mathbf{B}_i^{\mathbf{b}4} \end{bmatrix} \qquad (3.163)$$

$$\mathbf{B}_i^{b\alpha} = \begin{bmatrix} (N_i F_\alpha)_{,X}\,\mathsf{F}_{11} & (N_i F_\alpha)_{,X}\,\mathsf{F}_{21} \\ (N_i F_\alpha)_{,Y}\,\mathsf{F}_{12} & (N_i F_\alpha)_{,Y}\,\mathsf{F}_{22} \\ (N_i F_\alpha)_{,X}\,\mathsf{F}_{12} + (N_i F_\alpha)_{,Y}\,\mathsf{F}_{11} & (N_i F_\alpha)_{,X}\,\mathsf{F}_{22} + (N_i F_\alpha)_{,Y}\,\mathsf{F}_{21} \end{bmatrix} \qquad (3.164)$$

where the deformation gradient $\mathsf{F}_{il}$ is defined as the derivative of the current configuration $\mathbf{x}$ with respect to the initial configuration $\mathbf{X}$,

$$\mathsf{F}_{il} = \frac{\partial \mathbf{x}_i}{\partial \mathbf{X}_I} = \begin{bmatrix} \mathsf{F}_{11} & \mathsf{F}_{12} \\ \mathsf{F}_{21} & \mathsf{F}_{22} \end{bmatrix} \qquad (3.165)$$

To include the effects of interpolation, the following shifting amendments are required:

$$\mathbf{B}_i^{\mathbf{a}} = \begin{bmatrix} N_{i,X}\mathsf{F}_{11}\bar{H}_i & N_{i,X}\mathsf{F}_{21}\bar{H}_i \\ N_{i,Y}\mathsf{F}_{11}\bar{H}_i & N_{i,Y}\mathsf{F}_{22}\bar{H}_i \\ \left[N_{i,X}\mathsf{F}_{12} + N_{i,Y}\mathsf{F}_{11}\right]\bar{H}_i & \left[N_{i,X}\mathsf{F}_{22} + N_{i,Y}\mathsf{F}_{21}\right]\bar{H}_i \end{bmatrix} \qquad (3.166)$$

$$\mathbf{B}_i^{b\alpha} = \begin{bmatrix} \left(N_i \bar{F}_{\alpha i}\right)_{,X}\mathsf{F}_{11} & \left(N_i \bar{F}_{\alpha i}\right)_{,X}\mathsf{F}_{21} \\ \left(N_i \bar{F}_{\alpha i}\right)_{,Y}\mathsf{F}_{12} & \left(N_i \bar{F}_{\alpha i}\right)_{,Y}\mathsf{F}_{22} \\ \left(N_i \bar{F}_{\alpha i}\right)_{,X}\mathsf{F}_{12} + \left(N_i \bar{F}_{\alpha i}\right)_{,Y}\mathsf{F}_{11} & \left(N_i \bar{F}_{\alpha i}\right)_{,X}\mathsf{F}_{22} + \left(N_i \bar{F}_{\alpha i}\right)_{,Y}\mathsf{F}_{21} \end{bmatrix} \qquad (3.167)$$

where

$$\bar{H}_i = H(\xi) - H(\xi_i) \qquad (3.168)$$

$$\bar{F}_{\alpha i} = F_\alpha - F_{\alpha i}, \quad \alpha = 1 - 4 \qquad (3.169)$$

### 3.7.16   Automatic Enrichment

A number of studies have been devoted to the development of automatic or adaptive procedures for defining the appropriate crack-tip enrichment functions in different conditions. They are, however, in their earlier development stages and cannot be used for general problems nor can they compete with the existing enrichment functions.

## 3.8   Transition from Standard to Enriched Approximation

Application of the enrichment for near crack-tip analysis may lead to solution incompatibility and interior discontinuities, if it is not employed in the entire domain of consideration. The reason can be attributed to different orders of approximation for neighbouring domains while each domain follows a different basis function. As a result, different values may be obtained for the common nodes; an indication of the occurrence of internal discontinuities.

From a different point of view, finite elements used for modelling an entire domain may be classified into three categories: standard finite elements, elements with enriched nodes and partially enriched elements which consist of standard and enriched nodes (Figure 3.22). The first two categories are fully governed by either the classic FEM or XFEM approximations, whereas the third category (blending elements) is only partially involved with XFEM.

Three different types of blending elements are shown in Figure 3.22. The typical element A has three enriched nodes and one standard node, while the element B has two enriched and two standard nodes. Element C has the least number of enriched nodes in comparison to the number of standard nodes.

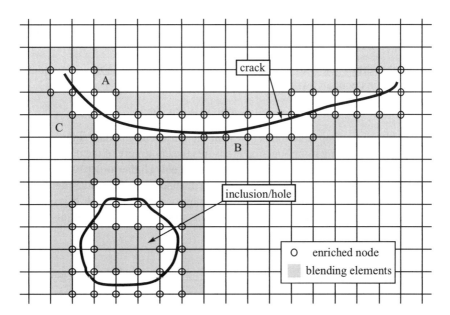

**Figure 3.22**   Standard, enriched and blending elements (Mohammadi, 2008). (Reproduced by permission of John Wiley & Sons, Ltd.)

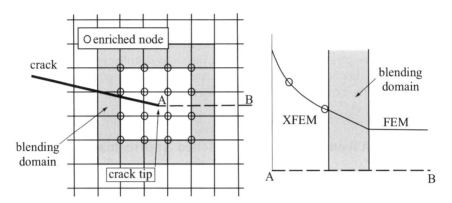

**Figure 3.23** A smooth transition between the enriched (XFEM) and standard (FEM) approximations (Mohammadi, 2008).

Now, consider a typical blending element B. Approximation of the displacement field for this element can be expressed as,

$$\mathbf{u}^h(\mathbf{x}) = \sum_{j=1}^{4} N_j(\mathbf{x})\mathbf{u}_j + \sum_{k=1}^{2} N_k(\mathbf{x})\psi(\mathbf{x})\mathbf{a}_k \tag{3.170}$$

The second part of Eq. (3.170) is no longer a partition of unity. Nevertheless, it has usually little direct effect on the approximation because such a blending element does not include any singularity at all.

### 3.8.1 Linear Blending

A remedy to this drawback is to design a blending procedure over a transition domain connecting the domains with and without enrichment (Figure 3.23):

$$\mathbf{u}^h(\mathbf{x}) = (1 - R)\mathbf{u}(\mathbf{x}) + R\mathbf{u}^{\text{enr}}(\mathbf{x}) \tag{3.171}$$

where $R$ is a blending ramp function set to 1 on the enriched boundary and 0 on the linear boundary. A linear blending ramp function $R$ ensures the continuity of the displacement field, while it cannot guarantee the continuity of the strain field. Higher order blending functions are, therefore, required to ensure continuous strain (displacement derivative) fields.

The same problem of internal discontinuities may occur in both intrinsic and extrinsic enrichments, if different types of approximation are to be used for modelling near and far fields.

### 3.8.2 Hierarchical Transition Domain

In this method, a new term, $\mathbf{u}^{\text{tra}}(\mathbf{x})$, is added to the original XFEM approximation (3.27) to overcome the incompatibility between the enriched and non-enriched domains by introduction of a transition (blending) zone.

$$\mathbf{u}^h(\mathbf{x}) = \mathbf{u}^{\text{XFEM}}(\mathbf{x}) + \mathbf{u}^{\text{tra}}(\mathbf{x}) \tag{3.172}$$

where $\mathbf{u}^{\text{XFEM}}$ is the original extended finite element method approximation (3.27) to calculate the displacement for a point $\mathbf{x}$ located within the domain (neglecting the weak discontinuity term),

$$\mathbf{u}^h(\mathbf{x}) = \mathbf{u}^h(\mathbf{x}) + \mathbf{u}^{\text{H}}(\mathbf{x}) + \mathbf{u}^{\text{tip}}(\mathbf{x}) + \mathbf{u}^{\text{tra}}(\mathbf{x}) \tag{3.173}$$

or more explicitly

$$\mathbf{u}^h(\mathbf{x}) = \left[ \sum_{j=1}^{n} N_j(\mathbf{x})\mathbf{u}_j \right] + \left[ \sum_{h=1}^{mh} N_h(\mathbf{x})H(\mathbf{x})\mathbf{a}_h \right]$$
$$+ \left[ \sum_{k=1}^{mt} N_k(\mathbf{x}) \left( \sum_{l=1}^{mf} F_l(\mathbf{x})\mathbf{b}_k^l \right) \right] + \left[ \sum_{m=1}^{mst} \bar{N}_m(\mathbf{x})\mathbf{c}_m + \sum_{n=1}^{msh} \bar{N}_n(\mathbf{x})H(\mathbf{x})\mathbf{d}_n \right] \tag{3.174}$$

where $\mathbf{u}_j$ is the displacement vector at node $j$, $\mathbf{b}_k^l$ are the added DOFs for tip enrichments $F_l$, $\mathbf{a}_h$ are the added DOFs for the crack discontinuity approximation by the Heaviside function $H$, and $\mathbf{c}_i$ and $\mathbf{d}_j$ are added DOFs which correspond to hierarchical nodes in the sets $mst$ and $msh$, respectively. $N_i$ are the standard finite element shape functions and $\bar{N}_i$ are the hierarchical shape functions for the transition domain.

Two types of hierarchical nodes exist: hierarchical nodes, such as $t_i$ in Figure 3.24, that correspond to edges connecting non-enriched and tip-enriched nodes (defined by $mst$), and the typical nodes $s_i$ in Figure 3.24, associated with those edges that connect tip and Heaviside-enriched nodes (defined by $mth$). Tarancon et al. (2009) have discussed the definition of hierarchical shape functions $\bar{N}_i$, as proposed by Szabo and Babuska (1991). For example, for

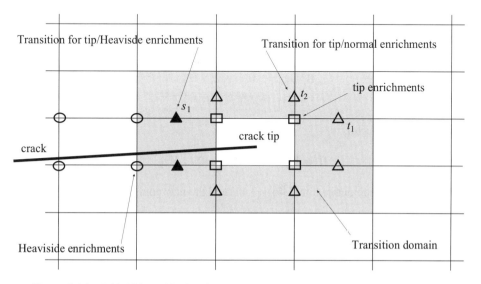

**Figure 3.24** Added hierarchical nodes in the blending elements (Tarancon et al., 2009).

nodes $t_1, t_2 \in msh$ in Figure 3.24, the hierarchical shape functions in terms of the isoparametric coordinates $\xi, \eta$ are defined as (Szabo and Babuska, 1991)

$$\bar{N}_{t_1}(\xi, \eta) = -\frac{1}{2}\sqrt{\frac{3}{2}}(1 - \xi^2)\frac{1 - \eta}{2} \tag{3.175}$$

$$\bar{N}_{t_2}(\xi, \eta) = -\frac{1}{2}\sqrt{\frac{3}{2}}(1 - \eta^2)\frac{1 - \xi}{2} \tag{3.176}$$

Discretization of Eq. (3.38) using the XFEM approximation (3.172) results in a discrete system of linear equilibrium equations:

$$\mathbf{K}\mathbf{u}^h = \mathbf{f} \tag{3.177}$$

where $\mathbf{u}^h$ is the vector of degrees of nodal freedom (for both classical and enriched ones),

$$\mathbf{u}^h = \{\mathbf{u} \quad \mathbf{a} \quad \mathbf{b}_\alpha \quad \mathbf{c} \quad \mathbf{d}\}^T \tag{3.178}$$

The global stiffness matrix $\mathbf{K}$ is assembled from the stiffness of each element $\mathbf{K}^e_{ij}$ with components $\mathbf{K}^{rs}_{ij}$

$$\mathbf{K}^{rs}_{ij} = \int_{\Omega^e} (\mathbf{B}^r_i)^T \mathbf{D}\mathbf{B}^s_j \, d\Omega \quad (r, s = \mathbf{u}, \mathbf{a}, \mathbf{b}, \mathbf{c}, \mathbf{d}) \tag{3.179}$$

where $\mathbf{B}^u_i$, $\mathbf{B}^a_i$ and $\mathbf{B}^b_i$ are derivatives of shape functions defined in Eqs. (3.40), (3.50) and (3.51), respectively, and $\mathbf{B}^c_i$ and $\mathbf{B}^d_i$ are defined as,

$$\mathbf{B}^c_i = \begin{bmatrix} \bar{N}_{i,x} & 0 \\ 0 & \bar{N}_{i,y} \\ \bar{N}_{i,y} & \bar{N}_{i,x} \end{bmatrix} \tag{3.180}$$

$$\mathbf{B}^d_i = \begin{bmatrix} (\bar{N}_i H)_{,x} & 0 \\ 0 & (\bar{N}_i H)_{,y} \\ (\bar{N}_i H)_{,y} & (\bar{N}_i H)_{,x} \end{bmatrix} \tag{3.181}$$

$\mathbf{f}$ is the nodal force vector,

$$\mathbf{f}^e_i = \{\mathbf{f}^u_i \quad \mathbf{f}^a_i \quad \mathbf{f}^{b_\alpha}_i \quad \mathbf{f}^c_i \quad \mathbf{f}^d_i\}^T \tag{3.182}$$

where $\mathbf{f}^u_i$, $\mathbf{f}^a_i$ and $\mathbf{f}^{b_\alpha}_i$ are defined in (3.46)–(3.47) and the new components are:

$$\mathbf{f}^c_i = \int_{\Gamma_t} \bar{N}_i \mathbf{f}^t \, d\Gamma + \int_{\Omega^e} \bar{N}_i \mathbf{f}^b \, d\Omega \tag{3.183}$$

$$\mathbf{f}^d_i = \int_{\Gamma_t} \bar{N}_i H \mathbf{f}^t \, d\Gamma + \int_{\Omega^e} \bar{N}_i H \mathbf{f}^b \, d\Omega \tag{3.184}$$

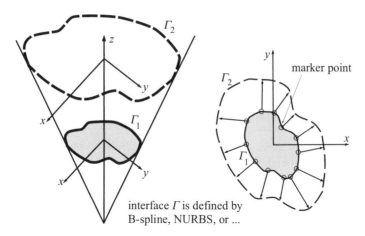

**Figure 3.25**   A simple description of LSM, including the original front projected on the *xy* plane and the level set function as the intersection of the surface and the *xy* plane (Mohammadi, 2008).

## 3.9   Tracking Moving Boundaries

One important aspect of problems with moving interfaces is to track them as they evolve. Most conventional numerical techniques attempt to follow moving boundaries by putting a collection of marker points on the evolving front. This is usually performed by the use of general B-spline or non-uniform rational B-spline (NURBS) functions (Figure 3.25) (Patrikalakis, 2003). The positions of the particles are then advanced in accordance with a set of finite difference approximations to the equations of motion. Such schemes usually become unstable around points of high curvature and cusps. The reason can be attributed to the fact that any small error in determining the position may produce large errors in the evaluation of the curvature (Sethian, 1987, 1996, 1999a, 1999b; Adalsteinsson and Sethian, 2003).

This idea of tracking evolving fronts with the equations of motion forms an important numerical approach based on partial differential equations. Accordingly, two different techniques have been developed: (i) a more general but slower general-purpose time-dependent level set method; (ii) a very efficient but limited-purpose fast marching method. Both methods are designed to handle problems in which the separating interfaces develop sharp corners and cusps, change topology, break apart and merge together. These techniques have a wide range of applications, including problems in fluid mechanics, combustion, manufacturing computer chips, computer animation, image processing, the structure of snowflakes, and the shape of soap bubbles (Sethian, 1987). In this section, a summarized review of the essential parts of algorithms for tracking moving boundaries is presented from Mohammadi (2008).

### 3.9.1   Level Set Method

A powerful tool for tracking interfaces is the level set method (LSM). Though it is not mandatory to use level sets in XFEM, many XFEM formulations take advantage of the level set method.

The level set approach, introduced by Osher and Sethian (1988), instead of following the curve interface itself, builds a surface from the original curve. A major property of this cone-shaped surface is that it intersects the *xy* plane exactly where the curve sits. Such a surface is called the level set function because it accepts a point in the plane and generates back its height (level) (Figure 3.25).

It seems odd to replace the problem of a moving curve with a moving surface. However, the level set function is well behaved and all the complicated problems of breaking and merging can be easily handled.

In LSM, the interface of interest is represented as the zero level set of a function $\phi(\mathbf{x})$. This function is one dimension higher than the dimension of the interface. The evolution equation for the interface can then be expressed as an equation for the evolution of $\phi$.

There are many advantages to using LSM for tracking interfaces. First, unlike many other interface tracking schemes, the motion of the interface is computed on a fixed Eulerian mesh. Second, the method handles changes in the topology of the interface naturally. Third, the method can be easily extended to higher dimensions. Finally, the geometric properties of the interface can be obtained from the level set function $\phi$.

One drawback of LSM is that the level set representation requires a function of a higher dimension than the original crack, potentially leading to higher storage and computational costs.

### 3.9.1.1 Definition of the Level Set Function

Consider a domain $\Omega$ divided into two non-overlapping subdomains, $\Omega_1$ and $\Omega_2$, sharing an interface $\Gamma$, as illustrated in Figure 3.26. The level set function $\phi(\mathbf{x})$ is defined as:

$$\phi(\mathbf{x}) = \begin{cases} > 0 & \mathbf{x} \in \Omega_1 \\ = 0 & \mathbf{x} \in \Gamma \\ < 0 & \mathbf{x} \in \Omega_2 \end{cases} \tag{3.185}$$

An interpretation of Eq. (3.185) is that the interface $\Gamma$ can be regarded as the zero level contour of the level set function $\phi(\mathbf{x})$.

One of the common choices for the level set function, $\phi(\mathbf{x})$, can then be simply defined in terms of the signed distance function:

$$\phi(\mathbf{x}) = \xi(\mathbf{x}) = \begin{cases} d & \mathbf{x} \in \Omega_1 \\ -d & \mathbf{x} \in \Omega_2 \end{cases} \tag{3.186}$$

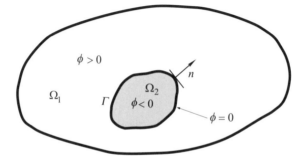

**Figure 3.26**  Definition of the level set function (Mohammadi, 2008).

where $d$ is the normal distance from a point $\mathbf{x}$ to the interface $\Gamma$. Using the definition (3.47) of the Heaviside function $H(\xi)$, domains $\Omega_1$ and $\Omega_2$ can then be defined as

$$\begin{cases} \Omega_1 = \{\mathbf{x} \in \Omega, H(\phi(\mathbf{x})) = 1\} \\ \Omega_2 = \{\mathbf{x} \in \Omega, H(-\phi(\mathbf{x})) = 1\} \end{cases} \tag{3.187}$$

The normal vector $\mathbf{n}$ to the interface $\Gamma$ at a point $\mathbf{x} \in \Gamma$ can then be defined as:

$$\mathbf{n} = \frac{\nabla \phi(\mathbf{x})}{\|\nabla \phi(\mathbf{x})\|} \tag{3.188}$$

Discretization of the level set allows for the evaluation of the level set at the element level based on the nodal level set values $\phi_j = \phi(\mathbf{x}_j)$ and known finite element shape functions $N_j(\mathbf{x})$:

$$\phi(\mathbf{x}) = \sum_{j=1}^{n} N_j(\mathbf{x})\phi_j \tag{3.189}$$

### 3.9.1.2   Evolving Surfaces

In the case of a moving surface, its position is no longer known *a priori*. From the fact that the level set is zero on the surface, its material time derivative has to vanish:

$$\frac{D\phi(\mathbf{x}, t)}{Dt} = 0 \tag{3.190}$$

which can be written in the form of the Hamilton–Jacobi equation of motion:

$$\frac{\partial \phi(\mathbf{x}, t)}{\partial t} + \nabla \phi(\mathbf{x}, t) \cdot \mathbf{v}(\mathbf{x}, t) = 0 \tag{3.191}$$

An appropriate incremental form using a first-order time integration scheme can be used to solve (3.191). This, however, is unlikely to be appropriate for crack propagation analyses, as the crack speed is not generally known.

### 3.9.1.3   Level Sets for a Crack

One of the main difficulties in the application of the level set method to crack problems is constraining the evolution of the signed distance function while the crack propagates so that the existing crack surface remains frozen. Since level sets are generally updated by the integration of the Hamilton–Jacobi equation, special techniques have to be adopted for cracks so the level sets describing the existing crack are not modified (Ventura, Budyn and Belytschko, 2003). Another reason for a new approach may be attributed to this fact that the level set functions are not updated with the speed of an interface in the direction normal to itself but with the speed at the crack fronts.

A different approach proposed by Ventura, Budyn and Belytschko (2003) is based on the vector level set formulation and avoids the difficulty mentioned above. In this method, the level set is only defined on a narrow band around the crack and the evolution of the level set function does not alter the previously formed crack. The method takes into consideration the effect of new points having a geometric closest point projection onto a segment when a crack advances. A simple updating procedure will then allow the inclusion of advancing cracks.

The previous definition of the level set for closed interfaces must be modified or altered if it is to be used for an open curve such as a crack. One level set $\phi$ is not generally sufficient to describe the crack, and another level set $\psi$ at the crack tip is required. A one-dimensional crack growth in a level set framework is modelled by representing the crack as the zero level set of a function $\psi(\mathbf{x}, t)$. An endpoint of the crack is represented as the intersection of the zero level set of $\psi(\mathbf{x}, t)$ with a zero level set of the function $\phi^k(\mathbf{x}, t)$, where $k$ is the number of tips on a given crack. The crack tip level set $\psi$ is generally assumed to be orthogonal to $\phi$

$$\phi_{,i}\psi_{,i} = 0 \tag{3.192}$$

The values of the level set functions are stored only at the nodes. The functions can be interpolated over the mesh by the same finite element shape functions (Stolarska and Chopp, 2003),

$$\phi^k(\mathbf{x}, t) = \sum_{j=1}^{n} N_j(\mathbf{x}) \left( \phi^k(\mathbf{x}, t) \right)_j \tag{3.193}$$

$$\psi(\mathbf{x}, t) = \sum N_j(\mathbf{x}) \psi_j(\mathbf{x}, t) \tag{3.194}$$

An important consideration is that, although the actual crack is embedded inside a domain, the zero level set of $\psi$ cuts through the entire domain. It is also assumed that once a part of a crack has formed, that part will no longer change shape or move.

## 3.9.2   Alternative Methods

### 3.9.2.1   Fast Marching Method

The fast marching method (FMM) was first introduced by Sethian (1996) and later improved by Sethian (1999a) and Chopp (2001). FMMs are designed to track a propagating interface and to find the first arrival of the interface as it passes a point. They are limited to problems in which the speed of propagation is isotropic; the speed function never changes sign, so that the front is always moving forward or backward. That speed can change from point to point, but there is no preferred direction. This allows the problem to be converted into a stationary formulation, which provides a tremendous speed. This is in contrast to level set methods that are designed for problems in which the speed function can be positive in some places and negative in others. As a result, the front may move forwards and backwards.

To explain the approach, consider a network in which a different cost has been assigned for reaching each node. In an optimal control, the cost of reaching a point depends on both where the present standing point is and the direction of movement. First, the starting point is placed in a set of accepted nodes. Grid points which are one link away are considered as neighbour or candidate nodes. Then the correct cost of reaching each of these candidates is computed. The node with smallest cost is removed from the set of candidates and added to the set of accepted nodes. The procedure continues by computing the cost and adding any new candidates that are not already accepted. The process terminates when all points are accepted. The algorithm is in fact a systematic ordering procedure for determining the cost of reaching points from a known starting point.

Fast marching methods have many desirable qualities. They do not require any iterative procedure and provide the solution in one pass, which allows a systematic ordering update of the points so that each point is touched only once. As a result, the method allows for very fast computation of order $O(n_t \log n_t)$, where $n_t$ is the total number of grid points. For a set of distributed finite element nodes, the methodology described by Sukumar, Chopp and Moran (2003) can be followed.

### 3.9.2.2 Ordered Upwind Method

In FMM, the solution is systematically updated from known values to unknown spots. At each step, one exploits the fact that the gradient of the front is in the direction from which information must come. This, however, is not true when the speed is not isotropic.

A possible solution is to keep track of the characteristic directions, defined as the ratio between the fastest and slowest speed at each point. Accordingly, the entire Dijkstra methodology can be held while modifying it to include anisotropic speeds. This method is called the ordered upwind method (OUM). This maintains the procedure of point ordering, while systematically computing the solution by relying on previously known computed information.

OUMs have been developed in both semi-Lagrangian and fully Eulerian versions. They use partial information about the characteristic directions, obtained by examining the anisotropy ratio between fastest and slowest speeds to decouple the large system of coupled nonlinear discretized equations, producing one pass algorithms of greatly reduced computational efforts of an order $O(n_t \log n_t)$ (Sethian, 2001).

Similar methodology has been reported as part of the fast marching method by a number of references. For further details on the fast marching and ordered upwind methods see Sukumar, Chopp and Moran (2003) and Sethian (2001).

## 3.10 Numerical Simulations

In this section several examples of various XFEM simulations of fracture mechanics problems are presented. A wide range of applications are covered to demonstrate the flexibility and efficiency of XFEM. Applications related to composites will be comprehensively presented in the corresponding chapters.

### 3.10.1 A Central Crack in an Infinite Tensile Plate

In order to verify the accuracy of XFEM, first a classical isotropic infinite plate with a central crack is considered (Figure 3.27a). Although, the crack length is not important in the infinite domain, the simulation is performed for the case of $2a = 200$. Figure 3.27b shows the typical crack-tip and Heaviside enrichment nodes. Elements that are fully cut by a crack are enriched by the Heaviside enrichment, whereas elements containing a crack tip are enriched by the crack-tip enrichment functions.

Instead of simulating the infinite domain, a finite domain ABCD is simulated, subjected to the analytical displacement field at its boundaries. The ABCD part is discretized by three structured quadrilateral finite element meshes $20 \times 20$, $40 \times 40$ and $80 \times 80$, as depicted in Figure 3.28.

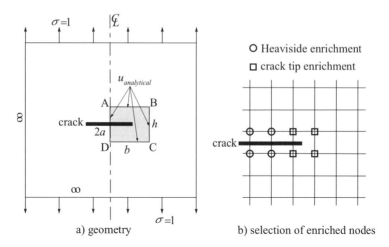

a) geometry

b) selection of enriched nodes

**Figure 3.27**  Geometry of the infinite tensile plate with a central crack.

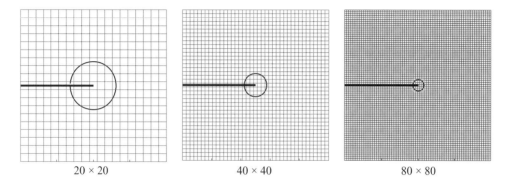

20 × 20                              40 × 40                              80 × 80

**Figure 3.28**  Contour domains for evaluation of the $J$ integral on each crack tip.

Evaluation of the stress intensity factor is performed by the interaction $J$ integral. Figure 3.28 illustrates the contours used for evaluation of the $J$ integral for different finite element meshes. Here, the same number of elements is used in each contour domain. In general, however, the optimum size of the contour integral should be obtained from a sensitivity analysis. Table 3.1 compares the predicted mode I normalized stress intensity factor $\bar{K}_I = K_I/(\sigma \sqrt{\pi a})$ for various meshes, which indicate a close and mesh independent result.

Figure 3.29 illustrates variations of stress components around the crack tip on the deformed shape of the model, magnified by a factor of 20. Clearly, the singular nature of the stress field can be captured by XFEM.

**Table 3.1**  Comparison of normalized stress intensity factors for different finite element meshes.

| Analytical | Mesh 20 × 20 | Mesh 40 × 40 | Mesh 80 × 80 |
| --- | --- | --- | --- |
| 17.725 | 17.363 | 17.281 | 17.242 |

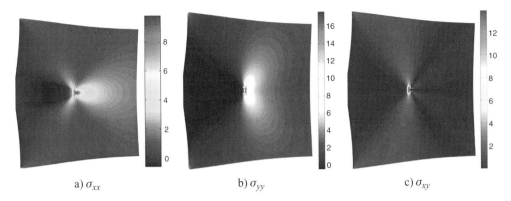

a) $\sigma_{xx}$                                b) $\sigma_{yy}$                                c) $\sigma_{xy}$

**Figure 3.29**   $\sigma_{xx}$, $\sigma_{yy}$ and $\sigma_{xy}$ stress contours on the deformed shape of the plate (magnified by 20).

### 3.10.2   An Edge Crack in a Finite Plate

The second example is a finite tensile plate with an edge crack, as depicted in Figure 3.30.
Uniform finite element meshes are used to simulate the problem for different crack lengths.

Table 3.2 compares the normalized mode I stress intensity factor $\bar{K}_I = K_I/(\sigma \sqrt{\pi a})$ for
various meshes and different crack lengths $a$. A close agreement is observed between the
numerical results and the analytical solution.

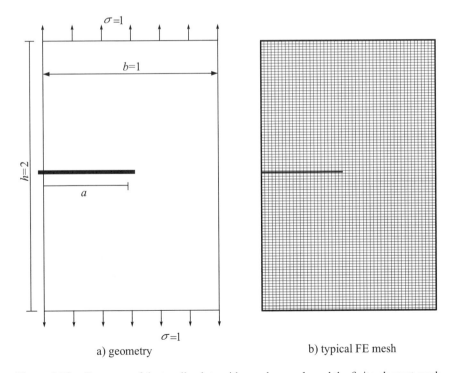

a) geometry                                                b) typical FE mesh

**Figure 3.30**   Geometry of the tensile plate with an edge crack, and the finite element mesh.

**Table 3.2**   Comparison of XFEM predictions for the stress intensity factor with the analytical solution for different crack sizes.

|  |  | XFEM | |
| --- | --- | --- | --- |
| $a$ | Analytical | Mesh $20 \times 40$ | Mesh $40 \times 80$ |
| 0.30 | 1.612 | 1.582 | 1.600 |
| 0.45 | 2.877 | 2.808 | 2.849 |
| 0.60 | 5.29 | 5.321 | 5.438 |

Distribution of the stress components $\sigma_{xx}$, $\sigma_{yy}$ and $\sigma_{xy}$ on the deformed shape of the plate (magnified by 20) are illustrated in Figure 3.31. Again, the singular nature of the stress field is clearly captured by XFEM.

### 3.10.3   Tensile Plate with a Central Inclined Crack

This test is dedicated to a classical mixed mode fracture problem. A tensile plate with a central inclined crack is simulated by a uniform structured finite element mesh, as depicted in Figure 3.32.

The fixed mesh is used to compute the stress intensity factors for all crack angles. Table 3.3 compares the normalized stress intensity factors, predicted by XFEM for various crack angles, with the analytical solution. A very good agreement is observed for all crack angles.

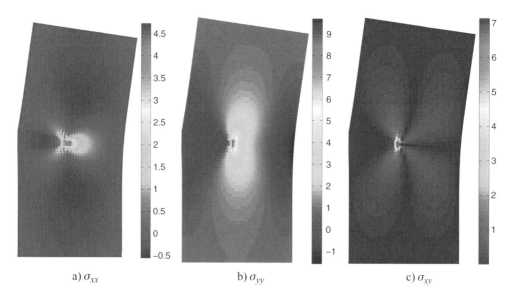

a) $\sigma_{xx}$                            b) $\sigma_{yy}$                            c) $\sigma_{xy}$

**Figure 3.31**   $\sigma_{xx}$, $\sigma_{yy}$ and $\sigma_{xy}$ stress contours on the deformed shape of the plate (magnified by 20).

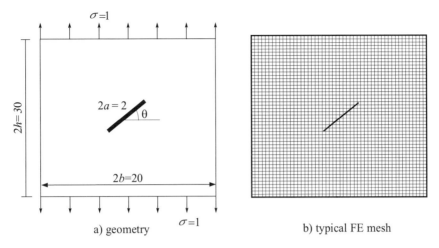

**Figure 3.32** Geometry and the finite element mesh for the mixed mode central crack problem.

### 3.10.4 A Bending Plate in Fracture Mode III

A clamped plate with an edge crack is subjected to two concentrated loads in opposite directions, as depicted in Figure 3.33. Bayesteh and Mohammadi (2011) designed this test to study the effect of rotational and out-of-plane tip enrichments. Elastic material properties are $E = 1000$ and $v = 0.3$. Structured finite element models are used to simulate the problem.

Figure 3.34 illustrates the contours of stress components $\sigma_{xz}$ and $\sigma_{yz}$, clearly indicating the discontinuity along the crack and the singularity at the crack tip.

Bayesteh and Mohammadi (2011) have illustrated that inclusion of out-of-plane enrichments is essential in reproducing the singular stress field. This is, in fact, a direct consequence of the effect of out-of-plane enrichments on out-of-plane displacements for computation of the relevant strain component. Accordingly, it is expected that the bending stress intensity factors $(K_{\theta\theta}, K_{r\theta}, K_{\theta3})$ will show different levels of sensitivity to various tip enrichment functions.

A similar conclusion can be made from fracture mechanics parameters. Table 3.4 compares the values of the stress intensity factor and the lateral displacement for different finite element meshes. While the lateral displacement of the point of loading is insensitive to the number

**Table 3.3** Comparison of normalized mode I and II stress intensity factors with the analytical solution for different crack angles.

| $\alpha°$ | $K_1/K_0$ | | $K_{II}/K_0$ | |
|---|---|---|---|---|
| | Exact | XFEM | Exact | XFEM |
| 0 | 1.000 | 1.001 | 0.000 | 0.000 |
| 15 | 0.933 | 0.935 | 0.250 | 0.250 |
| 30 | 0.750 | 0.752 | 0.433 | 0.433 |
| 45 | 0.500 | 0.502 | 0.500 | 0.500 |
| 60 | 0.250 | 0.251 | 0.433 | 0.434 |
| 75 | 0.067 | 0.067 | 0.250 | 0.250 |

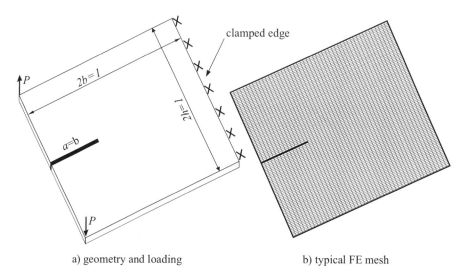

a) geometry and loading                          b) typical FE mesh

**Figure 3.33**   A bending plate in fracture mode III, and the typical finite element mesh (Bayesteh and Mohammadi, 2011).

of adopted enrichment functions, the stress intensity factor remains largely sensitive to those functions, with approximately 40% error when the number of enrichment functions is reduced from 5 to 1.

## 3.10.5   Crack Propagation in a Shell

Bayesteh and Mohammadi (2011) simulated a stiffened pressurized cylindrical shell with a longitudinal crack, as depicted in Figure 3.35, to study the efficiency of XFEM in the simulation of crack propagation in curved shells. The shell contains a longitudinal crack and the stiffening straps are meant to contain the crack propagation. Elastic material properties for the shell and stiffening straps are $E = 71422$ MPa and $v = 0.3$.

As the longitudinal crack propagates in a self-similar quasi-static procedure, it reaches the confining zone of stiffeners, causing the crack to turn to the circumferential direction.

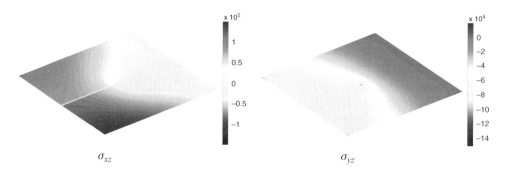

**Figure 3.34**   Contours of stress components (Bayesteh and Mohammadi, 2011).

**Table 3.4** Comparison of stress intensity factor and lateral displacement for different finite elements.

| $N_{elements}$ | $N_{nodes}$ | $N_{enrich.}$ | $N_{DOFs}$ | $K_{III}$ | $u_z$ |
|---|---|---|---|---|---|
| 289 | 936 | 5 | 5048 | 17.100 | 0.152 |
|  |  | 1 | 5016 | 11.284 | 0.153 |
| 1225 | 3816 | 5 | 19673 | 17.077 | 0.153 |
|  |  | 1 | 19641 | 11.662 | 0.153 |
| 2601 | 8008 | 5 | 40833 | 17.083 | 0.153 |
|  |  | 1 | 40801 | 11.770 | 0.153 |
| 4761 | 14560 | 5 | 73818 | 17.089 | 0.153 |
|  |  | 1 | 73786 | 11.833 | 0.153 |

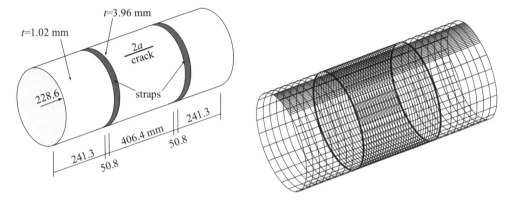

**Figure 3.35** Geometry of the stiffened cylinder and the finite element mesh.

Simulation of a steep turning path for crack propagation on a curved shell is very difficult to perform by a classical finite element approach, if even possible by special treatments such as adaptive refinements.

Conventional crack propagation criteria of fracture mechanics cannot be readily extended to crack propagation on curved surfaces. Therefore, the propagation state is assumed to be governed by the crack-tip opening angle (CTOA) criterion reaching a critical value of 5° (according to Keesecker *et al.*, 2003).

Variation of pressure as a function of half crack length is examined in Table 3.5. It is observed that by increasing the crack length, a very limited increase is expected in the critical pressure. Afterwards, the critical pressure virtually remains unchanged.

**Table 3.5** Comparison of various analyses performed for the stiffened cracked cylinder.

| $a$ (cm) | 4.73 | 6.83 | 8.02 | 10.40 | 12.38 | 14.50 | 16.62 | 18.73 |
|---|---|---|---|---|---|---|---|---|
| $P$ (MPa) | 0.50 | 0.30 | 0.25 | 0.18 | 0.15 | 0.12 | 0.11 | 0.11 |

Bayesteh and Mohammadi (2012) studied the effect of inclusion of crack-tip enrichment functions on the quality of local solution, and concluded that while the global solutions, such as the displacements, remain unaltered, inclusion of crack-tip enrichment functions is essential in reproducing the singular nature of the stress field near the crack tip, which directly influences the crack propagation criteria.

Figure 3.36 shows various stages of crack propagation in longitudinal and then circumferential directions with a clear illustration of the crack turning on the shell surface. The displacement contours are depicted on the deformed configurations of the cylinder.

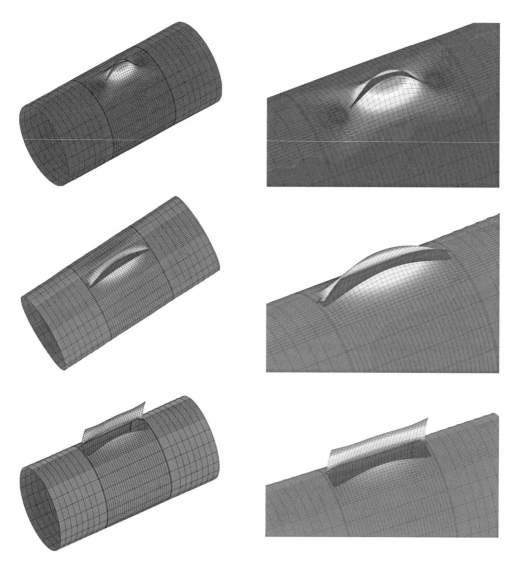

**Figure 3.36**  Crack propagation and displacement contours on the deformed shape of the shell (Bayesteh and Mohammadi, 2012). (Reproduced by permissions of Elsevier.)

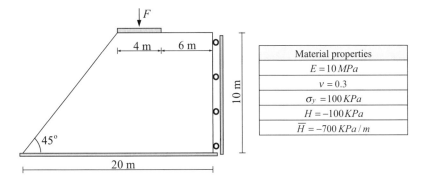

**Figure 3.37**   Geometry of the slope stability problem, and material properties.

## 3.10.6   Shear Band Simulation

A slope stability problem, previously studied by Oliver *et al.* (2006) and Armero and Linder (2008), is studied. Definition of the problem geometry and the boundary conditions are depicted in Figure 3.37. The material is assumed to follow the von Mises plastic criterion. $H$ is the slope of the linear softening curve, and $\bar{H} = H/h$ is the intrinsic hardening coefficient, which depends on the intrinsic shear band thickness $h$.

Figure 3.38 illustrates two different coarse and fine structured finite element meshes, which are used to capture the shear band creation process and to obtain the global force–displacement curve of the model (for the point of loading). Approximation (3.90) for a strong tangential discontinuity is used for XFEM simulation of elements that include a shear band. In addition, a normal cohesive stiffness model is adopted to better characterize the material behaviour in a shear band and to accurately represent the energy associated with plastic deformations. For details see Daneshyar (2012) and Daneshyar and Mohammadi (2012b).

Figure 3.39 compares the force–displacement response of the loading point of the model, predicted by the two models, with the available reference results. The close agreements indicate the accuracy of XFEM and the mesh independency of the results. Similar agreement is observed between the predicted shear band paths and the reference results in Figure 3.40.

Figure 3.41 depicts the deformed shape of the model, clearly showing the sliding behaviour, and the distribution of the effective plastic strain, which is concentrated, almost entirely, within the shear band.

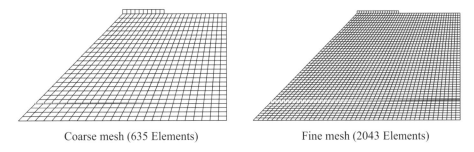

Coarse mesh (635 Elements)                                 Fine mesh (2043 Elements)

**Figure 3.38**   Coarse and fine structured finite element models (Daneshyar and Mohammadi, 2012b).

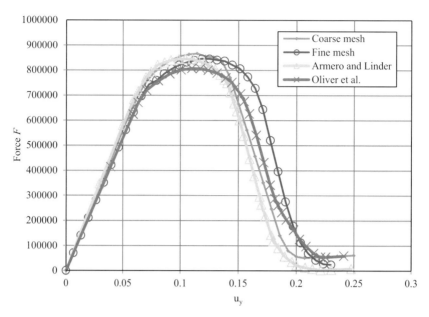

**Figure 3.39**  Comparison of predicted force–displacement response with the reference results (Daneshyar and Mohammadi, 2012b).

Finally, the shear stress contours for different loading increments are depicted in Figure 3.42. It also illustrates the trend of shear band creation and propagation.

### 3.10.7  Fault Simulation

A simple typical example of fault rupture, originally presented by Parchei, Mohammadi and Zafarani (2012), is examined. The problem can be regarded as a self-similarly propagating

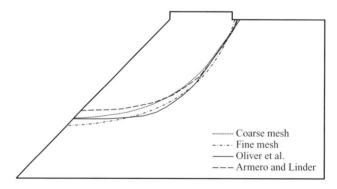

**Figure 3.40**  Comparison of the shear band path with the reference results (Daneshyar and Mohammadi, 2012b).

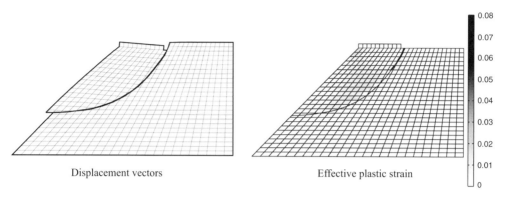

Displacement vectors                    Effective plastic strain

**Figure 3.41** Deformed shape of the slope (magnified by a factor of 4) and the effective plastic strain contour (Daneshyar and Mohammadi, 2012b).

mode II crack under uniform shear tractions (shear stress drop) on crack faces in an infinite, homogeneous and isotropic elastic medium (Figure 3.43).

The fault zone is assumed to have shear wave speed $c_s = 1$, dilatational wave speed $c_d = \sqrt{3}$ and the density and shear modulus, $\rho = 1$ and $\mu = 1$, respectively. The crack is assumed to suddenly appear with zero length and propagate self-similarly with a constant sub-Rayleigh velocity. The size of the domain is selected large enough such that no reflected waves would reach back to the crack.

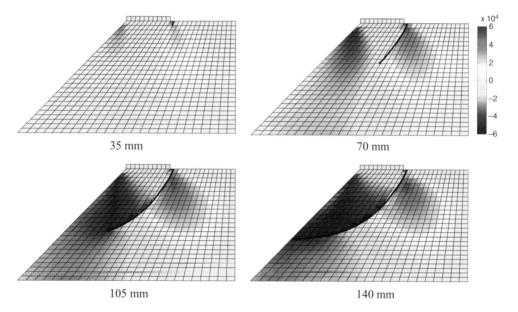

35 mm                                    70 mm

105 mm                                   140 mm

**Figure 3.42** Shear stress contours for different stages of the loading (Daneshyar and Mohammadi, 2012b).

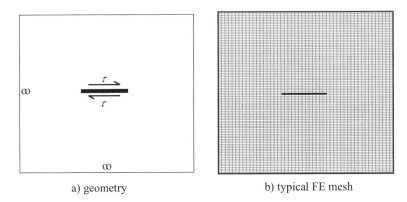

a) geometry                                          b) typical FE mesh

**Figure 3.43**   The fault rupture problem, and the finite element model.

This domain is discretized by a $90 \times 140$ grid of four-node quadrilateral elements; finer around the initial crack and coarser further from the centre of the domain. The problem is solved for several propagation speeds and the amount of slip at the crack centre is computed at the time when the crack length reaches 20 units. Figure 3.44 compares the XFEM predictions of slips (by Parchei, Mohammadi and Zafarani, 2012) in terms of the ratio $v_c/c_s$ with the reference analytical solution from Freund (1990) and numerical BIEM results (Tada and Yamashita, 1997).

In numerical analysis of crack propagation, a sudden growth of the crack at each time step usually results in an undesirable noise generation that affects the accuracy of the results. Such noise, in many cases, may cause instabilities in the numerical results and is mainly suppressed by means of an artificial damping coefficient, $C$, added to the numerical method (Tada and

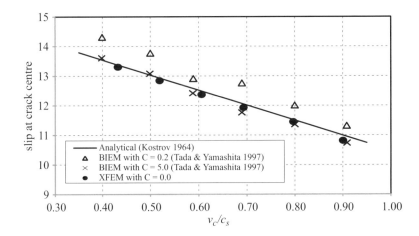

**Figure 3.44**   Comparison of XFEM predictions of slips for different ratios of $v_c/c_s$ (Parchei, Mohammadi and Zafarani, 2012) with the reference analytical (Freund, 1990) and numerical BIEM results (Tada and Yamashita, 1997). (Reproduced by International Institute of Earthquake Engineering and Seismology, Iran. In press.)

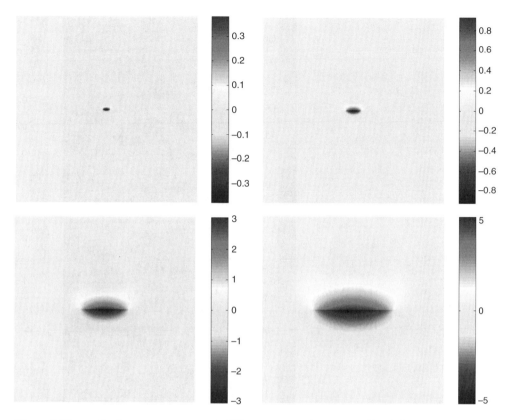

**Figure 3.45** $u_x$ displacement contours at different propagation stages (Parchei, Mohammadi and Zafarani, 2012). (Reproduced by International Institute of Earthquake Engineering and Seismology, Iran. In press.)

Yamashita, 1997; Kame and Yamashita, 1999). In the present simulation of the dynamic fault rupture with the extended finite element method, these noises have negligible effects on the resulting displacements and slips, as clearly observed in Figure 3.44. Therefore, the analysis can be performed without addition of artificial damping (Parchei, Mohammadi and Zafarani, 2012).

Figure 3.45 illustrates the $u_x$ displacement contours during the crack propagations. In addition, contours of $\sigma_{yy}$ and $\sigma_{xy}$ stress contours are depicted in Figure 3.46.

## 3.10.8   Sliding Contact Stress Singularity by PUFEM

The existence of stress singularity at the corner of a contact wedge has been studied by several researchers for various free/stick/slip edges and wedge angles (Barber, 2004; Lee and Barber, 2006; Dini et al., 2008; Giner et al., 2008; Wu and Liu, 2010). Recently, Ebrahimi (2012) and Ebrahimi, Mohammadi and Kani (2012) have comprehensively studied the same problem and proposed an analytical/numerical solution for evaluation of the order of stress singularity at the corner of a contact problem. The characteristic equation (3.153) may be solved numerically

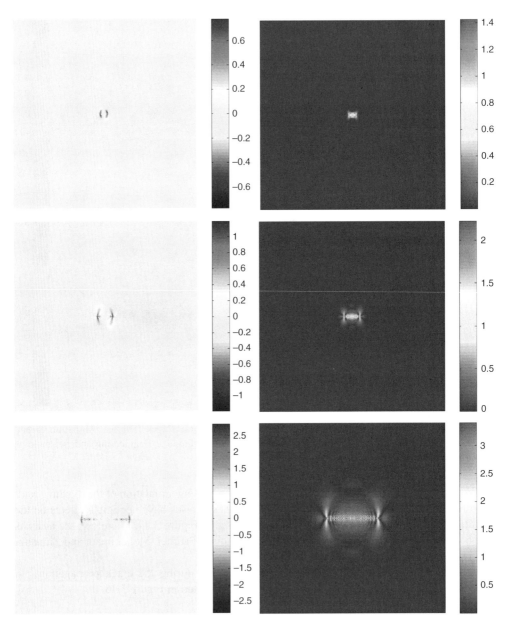

**Figure 3.46** Contours of $\sigma_{yy}$ and $\sigma_{xy}$ stress contours at different propagation stages (Parchei, Mohammadi and Zafarani, 2012). (Reproduced by International Institute of Earthquake Engineering and Seismology, Iran. In press.)

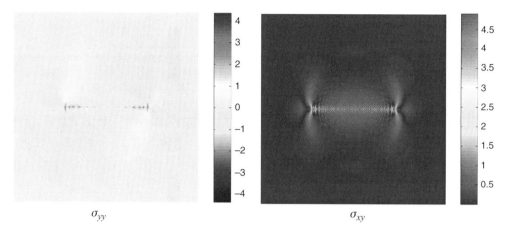

$\sigma_{yy}$                                                                       $\sigma_{xy}$

**Figure 3.46**   (*Continued*)

and the results will include real and imaginary parts of the characteristic determinant and their parametric derivatives with respect to real and imaginary parts of the eigenvalue $\lambda$.

A contact problem, depicted in Figure 3.47, is now considered by the PUFEM based on the methodology proposed by Ebrahimi, Mohammadi and Kani (2012) for determining the order of contact stress singularity. The material properties are $E = 210$ MPa and $\nu = 0.3$.

An unstructured finite element mesh, consisting of 1316 nodes and 1220 elements, is adaptively refined around the sliding interface to improve the accuracy of discretization. The asymptotic displacement field with $K_{GSIF} = -0.1$ is applied on the slave and master

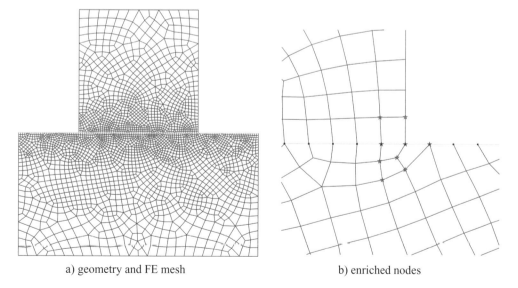

a) geometry and FE mesh                                b) enriched nodes

**Figure 3.47**   Contact problem, finite element mesh and the enrichment nodes for the PUFEM framework (Ebrahimi, Mohammadi and Kani, 2012). (Reproduced by permissions of Elsevier.)

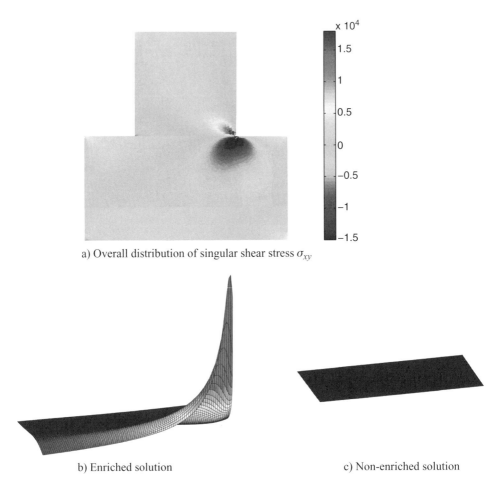

a) Overall distribution of singular shear stress $\sigma_{xy}$

b) Enriched solution           c) Non-enriched solution

**Figure 3.48** Distribution of singular shear stress $\sigma_{xy}$, (a) overall view, (b) local enriched solution around the contact corner, (c) local non-enriched solution around the contact corner (Ebrahimi, Mohammadi and Kani, 2012). (Reproduced by permissions of Elsevier.)

boundaries. The nodes marked near the corner vertex node indicate the topological enrichment nodes within the partition of unity finite element approach. In total, 1 slave element and 3 master elements are enriched (see Figure 3.47).

Figure 3.48a illustrates the shear stress $\sigma_{xy}$ contour, with a clear indication of stress singularity at the contact corner. This is better demonstrated in Figure 3.48b,c where the stress contour is depicted over the singular slave element for two cases of standard finite element solution and the present enriched PUFEM. Clearly, in contrast to conventional FEM solution, PUFEM is capable of reproducing the singular stress field around the contact corner.

## 3.10.9 Hydraulic Fracture

A hydraulic fracture problem is now simulated to assess the performance of XFEM in modelling a crack propagation problem under high pressure gas inside a borehole.

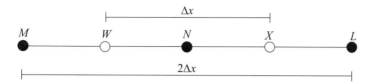

**Figure 3.49**   The finite difference model for solution of gas-dynamics.

The method proposed by Nilson, Proffer and Duff (1985) can be used to formulate the gas flow in cracks. It includes the thermal effects and permeability of solid material. The one-dimensional compressible adiabatic gas flow in a 1D crack can be written as,

$$\frac{\partial(\rho h)}{\partial t} + \frac{\partial(\rho v h)}{\partial t} = 0 \tag{3.195}$$

$$\frac{\partial(\rho v h)}{\partial t} + \frac{\partial(\rho v^2 h)}{\partial x} = -\rho h \left( \frac{1}{\rho} \frac{\partial P}{\partial x} + \psi \right) \tag{3.196}$$

where $\psi$ is the friction coefficient:

$$\psi = a \left( \frac{\varepsilon}{h} \right)^b \frac{v^2}{h} \tag{3.197}$$

It is well accepted within the gas dynamics literature that for a high pressure hot gas, $a = 0.1, b = 0.5$ and (3.196) is transformed into

$$\frac{\partial(\rho h)}{\partial t} + \frac{\partial}{\partial x} \left( h \sqrt{-\rho \frac{h}{f} \frac{\partial P}{\partial x}} \right) = 0, \quad f = a \left( \frac{\varepsilon}{h} \right)^b \tag{3.198}$$

A finite difference scheme is then used to discretize the equations on a 1D gas-mass model, as depicted in Figure 3.49:

$$\rho_N^{t+\Delta t} - \rho_N^t = \frac{-2\Delta t}{\Delta x(h_w + h_x)} \left( h_x \sqrt{-\rho_x \frac{h_x}{f} \frac{(P_L - P_N)}{\Delta x}} - h_w \sqrt{-\rho_w \frac{h_w}{f} \frac{(P_N - P_M)}{\Delta x}} \right) \tag{3.199}$$

It is important to note that the enriched parts of the force vector in (3.39) should now include the internal gas pressure on crack faces (see Figure 3.50), (QingWen, Yuwen and Tian Tang, 2009)

$$\mathbf{f}_i^a = \int_{\Gamma_t} N_i H \mathbf{f}^t \, d\Gamma + \int_{\Omega^e} N_i H \mathbf{f}^b \, d\Omega + 2 \int_{\Gamma_c} n N_i p \, d\Gamma \tag{3.200}$$

$$\mathbf{f}_i^{b\alpha} = \int_{\Gamma_t} N_i F_\alpha \mathbf{f}^t \, d\Gamma + \int_{\Omega^e} N_i F_\alpha \mathbf{f}^b \, d\Omega + 2 \int_{\Gamma_c} n \sqrt{r} N_i p \, d\Gamma \quad (\alpha = 1,2,3 \text{ and } 4) \tag{3.201}$$

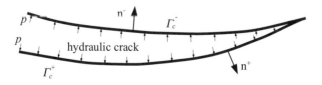

**Figure 3.50**   A hydraulic fracture.

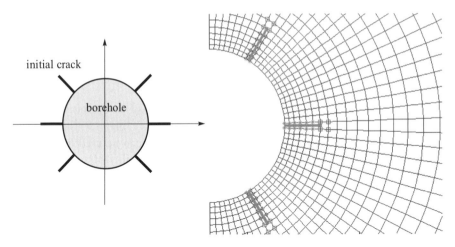

**Figure 3.51** Definitions of the gas-driven hydraulic fracture problem, and half of the finite element model around the hole for the case of six initial cracks.

In order to verify the process, a polar XFEM model is used to predict the analytical solution for the stress intensity factor of a hydraulic crack emanated from a gas bore hole with pressure $p$ can be defined as (Saouma, 2000):

$$K = \beta_1 p \sqrt{\pi a} \tag{3.202}$$

where $\beta_1$ is a constant depending on the ratio of the crack-tip distance to the hole diameter. For a specific case in which the bore diameter is 5 cm, the crack length is 0.15 cm and the bore pressure is 1 Mpa, $\beta_1 = 0.9976$. The XFEM prediction, $7.69 \times 10^6$ is only about 2% different from the analytical solution.

A series of tests by Nilson, Proffer and Duff (1985) on gas-driven rock fracture is now simulated by XFEM. Geometry and the typical finite element model are defined in Figure 3.51 and the material properties are defined in Table 3.6. $\beta$ is the crack surface roughness.

Since the crack propagation length directly affects the gas pressure, a sensitivity analysis is performed to evaluate the proper value of the crack propagation velocity. Figure 3.52 compares the variations in gas pressure for different values of crack propagation velocity with the experimental results. Clearly, a good agreement is obtained with the propagation velocity of $50 \, \text{m s}^{-1}$.

The same mesh is used for a different number of initial cracks. Three cases of 2, 4 and 6 initial cracks of length 0.5 m, subjected to pressure $p = 40 \, \text{MPa}$, are analysed. The predicted stress

**Table 3.6** Material properties for the gas-driven rock fracture.

| $G$ | $v$ | $\beta$ | $K_c$ | $c_s$ |
|---|---|---|---|---|
| 3 GPa | 0.3 | 0.4 mm | 0.5 MPa m$^{1/2}$ | 1200 m s$^{-1}$ |

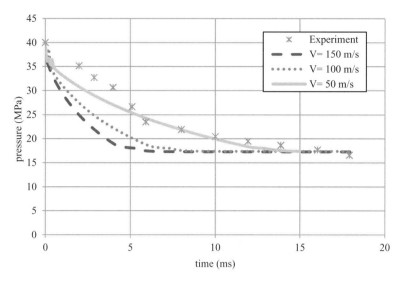

**Figure 3.52** Sensitivity analysis of gas pressure with respect to time for various crack propagation velocities.

intensity factors for the three cases are, 1.79, 1.12 and 0.824, respectively. Also, Figure 3.53 depicts the variations of crack opening in terms of the crack lengths for different numbers of initial cracks. It is observed that by increasing the number of cracks around the holes, both the crack opening and the crack stress intensity factor are decreased.

Figure 3.54 illustrates the radial $\sigma_{rr}$ and circumferential $\sigma_{\theta\theta}$ stress contours for the six-crack model at time $t = 16\,\mu s$.

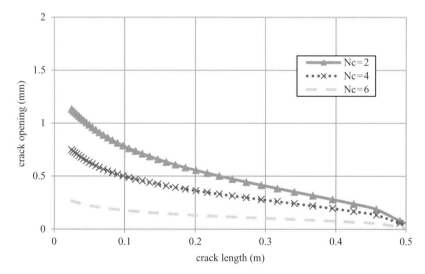

**Figure 3.53** Variations of the crack opening with respect to the crack length for different numbers of initial cracks.

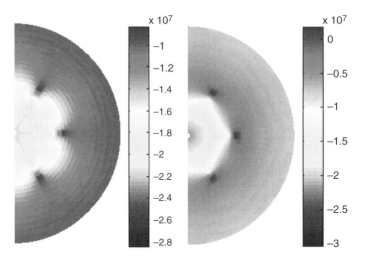

**Figure 3.54** $\sigma_{rr}$ and $\sigma_{\theta\theta}$ stress contours for the six-crack model at time $t = 16\mu s$.

### 3.10.10   Dislocation Dynamics

Three different dislocation problems, depicted in Figure 3.55, are examined by a fixed $25 \times 25$ structured mesh of finite elements, to illustrate the application of XFEM to various dislocation problems, where a predefined discontinuous slip across the glide plane is present. No external loading is present and the models are analysed for dislocation self-stress states.

The first example is to simulate the square ABCD part around a straight edge dislocation near the free edge of a semi-infinite domain (Figure 3.55a). The edge dislocation is placed at $l = 0.5\,\mu m$ and the slip plane is normal to the free edge .The square model is then subjected

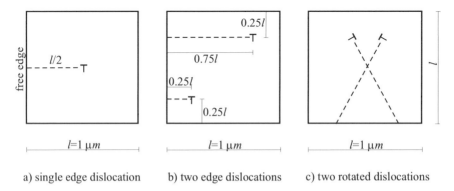

a) single edge dislocation      b) two edge dislocations      c) two rotated dislocations

**Figure 3.55**   Comparison of analytical and XFEM displacements for the sliding system $s_1$ (Malekafzali, 2010).

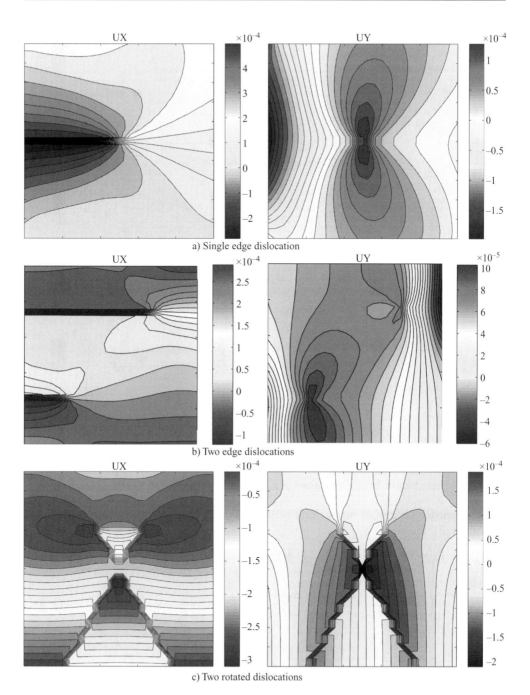

a) Single edge dislocation

b) Two edge dislocations

c) Two rotated dislocations

**Figure 3.56** $u_x$ and $u_y$ displacement contours (Malekafzali, 2010).

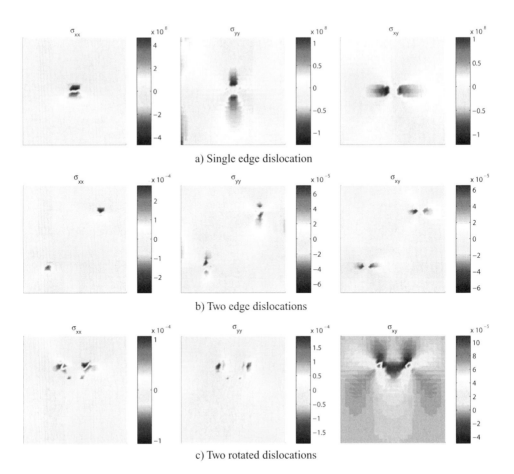

**Figure 3.57** Comparison of stress contours (Malekafzali, 2010).

to the boundary conditions in the form of an analytical stress field to represent the ideal semi-infinite domain,

$$\sigma_{xx} = D \left( -\frac{y(3(x-l)^2 + y^2)}{((x-l)^2 + y^2)^2} + \frac{y(3(x+l)^2 + y^2)}{((x+l)^2 + y^2)^2} + 4lxy\frac{(3(x-l)^2 - y^2)}{((x+l)^2 + y^2)^3} \right) \quad (3.203)$$

$$\sigma_{yy} = D \left( \frac{y((x-l)^2 - y^2)}{((x-l)^2 + y^2)^2} + \frac{y(3(x+l)^2 + y^2)}{((x+l)^2 + y^2)^2} + 4lxy\frac{(3(x-l)^2 - y^2)}{((x+l)^2 + y^2)^3} \right) \quad (3.204)$$

$$\sigma_{xy} = D \left( -\frac{(x-l)((x-l)^2 - y^2)}{((x-l)^2 + y^2)^2} - \frac{(x+l)((x+l)^2 - y^2)}{((x+l)^2 + y^2)^2} \right.$$
$$\left. + 2l\frac{((l-x)(x+l)^3 + 6x(x+l)y^2 - y^4)}{((x+l)^2 + y^2)^3} \right) \quad (3.205)$$

where

$$D = \frac{Eb}{4\pi(1-\nu^2)} \quad (3.206)$$

The second example is to analyse a finite domain of $1\,\mu m \times 1\,\mu m$ which includes two straight edge dislocations at $(0.75, 0.75)\mu m$ and $(0.25, 0.25)\mu m$, respectively, as depicted in Figure 3.55b. Similar material properties are adopted and the same structured mesh is used for XFEM simulation.

The third model, illustrated in Figure 3.55c, is related to the same domain that includes two simultaneous rotated edge dislocations ($60°$ and $120°$). The glide planes are defined as $(0.4226, 0) - (0.3125 - 1.43330)\mu m$ and $(1.5774, 0) - (0.3750 - 1.43330)\mu m$, respectively, and they both pass through the centre of the square domain.

The mechanical properties for all three cases are assumed as: $E = 0.12141\,N\,\mu m^{-2}$, $v = 0.34$ and $b = 2.551 \times 10^{-4}\,\mu m$ (Malekafzali, 2010).

Figure 3.56 depicts the $x$ and $y$ displacement contours for the three dislocation problems. Clearly, discontinuous tangential displacement fields across the slip plane are observed, whereas the displacement component normal to the slip plane remains continuous.

Figure 3.57a depicts the $\sigma_{xx}$, $\sigma_{yy}$ and $\sigma_{xy}$ stress contours for the single edge dislocation in the semi-infinite domain, which match the analytical solution (Malekafzali, 2010). The stress patterns for the two other cases seem logical, but there is no analytical solution available for comparison.

# 4

# Static Fracture Analysis of Composites

## 4.1 Introduction

Composite materials are used extensively in engineering and industrial applications; from traditional structural strengthening (Meier, 1995; Mohammadi, 2003b; Rizkalla, Hassan and Hassan, 2003; Lau and Zhou, 2001; Teng *et al.*, 2002) to advanced aerospace and defence systems, and from nanoscale carbon nanotubes (CNTs) (Eftekhari and Mohammadi, 2012) to large scale turbines and power plants, and so on. Their highly flexible design allows prescribed tailoring of material properties fitted to specific engineering requirements. These include a wide variety of properties across various length scales, including nano and micro-mechanical structural needs, thermo-mechanical specifications and even electro-magneto-mechanical characteristics.

Despite enormous advantages, composites suffer from a number of shortcomings, such as high thermal and residual stresses, poor interfacial bonding strength and low toughness, which may facilitate the process of unstable cracking under different circumstances, such as initial weakness in material strength, fatigue, yielding and imperfection in production procedure. These failures can cause extensive damage accompanied by substantial reduction in stiffness and load bearing capacity, lower ductility and potential abrupt fracture mechanisms. The problem becomes more important in intensive concentrated loading conditions such as impact and explosion (Mohammadi and Moosavi, 2007; Mohammadi and Forouzan-Sepehr, 2003).

Fracture mechanics of composite structures has been studied by many researchers through either analytical solutions or numerical methodologies. Several analytical investigations are available on the fracture behaviour of composite materials. Apart from the pioneering work by Muskhelishvili (1953) on isotropic elastic materials, sometimes used for composites, Sih, Paris and Irwin (1965), Bogy (1972), Bowie and Freese (1972), Barnett and Asaro (1972), Kuo and Bogy (1974), Tupholme (1974), Atluri, Kobayashi and Nakagaki (1975a, 1975b), Forschi and Barret (1976), Boone, Wawrzynek and Ingraffea (1987), Viola, Piva and Radi (1989) and, more recently, Lim, Choi and Sankar (2001), Carloni and Nobile (2002), Carloni, Piva and Viola (2003) and Nobile and Carloni (2005) have proposed solutions for various anisotropic elastostatic crack problems.

*XFEM Fracture Analysis of Composites*, First Edition. Soheil Mohammadi.
© 2012 John Wiley & Sons, Ltd. Published 2012 by John Wiley & Sons, Ltd.

On the other hand, several researchers have contributed in finding the elastodynamic fields around a propagating crack within an anisotropic medium, including Achenbach and Bazant (1975), Arcisz and Sih (1984), Piva and Viola (1988), Viola, Piva and Radi (1989), Shindo and Hiroaki (1990), De and Patra (1992), Gentilini, Piva and Viola (2004), Kasmalkar (1996), Chen and Erdogan (1996), Lee, Hawong and Choi (1996), Gu and Asaro (1997), Rubio-Gonzales and Mason (1998), Broberg (1999), Lim, Choi and Sankar (2001), Federici *et al.* (2001), Nobile and Carloni (2005), Piva, Viola and Tornabene (2005) Sethi *et al.* (2011), and Abd-Alla *et al.* (2011), among others.

The research has not been limited to single layer orthotropic homogeneous composites. Analytical solutions for delamination in multilayer composites have also been investigated comprehensively. The first attempt was probably by Williams (1959) who discovered the oscillatory near-tip behaviour for a traction-free interface crack between two dissimilar isotropic elastic materials, followed by several others, such as Erdogan (1963), Rice and Sih (1965), Malysev and Salganik (1965), England (1965), Comninou (1977), Comninou and Schmuser (1979), Sun and Jih (1987), Hutchinson, Mear and Rice (1987) and Rice (1988), among others. The study of interface cracks between two anisotropic materials was performed by Gotoh (1967), Clements (1971), Willis (1971) and followed by Wang and Choi (1983a, 1983b), Ting (1986), Tewary, Wagoner and Hirth (1989), Wu (1990), Gao, Abbudi and Barnett (1992), Hwu (1993a, 1993b), Bassani and Qu (1989), Sun and Manoharan (1989), Suo (1990), Yang, Sou and Shih (1991), Qian and Sun (1998), Lee (2000) and Hemanth *et al.* (2005).

With the development of functionally graded materials (FGMs), several researchers have concentrated on their fracture behaviour. For instance, Yamanouchi *et al.* (1990), Holt *et al.* (1993), Ilschner and Cherradi (1995), Nadeau and Ferrari (1999), Takahashi *et al.* (1993) and Pipes and Pagano (1970, 1974), Pagano (1974), Kurihara, Sasaki and Kawarada (1990), Niino and Maeda (1990), Sampath *et al.* (1995), Kaysser and Ilschner (1995), Erdogan (1995) and Lee and Erdogan (1995) have studied various aspects of FGM properties. Despite material inhomogeneity, Sih and Chen (1980), Eischen (1983) and Delale and Erdogan (1983) have shown that the asymptotic crack-tip stress and displacement fields for certain classes of FGMs follow the general form of homogeneous materials and Ozturk and Erdogan (1997) and Konda and Erdogan (1994) analytically solved for crack-tip fields in inhomogeneous orthotropic infinite FGM problems. Evaluation of the $J$ integral for determining the mixed-mode stress intensity factors in general FGM problems was studied by Gu and Asaro (1997), Gu, Dao and Asaro (1999) and Anlas, Santare and Lambros (2000). Also, Kim and Paulino (2002a, 2002b, 2002c, 2003a, 2003b, 2005) examined various forms of the $J$ integral and developed three independent formulations: non-equilibrium, incompatibility and constant-constitutive-tensor.

Due to the limitations of analytical methods in handling general crack propagations and arbitrary complex geometries and boundary conditions, several numerical techniques have been developed for solving composite fracture mechanics problems. Cruse (1988), Aliabadi and Sollero (1998) and García-Sánchez, Zhang and Sáez (2008) developed boundary element solutions for quasi-static crack propagation and dynamic analysis of cracks in orthotropic media. The boundary element methods, regardless of all the benefits, are barely capable of being employed in nonlinear systems and are not suited to general crack propagation problems.

Alternative methods are also available, including the discrete element method (DEM) (Mohammadi, 2003a), meshless methods, such as the element-free Galerkin method (EFG) (Belytschko, Organ and Krongauz, 1995; Belytschko and Tabbara, 1996; Ghorashi, Mohammadi and Sabbagh-Yazdi, 2011), the extended finite element method (XFEM) and the extended isogeometric analysis (XIGA) (Ghorashi, Valizadeh and Mohammadi, 2012). Despite higher accuracy and flexible adaptive schemes, many meshless methods are yet to be user friendly and need sensitivity analysis, calibration and difficult stabilization schemes (Motamedi and Mohammadi, 2012).

In contrast, the finite element method is well developed into almost every possible engineering application, including nonlinear, inhomogeneous, anisotropic, multilayer, large deformation and dynamic problems. Since the earlier application of FEM to crack analysis of composites by Swenson and Ingraffea (1988), various techniques have been developed for crack analysis of composites within the FEM framework. For example, quarter-point singular elements are very efficient in reproducing the singular stress field at a crack tip. Alternatively, the adaptive remeshing technique within the finite element method has been adopted for simulation of progressive crack propagation in composites under quasi-static and dynamic loadings (Moosavi and Mohammadi, 2007). They mainly include continuous smeared crack models, discrete inter-element crack models, cracked interface elements and the discrete element based approach, which may use general contact mechanics algorithms to simulate progressive delamination and fracture problems (Mohammadi, Owen and Peric, 1997; Sprenger, Gruttmann and Wagner, 2000; Wu, Yuan and Niu, 2002; Mohammadi and Forouzan-Sepehr, 2003; Wong and Vecchio, 2003; Wu and Yin, 2003; Lu et al., 2005; Wang, 2006; Teng, Yuan and Chen, 2006; Mohammadi and Moosavi, 2007; Mohammadi, 2008; Rabinovitch, 2008).

The natural extension of FEM into XFEM allows new capabilities while preserving the finite element original advantages. In the past decade, development of XFEM has substantially contributed to new studies of fracture analysis of various types of composite materials. The first application of XFEM for composites was reported by Remmers, Wells and de Borst (2003) and Nagashima and Suemasu (2004, 2006) for simulation of thin layered composites, concluding that the isotropic enrichment functions cannot represent the asymptotic solution for a crack in an orthotropic material. Then, Sukumar et al. (2004) developed partition of unity-based enrichment functions for bimaterial interface cracks between two isotropic media. Later, Hettich and Ramm (2006) simulated the delamination crack as a jump in the displacement field without using any crack-tip enrichment. Also, a number of XFEM simulations have focused on thermo-mechanical analysis of orthotropic FGMs (Dag, Yildirim and Sarikaya, 2007), 3D isotropic FGMs (Ayhan, 2009; Zhang et al., 2011; Moghaddam, Ghajar and Alfano, 2011) and frequency analysis of cracked isotropic FGMs (Natarajan et al., 2011).

Development of orthotropic crack tip enrichment functions was reported in a series of papers by Asadpoure, Mohammadi and Vafai (2006, 2007), Asadpoure and Mohammadi (2007) and Mohammadi and Asadpoure (2006). Later, Motamedi (2008) and Motamedi and Mohammadi (2010a, 2010b, 2012) studied the dynamic crack stability and propagation in composites based on static and dynamic orthotropic enrichment functions and Esna Ashari (2009) and Esna Ashari and Mohammadi (2009, 2010a, 2010b, 2011a, 2011b, 2012) have further extended the method for orthotropic bimaterial interfaces. Recently, Bayesteh and Mohammadi (2012) have used XFEM with orthotropic crack-tip enrichment functions to analyse several FGM crack stability and propagation problems.

This chapter, which is a modified edition of a similar chapter in Mohammadi (2008), begins with a review of anisotropic and orthotropic elasticity. It is followed by a comprehensive discussion on available analytical solutions for near crack-tip fields in orthotropic materials. These fields are then used to develop the orthotropic enrichment functions for the XFEM formulation. A section is then devoted to orthotropic mixed mode fracture mechanics, which discusses orthotropic mixed mode criteria, the $J$ integral, crack propagation criteria for orthotropic media and other related issues. Finally, a number of numerical simulations are provided to illustrate the validity, robustness and efficiency of the proposed approach for evaluation of mixed mode stress intensity factors in homogeneous composites.

## 4.2  Anisotropic Elasticity

### 4.2.1  Elasticity Solution

The general form of an anisotropic stress–strain relationship can be defined in terms of the 4th order material modulus and compliance tensors, **D** and **C**, respectively,

$$\sigma = \mathbf{D} : \varepsilon \tag{4.1}$$

$$\varepsilon = \mathbf{C} : \sigma \tag{4.2}$$

or in the component forms of

$$\sigma_{ij} = \mathbf{D}_{ijkl}\varepsilon_{kl} \quad i, j = 1, 2, 3 \tag{4.3}$$

$$\varepsilon_{ij} = \mathbf{C}_{ijkl}\sigma_{kl} \quad i, j = 1, 2, 3 \tag{4.4}$$

Equations (4.3) and (4.4) can also be written in a more appropriate component form,

$$\sigma_i = d_{ij}\varepsilon_j \quad i, j = 1, 2, 3, \ldots, 6 \tag{4.5}$$

$$\varepsilon_i = c_{ij}\sigma_j \quad i, j = 1, 2, 3, \ldots, 6 \tag{4.6}$$

For an orthotropic material which has three mutually orthogonal planes of elastic symmetry, Equations (4.5) and (4.6) are reduced to:

$$\begin{Bmatrix} \sigma_1 \\ \sigma_2 \\ \sigma_3 \\ \sigma_4 \\ \sigma_5 \\ \sigma_6 \end{Bmatrix} = \begin{bmatrix} d_{11} & d_{12} & d_{13} & 0 & 0 & 0 \\ d_{12} & d_{22} & d_{23} & 0 & 0 & 0 \\ d_{13} & d_{23} & d_{33} & 0 & 0 & 0 \\ 0 & 0 & 0 & d_{44} & 0 & 0 \\ 0 & 0 & 0 & 0 & d_{55} & 0 \\ 0 & 0 & 0 & 0 & 0 & d_{66} \end{bmatrix} \begin{Bmatrix} \varepsilon_1 \\ \varepsilon_2 \\ \varepsilon_3 \\ \varepsilon_4 \\ \varepsilon_5 \\ \varepsilon_6 \end{Bmatrix} \tag{4.7}$$

$$\begin{Bmatrix} \varepsilon_1 \\ \varepsilon_2 \\ \varepsilon_3 \\ \varepsilon_4 \\ \varepsilon_5 \\ \varepsilon_6 \end{Bmatrix} = \begin{bmatrix} c_{11} & c_{12} & c_{13} & 0 & 0 & 0 \\ c_{12} & c_{22} & c_{23} & 0 & 0 & 0 \\ c_{13} & c_{23} & c_{33} & 0 & 0 & 0 \\ 0 & 0 & 0 & c_{44} & 0 & 0 \\ 0 & 0 & 0 & 0 & c_{55} & 0 \\ 0 & 0 & 0 & 0 & 0 & c_{66} \end{bmatrix} \begin{Bmatrix} \sigma_1 \\ \sigma_2 \\ \sigma_3 \\ \sigma_4 \\ \sigma_5 \\ \sigma_6 \end{Bmatrix} \tag{4.8}$$

where

$$
\begin{Bmatrix} \sigma_1 \\ \sigma_2 \\ \sigma_3 \\ \sigma_4 \\ \sigma_5 \\ \sigma_6 \end{Bmatrix} = \begin{Bmatrix} \sigma_{11} \\ \sigma_{22} \\ \sigma_{33} \\ \sigma_{23} \\ \sigma_{13} \\ \sigma_{12} \end{Bmatrix}, \quad \begin{Bmatrix} \varepsilon_1 \\ \varepsilon_2 \\ \varepsilon_3 \\ \varepsilon_4 \\ \varepsilon_5 \\ \varepsilon_6 \end{Bmatrix} = \begin{Bmatrix} \varepsilon_{11} \\ \varepsilon_{22} \\ \varepsilon_{33} \\ 2\varepsilon_{23} \\ 2\varepsilon_{13} \\ 2\varepsilon_{12} \end{Bmatrix} \tag{4.9}
$$

Equations (4.7) and (4.8) are valid for arbitrarily selected coordinate systems. Further simplification is obtained for the principal directions of orthotropy. The components $d_{ij}$ of Eq. (4.7) can then be defined as (Bower, 2012),

$$
\begin{bmatrix}
E_1(1-v_{23}v_{32})\Delta & E_1(v_{12}+v_{31}v_{23})\Delta & E_1(v_{31}+v_{21}v_{31})\Delta & 0 & 0 & 0 \\
E_1(v_{12}+v_{31}v_{23})\Delta & E_2(1-v_{13}v_{31})\Delta & E_2(v_{32}+v_{12}v_{31})\Delta & 0 & 0 & 0 \\
E_1(v_{31}+v_{21}v_{31})\Delta & E_2(v_{32}+v_{12}v_{31})\Delta & E_3(1-v_{12}v_{21})\Delta & 0 & 0 & 0 \\
0 & 0 & 0 & \mu_{23} & 0 & 0 \\
0 & 0 & 0 & 0 & \mu_{13} & 0 \\
0 & 0 & 0 & 0 & 0 & \mu_{12}
\end{bmatrix} \tag{4.10}
$$

where

$$
\frac{1}{\Delta} = 1 - v_{12}v_{21} - v_{23}v_{32} - v_{31}v_{13} - 2v_{21}v_{32}v_{13} \tag{4.11}
$$

and the components $c_{ij}$ of Eq. (4.8) can be written as,

$$
\begin{bmatrix}
\dfrac{1}{E_1} & -\dfrac{v_{21}}{E_2} & -\dfrac{v_{31}}{E_3} & 0 & 0 & 0 \\[2mm]
-\dfrac{v_{12}}{E_1} & \dfrac{1}{E_2} & -\dfrac{v_{32}}{E_3} & 0 & 0 & 0 \\[2mm]
-\dfrac{v_{13}}{E_1} & -\dfrac{v_{23}}{E_2} & \dfrac{1}{E_3} & 0 & 0 & 0 \\[2mm]
0 & 0 & 0 & \dfrac{1}{\mu_{23}} & 0 & 0 \\[2mm]
0 & 0 & 0 & 0 & \dfrac{1}{\mu_{31}} & 0 \\[2mm]
0 & 0 & 0 & 0 & 0 & \dfrac{1}{\mu_{12}}
\end{bmatrix} \tag{4.12}
$$

with the following additional relations to keep the number of independent constants unchanged:

$$
E_1 v_{21} = E_2 v_{12} \tag{4.13}
$$

$$
E_2 v_{32} = E_3 v_{23} \tag{4.14}
$$

$$
E_3 v_{13} = E_1 v_{31} \tag{4.15}
$$

These components can be further simplified for plane stress and strain states. The components $d_{ij}$ for a generalized plane stress problem are directly computed from (4.10)

$$
\begin{bmatrix} d_{11} & d_{12} & 0 \\ d_{12} & d_{22} & 0 \\ 0 & 0 & d_{66} \end{bmatrix} = \begin{bmatrix} \dfrac{E_1}{1 - \nu_{12}\nu_{21}} & \dfrac{\nu_{12}E_2}{1 - \nu_{12}\nu_{21}} & 0 \\ \dfrac{\nu_{12}E_2}{1 - \nu_{12}\nu_{21}} & \dfrac{E_2}{1 - \nu_{12}\nu_{21}} & 0 \\ 0 & 0 & \mu_{12} \end{bmatrix}
\tag{4.16}
$$

and for generalized plane strains, (García-Sánchez, Zhang and Sáez, 2008)

$$
\begin{bmatrix} d_{11} & d_{12} & 0 \\ d_{12} & d_{22} & 0 \\ 0 & 0 & d_{66} \end{bmatrix} = \begin{bmatrix} E_1\Delta(1 - \nu_{12}\nu_{21}) & E_1\Delta\left(\nu_{21} + \dfrac{E_2}{E_1}\nu_{13}\nu_{32}\right) & 0 \\ E_1\Delta\left(\nu_{21} + \dfrac{E_2}{E_1}\nu_{13}\nu_{32}\right) & E_2\Delta(1 - \nu_{13}\nu_{31}) & 0 \\ 0 & 0 & \mu_{12} \end{bmatrix}
\tag{4.17}
$$

The components of the compliance matrix $c_{ij}$ for plane stress problems are also directly determined from (4.12). A simple rule for change of parameters can then be used for the plane stress problems (Ghorashi, Mohammadi and Sabbagh-Yazdi, 2011),

$$
c_{ij} = \begin{cases} c_{ij} & i, j = 1, 2, 6 \quad \text{plane stress} \\ c_{ij} - \dfrac{c_{i3}.c_{j3}}{c_{33}} & i, j = 1, 2, 6 \quad \text{plane strain} \end{cases}
\tag{4.18}
$$

### 4.2.2 Anisotropic Stress Functions

The Airy stress function is limited to isotropic problems. For an extension to more complex problems, including anisotropic problems, the stress function $\Phi(x, y)$ can be written as

$$
\Phi(x, y) = 2\,\text{Re}\,[\Phi_1(z_1) + \Phi_2(z_2)]
\tag{4.19}
$$

where $\Phi_1(z_1)$ and $\Phi_2(z_2)$ are arbitrary functions of $z_1 = x + s_1 y$ and $z_2 = x + s_2 y$, respectively. Combining the definition of stress components from the Airy stress function and satisfying the compatibility equation, the following relation is obtained for anisotropic solids in the absence of body forces:

$$
c_{22}\frac{\partial^4 \Phi}{\partial x^4} - 2c_{26}\frac{\partial^4 \Phi}{\partial x^3 \partial y} + (2c_{12} + c_{66})\frac{\partial^4 \Phi}{\partial x^2 \partial y^2} - 2c_{16}\frac{\partial^4 \Phi}{\partial x \partial y^3} + c_{22}\frac{\partial^4 \Phi}{\partial y^4} = 0
\tag{4.20}
$$

where coordinates $(x, y)$ coincide with the material axes $(1, 2)$ or $(x_1, y_1)$. Eq. (4.20) reduces to the following simplified equation for isotropic problems,

$$
\frac{\partial^4 \Phi}{\partial x^4} + 2\frac{\partial^4 \Phi}{\partial x^2 \partial y^2} + \frac{\partial^4 \Phi}{\partial y^4} = \nabla^2(\nabla^2 \Phi) = 0
\tag{4.21}
$$

The characteristic equation of the homogenous partial differential equation (4.20) is

$$
c_{11}s^4 - 2c_{16}s^3 + (2c_{12} + c_{66})s^2 - 2c_{26}s + c_{22} = 0
\tag{4.22}
$$

Lekhnitskii (1968) discussed the availability and conditions for the roots of Eq. (4.22). Here, only the two isotropic and orthotropic cases are considered. For an isotropic case, the roots are $s_1 = s_2 = i$ and $\bar{s}_1 = \bar{s}_2 = -i$, whereas for an orthotropic material with axes of orthotropy $(1, 2)$ coinciding Cartesian $(x, y)$ axes, $c_{16} = c_{26} = 0$, the characteristic Eq. (4.22) is reduced to:

$$c_{11}s^4 + (2c_{12} + c_{66})s^2 + c_{22} = 0 \tag{4.23}$$

with pure imaginary roots $s_1$ and $s_2$(Sadd, 2005). Finally, the stress components are defined from the second derivatives of the complex stress function $\Phi_i''$:

$$\sigma_x = 2\,\mathrm{Re}\left[s_1^2\Phi_1''(z_1) + s_2^2\Phi_2''(z_2)\right] \tag{4.24}$$

$$\sigma_y = 2\,\mathrm{Re}\left[\Phi_1''(z_1) + \Phi_2''(z_2)\right] \tag{4.25}$$

$$\sigma_{xy} = -2\,\mathrm{Re}\left[s_1\Phi_1''(z_1) + s_2\Phi_2''(z_2)\right] \tag{4.26}$$

and the displacements are obtained from the first derivatives of the complex stress function, $\Phi_i'$:

$$u_x = 2\,\mathrm{Re}\left[p_1\Phi_1'(z_1) + p_2\Phi_2'(z_2)\right] \tag{4.27}$$

$$u_y = 2\,\mathrm{Re}\left[q_1\Phi_1'(z_1) + q_2\Phi_2'(z_2)\right] \tag{4.28}$$

where

$$p_i = c_{11}s_i^2 + c_{12} - c_{16}s_i \quad i = 1, 2 \tag{4.29}$$

$$q_i = c_{12}s_i + \frac{c_{22}}{s_i} - c_{26} \quad i = 1, 2 \tag{4.30}$$

## 4.3    Analytical Solutions for Near Crack Tip

In the extended finite element method, near tip displacement fields are required to derive a basis for enrichment functions. Several analytical solutions for near crack-tip fields in orthotropic materials have been proposed. Some of them can only be used with specific applications, while others can be applied to general orthotropic media.

In this section, first the general solutions of displacement and stress fields for near a crack tip in an orthotropic medium are discussed. Then, special solutions for different types of composites are briefly reviewed.

Assume an anisotropic body containing a crack is subjected to arbitrary forces with general displacement and traction boundary conditions. Global Cartesian coordinates are $(X, Y)$ and the local Cartesian coordinate $(x, y)$ and local polar coordinate $(r, \theta)$ are defined on the crack tip, as illustrated in Figure 4.1.

### 4.3.1    The General Solution

The stress–strain equations for an orthotropic medium, with axes of elastic symmetry $(1, 2) = (x_1, x_2)$ coincident with the Cartesian coordinates axes $(x_1, x_2) = (x, y)$, can be defined in

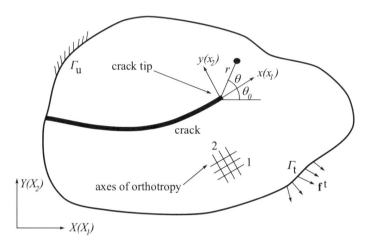

**Figure 4.1**  Local coordinates at the crack tip.

terms of the material modulus coefficients $d_{ij}$:

$$\sigma_x = d_{11}\frac{\partial u_x}{\partial x} + d_{12}\frac{\partial u_y}{\partial y} \tag{4.31}$$

$$\sigma_y = d_{12}\frac{\partial u_x}{\partial x} + d_{22}\frac{\partial u_y}{\partial y} \tag{4.32}$$

$$\sigma_{xy} = d_{66}\left(\frac{\partial u_x}{\partial y} + \frac{\partial u_y}{\partial x}\right) \tag{4.33}$$

The set of equations for an inplane elastostatic problem can be expressed as:

$$d_{11}\frac{\partial^2 u_x}{\partial x^2} + d_{66}\frac{\partial^2 u_x}{\partial y^2} + (d_{12} + d_{66})\frac{\partial^2 u_y}{\partial x \partial y} = 0 \tag{4.34}$$

$$d_{66}\frac{\partial^2 u_y}{\partial x^2} + d_{22}\frac{\partial^2 u_y}{\partial y^2} + (d_{12} + d_{66})\frac{\partial^2 u_x}{\partial x \partial y} = 0 \tag{4.35}$$

Now recalling the characteristic equation (4.22) of the governing fourth-order partial differential equation for the general case of Figure 4.1, the roots of Eq. (4.22) are always complex or purely imaginary ($s_k = s_{kx} + i s_{ky}$, $k = 1, 2$) and occur in conjugate pairs as $s_1$, $\bar{s}_1$ and $s_2$, $\bar{s}_2$ (Lekhnitskii, 1968).

Sih, Paris and Irwin (1965) derived the two-dimensional displacement and stress fields in the vicinity of the crack tip in an infinite anisotropic plate by means of the stress function

(4.19), analytical functions and complex variables, $z_k = x + s_k y$, $k = 1, 2$. The displacement components for pure mode I are defined as,

$$u_1^I = K_I \sqrt{\frac{2r}{\pi}} \operatorname{Re} \left\{ \frac{1}{s_1 - s_2} \left[ s_1 p_2 (\cos\theta + s_2 \sin\theta)^{\frac{1}{2}} - s_2 p_1 (\cos\theta + s_1 \sin\theta)^{\frac{1}{2}} \right] \right\} \quad (4.36)$$

$$u_2^I = K_I \sqrt{\frac{2r}{\pi}} \operatorname{Re} \left\{ \frac{1}{s_1 - s_2} \left[ s_1 q_2 (\cos\theta + s_2 \sin\theta)^{\frac{1}{2}} - s_2 q_1 (\cos\theta + s_1 \sin\theta)^{\frac{1}{2}} \right] \right\} \quad (4.37)$$

and the stress components are

$$\sigma_{11}^I = \frac{K_I}{\sqrt{2\pi r}} \operatorname{Re} \left[ \frac{s_1 s_2}{s_1 - s_2} \left( \frac{s_2}{(\cos\theta + s_2 \sin\theta)^{1/2}} - \frac{s_1}{(\cos\theta + s_1 \sin\theta)^{1/2}} \right) \right] \quad (4.38)$$

$$\sigma_{22}^I = \frac{K_I}{\sqrt{2\pi r}} \operatorname{Re} \left[ \frac{1}{s_1 - s_2} \left( \frac{s_1}{(\cos\theta + s_2 \sin\theta)^{1/2}} - \frac{s_2}{(\cos\theta + s_1 \sin\theta)^{1/2}} \right) \right] \quad (4.39)$$

$$\sigma_{12}^I = \frac{K_I}{\sqrt{2\pi r}} \operatorname{Re} \left[ \frac{s_1 s_2}{s_1 - s_2} \left( \frac{-1}{(\cos\theta + s_2 \sin\theta)^{1/2}} + \frac{1}{(\cos\theta + s_1 \sin\theta)^{1/2}} \right) \right] \quad (4.40)$$

and, in the same way, for pure mode II (skew-symmetric loading), the displacement and stress fields are defined as:

$$u_1^{II} = K_{II} \sqrt{\frac{2r}{\pi}} \operatorname{Re} \left\{ \frac{1}{s_1 - s_2} \left[ p_2 (\cos\theta + s_2 \sin\theta)^{\frac{1}{2}} - p_1 (\cos\theta + s_1 \sin\theta)^{\frac{1}{2}} \right] \right\} \quad (4.41)$$

$$u_2^{II} = K_{II} \sqrt{\frac{2r}{\pi}} \operatorname{Re} \left\{ \frac{1}{s_1 - s_2} \left[ q_2 (\cos\theta + s_2 \sin\theta)^{\frac{1}{2}} - q_1 (\cos\theta + s_1 \sin\theta)^{\frac{1}{2}} \right] \right\} \quad (4.42)$$

and the stress components are

$$\sigma_{11}^{II} = \frac{K_{II}}{\sqrt{2\pi r}} \operatorname{Re} \left[ \frac{1}{s_1 - s_2} \left( \frac{s_2^2}{(\cos\theta + s_2 \sin\theta)^{1/2}} - \frac{s_1^2}{(\cos\theta + s_1 \sin\theta)^{1/2}} \right) \right] \quad (4.43)$$

$$\sigma_{22}^{II} = \frac{K_{II}}{\sqrt{2\pi r}} \operatorname{Re} \left[ \frac{1}{s_1 - s_2} \left( \frac{1}{(\cos\theta + s_2 \sin\theta)^{1/2}} - \frac{1}{(\cos\theta + s_1 \sin\theta)^{1/2}} \right) \right] \quad (4.44)$$

$$\sigma_{12}^{II} = \frac{K_{II}}{\sqrt{2\pi r}} \operatorname{Re} \left[ \frac{1}{s_1 - s_2} \left( \frac{-s_2}{(\cos\theta + s_2 \sin\theta)^{1/2}} + \frac{s_1}{(\cos\theta + s_1 \sin\theta)^{1/2}} \right) \right] \quad (4.45)$$

where $K_I$ and $K_{II}$ are the stress intensity factors for modes I and II, respectively, and $p_k$ and $q_k$ are defined in Eqs. (4.29) and (4.30).

The original assumption of coincident material and crack-tip local axes in deriving the analytical solutions can be generalized to non-coincident axes for developing the basis functions for the orthotropic crack-tip enrichment function. The crack-tip enrichment functions in the polar coordinate system are then obtained from Eqs. (4.36), (4.37), (4.41) and (4.42), as discussed in Section 4.5.3 (Asadpoure and Mohammadi, 2007).

## 4.3.2 Special Solutions for Different Types of Composites

Two types of composites can be distinguished from the roots of the characteristic equation (4.22). Defining

$$a_1 = \frac{(\alpha_1 + \alpha_2 - 4\beta_1\beta_2)}{2} \tag{4.46}$$

$$a_2 = \alpha_1\alpha_2 \tag{4.47}$$

with

$$\alpha_1 = \frac{d_{66}}{d_{11}} \quad \alpha_2 = \frac{d_{22}}{d_{66}} \quad \beta_1 = \frac{d_{12} + d_{66}}{2d_{11}} \quad \beta_2 = \frac{d_{12} + d_{66}}{2d_{66}} \tag{4.48}$$

Then, two different types of orthotropic materials can be defined based on the existence of the real part of the solution: Type I ($a_1 > \sqrt{a_2}, a_2 > 0$) where the characteristic equation has only imaginary roots, and Type II ($|a_1| < \sqrt{a_2}, a_2 > 0$) with complex roots for (4.22), as discussed by Piva and Viola (1988) and Gentilini, Piva and Viola (2004), among others.

The first type was solved by Carloni and Nobile (2002), Carloni, Piva and Viola (2003), Nobile and Carloni (2005),

$$u_1 = \frac{2m_2}{d_{66}(e_1 - e_2)}\sqrt{2lr}\left\{\sigma_{02}\left[\frac{e_2\sqrt{g_2(\theta)}}{m_4(m_1 - e_2^2)}\cos\frac{\theta_2}{2} - \frac{e_1\sqrt{g_1(\theta)}}{m_3(m_1 - e_1^2)}\cos\frac{\theta_1}{2}\right]\right.$$

$$+ e_1e_2\sigma_{03}\left[\frac{\sqrt{g_2(\theta)}}{m_4(m_1 - e_2^2)}\sin\frac{\theta_2}{2} - \frac{\sqrt{g_1(\theta)}}{m_3(m_1 - e_1^2)}\sin\frac{\theta_1}{2}\right]\right\}$$

$$- \frac{2m_2e_1e_2\,(\sigma_{02} - e_1e_2\sigma_{01})}{d_{66}m_3m_4(m_1 - e_1^2)(m_1 - e_2^2)}(l + r\cos\theta) - \frac{m_2\sigma_{03}\,(e_1 + e_2)^2}{d_{66}m_3m_4(m_1 - e_1^2)(m_1 - e_2^2)}r\sin\theta \tag{4.49}$$

$$u_2 = \frac{1}{d_{66}(e_1 - e_2)}\frac{\sqrt{2lr}}{m_3m_4}\left\{\sigma_{02}\left[m_3\sqrt{g_2(\theta)}\sin\frac{\theta_2}{2} - m_4\sqrt{g_1(\theta)}\sin\frac{\theta_1}{2}\right]\right.$$

$$+ \sigma_{03}\left[m_4e_2\sqrt{g_1(\theta)}\cos\frac{\theta_1}{2} - m_3e_1\sqrt{g_2(\theta)}\cos\frac{\theta_2}{2}\right]\right\}$$

$$+ \frac{\sigma_{03}\,(e_1 + e_2)\,(m_3 - m_4)}{2d_{66}m_3m_4\,(e_1 - e_2)}(l + r\cos\theta)$$

$$+ \frac{(\sigma_{02} - e_1e_2\sigma_{01})}{d_{66}(e_1^2 - e_2^2)}\left(\frac{e_2}{m_3e_1} - \frac{e_1}{m_4e_2}\right)\frac{m_2\sigma_{03}\,(e_1 + e_2)^2}{d_{66}m_3m_4(m_1 - e_1^2)(m_1 - e_2^2)}r\sin\theta \tag{4.50}$$

where $m_1$, $m_2$, $m_3$ and $m_4$ are coefficients related to material properties (see Carloni, Piva and Viola, 2003; Mohammadi, 2008), and $e_1$ and $e_2$ are defined as (Carloni, Piva and Viola, 2003),

$$
e_1 = \left\{ \frac{1}{2} \left[ \frac{d_{66}}{d_{11}} + \frac{d_{22}}{d_{66}} - \frac{(d_{12} + d_{66})^2}{d_{11} d_{66}} \right] - \left( \frac{1}{4} \left[ \frac{d_{66}}{d_{11}} + \frac{d_{22}}{d_{66}} - \frac{(d_{12} + d_{66})^2}{d_{11} d_{66}} \right]^2 - \frac{d_{22}}{d_{11}} \right)^{1/2} \right\}^{1/2}
$$

$$(4.51)$$

$$
e_2 = \left\{ \frac{1}{2} \left[ \frac{d_{66}}{d_{11}} + \frac{d_{22}}{d_{66}} - \frac{(d_{12} + d_{66})^2}{d_{11} d_{66}} \right] + \left( \frac{1}{4} \left[ \frac{d_{66}}{d_{11}} + \frac{d_{22}}{d_{66}} - \frac{(d_{12} + d_{66})^2}{d_{11} d_{66}} \right]^2 - \frac{d_{22}}{d_{11}} \right)^{1/2} \right\}^{1/2}
$$

$$(4.52)$$

Equations (4.51) and (4.52) were used by Asadpoure (2006) and Asadpoure, Mohammadi and Vafai (2007) to derive XFEM crack-tip enrichment functions (See Section 4.5.3).

The same problem with the condition of $|a_1| < \sqrt{a_2}$ (type II) was considered by Viola, Piva and Radi (1989). Accordingly, Asadpoure, Mohammadi and Vafai (2006) proposed the following enrichment functions for this class of orthotropic composites:

$$
\begin{aligned}
u_1 = -2\beta_1 t_5 + \frac{\beta \sigma_0}{d_{66} k_7} & \left\{ t_1 \left[ 2(a + r\cos\theta) - \sqrt{2ar} \left( \sqrt{g_1(\theta)} \cos\frac{\theta_1}{2} + \sqrt{g_2(\theta)} \cos\frac{\theta_2}{2} \right) \right] \right. \\
& \left. - t_2 \sqrt{2ar} \left( \sqrt{g_1(\theta)} \sin\frac{\theta_1}{2} - \sqrt{g_2(\theta)} \sin\frac{\theta_2}{2} \right) \right\} \\
+ \frac{\beta_1 \tau_0}{d_{66} k_7} & \left\{ t_3 \left[ X_1 - X_2 + \sqrt{2ar} \left( \sqrt{g_2(\theta)} \cos\frac{\theta_2}{2} - \sqrt{g_1(\theta)} \cos\frac{\theta_1}{2} \right) \right] \right. \\
& \left. - t_4 \left[ 2Y_1 - \sqrt{2ar} \left( \sqrt{g_1(\theta)} \sin\frac{\theta_1}{2} + \sqrt{g_2(\theta)} \sin\frac{\theta_2}{2} \right) \right] \right\}
\end{aligned}
$$

$$(4.53)$$

$$
\begin{aligned}
u_2 = -h_5 + \frac{\sigma_0}{2 d_{66} k_7} & \left\{ h_1 \left[ (X_1 - X_2) + \sqrt{2ar} \left( \sqrt{g_1(\theta)} \cos\frac{\theta_1}{2} - \sqrt{2ar} \cos\frac{\theta_2}{2} \right) \right] \right. \\
& \left. + h_2 \left[ 2Y_1 - \sqrt{2ar} \left( \sqrt{g_1(\theta)} \sin\frac{\theta_1}{2} + \sqrt{g_2(\theta)} \sin\frac{\theta_2}{2} \right) \right] \right\} \\
+ \frac{\tau_0}{2 d_{66} k_7} & \left\{ h_3 \left[ 2(a + r\cos\theta) - \sqrt{2ar} \left( \sqrt{g_1(\theta)} \cos\frac{\theta_1}{2} + \sqrt{g_2(\theta)} \cos\frac{\theta_2}{2} \right) \right] \right. \\
& \left. + h_4 \sqrt{2ar} \left( \sqrt{g_1(\theta)} \sin\frac{\theta_1}{2} + \sqrt{g_2(\theta)} \sin\frac{\theta_2}{2} \right) \right\}
\end{aligned}
$$

$$(4.54)$$

where

$$\gamma_1 = \left[\frac{1}{2} + \left(\sqrt{a_2} + a_1\right)\right]^{1/2} \tag{4.55}$$

$$\gamma_2 = \left[\frac{1}{2} + \left(\sqrt{a_2} - a_1\right)\right]^{1/2} \tag{4.56}$$

$$l^2 = \left(\gamma_1^2 + \gamma_2^2\right)^{-1} \quad j = 1, 2 \tag{4.57}$$

For definition of other parameters see Viola, Piva and Radi (1989) and Asadpoure, Moham-madi and Vafai (2006).

## 4.4  Orthotropic Mixed Mode Fracture

An orthotropic composite material behaviour can be further idealized in order to simplify the process of crack modelling; a fibre composite structure is assumed as a homogeneous orthotropic continuum, where the crack growth takes place in an idealized material with anisotropic constituents. In this approach, the details of local failures of the composite, such as broken fibres or cracked matrix, are not considered and an equivalent orthotropic continuum is adopted.

### 4.4.1  Energy Release Rate for Anisotropic Materials

Sih, Paris and Irwin (1965) extended Eqs. (2.136)–(2.137) for anisotropic materials:

$$G_{\mathrm{I}} = -\frac{\pi}{2}K_{\mathrm{I}}c_{22}\mathrm{Im}\left[\frac{K_{\mathrm{I}}(s_1 + s_2) + K_{\mathrm{II}}}{s_1 s_2}\right] \tag{4.58}$$

$$G_{\mathrm{II}} = \frac{\pi}{2}K_{\mathrm{II}}c_{11}\mathrm{Im}\left[K_{\mathrm{II}}(s_1 + s_2) + K_{\mathrm{I}}s_1 s_2\right] \tag{4.59}$$

Eqs. (4.58) and (4.59) can be conveniently simplified for orthotropic problems with the crack on one plane of material symmetry (Sih, Paris and Irwin, 1965):

$$G_{\mathrm{I}} = \pi K_{\mathrm{I}}^2\sqrt{\frac{c_{11}c_{22}}{2}}\left[\sqrt{\frac{c_{22}}{c_{11}}} + \frac{2c_{12} + c_{66}}{2c_{11}}\right]^{\frac{1}{2}} \tag{4.60}$$

$$G_{\mathrm{II}} = \pi K_{\mathrm{II}}^2\frac{c_{11}}{\sqrt{2}}\left[\sqrt{\frac{c_{22}}{c_{11}}} + \frac{2c_{12} + c_{66}}{2c_{11}}\right]^{\frac{1}{2}} \tag{4.61}$$

### 4.4.2  Anisotropic Singular Elements

The same idea of singular quarter point finite elements can be extended to anisotropic problems. Saouma and Sikiotis (1986) proposed the following procedure for anisotropic materials in the

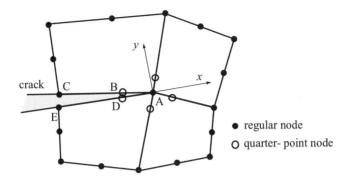

**Figure 4.2**  Displacement field around the crack tip (Mohammadi, 2008).

form of:

$$\begin{Bmatrix} K_I \\ K_{II} \end{Bmatrix} = \frac{1}{\Delta'} \mathbf{B} \begin{Bmatrix} 4(u_{xB} - u_{xD}) - (u_{xC} - u_{xE}) \\ 4(u_{yB} - u_{yD}) - (u_{yC} - u_{yE}) \end{Bmatrix} \sqrt{\frac{\pi}{2L}} \tag{4.62}$$

where $u_{ia}$ is the $i$-th local crack tip displacement component $(x, y)$ of nodes $a = B, C, D, E$ of the singular quarter point element, as shown in Figure 4.2, and

$$\mathbf{B} = \begin{bmatrix} \mathrm{Re}\left[\dfrac{i}{s_1 - s_2}(q_2 - q_1)\right] & \mathrm{Re}\left[\dfrac{-i}{s_1 - s_2}(p_2 - p_1)\right] \\[2ex] \mathrm{Re}\left[\dfrac{-i}{s_1 - s_2}(s_1 q_2 - s_2 q_1)\right] & \mathrm{Re}\left[\dfrac{i}{s_1 - s_2}(s_1 p_2 - s_2 p_1)\right] \end{bmatrix} \tag{4.63}$$

$$\Delta' = \det \begin{bmatrix} \mathrm{Re}\left[\dfrac{i}{s_1 - s_2}(s_1 p_2 - s_2 p_1)\right] & \mathrm{Re}\left[\dfrac{i}{s_1 - s_2}(p_2 - p_1)\right] \\[2ex] \mathrm{Re}\left[\dfrac{i}{s_1 - s_2}(s_1 q_2 - s_2 q_1)\right] & \mathrm{Re}\left[\dfrac{i}{s_1 - s_2}(q_2 - q_1)\right] \end{bmatrix} \tag{4.64}$$

and $L$ is the side length of the singular element. Further investigation, however, is required to assess the accuracy of such a complex model.

### 4.4.3   SIF Calculation by Interaction Integral

The stress intensity factor (SIF) is one of the important parameters representing fracture properties of a crack tip. Here, the domain integral method, proposed by Kim and Paulino (2002a) is adopted to evaluate the mixed mode stress intensity factors in homogenous orthotropic media.

Figure 4.3 shows an arbitrary area surrounding a crack tip. The standard path independent $J$ integral for the cracked body is defined as (Rice and Rosengren, 1968):

$$J = \int_\Gamma \left( w_s \delta_{1j} - \sigma_{ij} \frac{\partial u_i}{\partial x_1} \right) \mathbf{n}_j d\Gamma \tag{4.65}$$

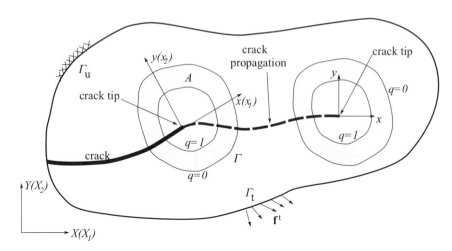

**Figure 4.3**  Definitions of the $J$ integral computations for a propagating crack.

where $\mathbf{n}_j$ is the $j^{th}$ component of the outward unit normal to $\Gamma$, $\delta_{1j}$ is the Kronecker delta, $w_s$ is the strain energy density for linear elastic material, defined in (2.154), and $\Gamma$ is an arbitrary contour around the crack tip which encloses no other cracks or discontinuities.

Eq. (4.66) is not well suited for the finite element solutions, and an equivalent form of the $J$ integral can be obtained by exploiting the divergence theorem in the form of the domain integral approach (Kim and Paulino, 2003d)

$$J = \int_A \left( \sigma_{ij} \frac{\partial u_i}{\partial x_1} - w_s \delta_{1j} \right) \frac{\partial q}{\partial x_j} dA \qquad (4.66)$$

where $A$ is an area surrounding the crack tip (the interior region of $\Gamma$) and $q$ is a smoothly varying function. $\Gamma$ is usually assumed as a circular or rectangular area whose centre locates on the crack tip.

As proposed by Dolbow (1999) and Moës, Dolbow and Belytschko (1999), a simple function $q$, varying linearly from $q = 1$ at the crack tip to $q = 0$ at the exterior boundary $\Gamma$, is used in the finite element model (Figure 4.3). Therefore, the elements away from the boundary can be neglected. It is worth noting that the value within the parentheses in Eq. (4.60) need not be evaluated in the area where $q$ is constant (and so its gradient vanishes).

From a numerical aspect, in spite of the fact that the stress gradient in elements containing a crack tip is usually very high, it is more appropriate to avoid locating the contour on the elements including a crack tip (Asadpoure, Mohammadi and Vafai, 2007).

Evaluation of (4.65) can be implemented by the Gauss integration rule along the path $\Gamma$,

$$J = \sum_{g=1}^{ng_\Gamma} W_g^\Gamma J(\xi_g, \eta_g) \qquad (4.67)$$

where $W_g^\Gamma$ is the Gauss weighting factor, $ng_\Gamma$ is the order of integration and $(\xi_g, \eta_g)$ are the Gauss points along the contour $\Gamma$.

Similar evaluation of the domain $J$ integral (4.66) is usually performed by the Gauss integration rule on the modified form of the $J$ integral. Discretization of $q$ in terms of its nodal values by the finite element shape functions $N_i$,

$$q(\mathbf{x}) = \sum N_i(\mathbf{x}) q_i \tag{4.68}$$

leads to the following numerical form of the domain integral

$$J = \sum_{\substack{\text{elements} \\ \text{in } A}} \left( \sum_{g=1}^{ng_A} \left\{ \left[ \left( \sigma_{ij} \frac{\partial u_i}{\partial x_1} - w_s \delta_{1i} \right) \frac{\partial q}{\partial x_j} \right] \det \mathbf{J} \right\}_g W_g^A \right) \tag{4.69}$$

where $W_g^A$ is the Gauss weighting factor, $ng_A$ is the number of integration points for the domain $A$ and $\mathbf{J}$ is the finite element Jacobian matrix.

The domain integral can also be computed using a set of additional integration cells rich in Gauss points. Such a domain can be chosen independent of the finite element mesh and follows the crack tip and its orientations during the crack growth (Gosz, Dolbow and Moran, 1998; Gregoire *et al.*, 2007).

The method of interaction integral (Section 2.7.5), based on the definition of an auxiliary state, is used to extract mixed mode stress intensity factors. By combining the actual and auxiliary states for obtaining the $J$ integral, one can write:

$$J = J^{\text{act}} + J^{\text{aux}} + M \tag{4.70}$$

where $J$ corresponds to the superposition state, and $J^{\text{act}}$ and $J^{\text{aux}}$ are the actual and auxiliary states $J$ integrals, respectively, defined in (2.164)–(2.165), and the interaction integral $M$ is defined in (2.166):

$$M = \int_A \left[ \sigma_{ij} \frac{\partial u_i^{\text{aux}}}{\partial x_1} + \sigma_{ij}^{\text{aux}} \frac{\partial u_i}{\partial x_1} - w^{\text{M}} \delta_{1j} \right] \frac{\partial q}{\partial x_j} \, \mathrm{d}A \tag{4.71}$$

where $w^{\text{M}} = 0.5(\sigma_{ij} \varepsilon_{ij}^{\text{aux}} + \sigma_{ij}^{\text{aux}} \varepsilon_{ij})$ for linear elastic conditions (see Eq. (2.169)).

The strain of the auxiliary field can be chosen by either the strain–stress relationship,

$$\varepsilon_{ij}^{\text{aux}} = \mathsf{C}_{ijkl} \sigma_{kl}^{\text{aux}} \tag{4.72}$$

or as the symmetric part of the displacement gradient,

$$\varepsilon_{ij}^{\text{aux}} = 0.5 \left( u_{i,j}^{\text{aux}} + u_{j,i}^{\text{aux}} \right) \tag{4.73}$$

these options are compatible with each other as long as the material is homogeneous.

Auxiliary stress and strain states should be chosen so as to satisfy both the equilibrium equation and the traction free boundary condition on the crack surface in the area $A$. One of the choices is the displacement and stress fields in the vicinity of the crack tip provided by

Sih, Paris and Irwin (1965) and Asadpoure, Mohammadi and Vafai (2007) in local crack tip coordinates $(x, y)$ (see Section 4.3.1),

$$u_1^{\text{aux}} = K_{\text{I}}^{\text{aux}} \sqrt{\frac{2r}{\pi}} \, \text{Re} \left[ \frac{1}{s_1 - s_2} \left\{ s_1 p_2 \sqrt{\cos\theta + s_2 \sin\theta} - s_2 p_1 \sqrt{\cos\theta + s_1 \sin\theta} \right\} \right]$$

$$+ K_{\text{II}}^{\text{aux}} \sqrt{\frac{2r}{\pi}} \, \text{Re} \left[ \frac{1}{s_1 - s_2} \left\{ p_2 \sqrt{\cos\theta + s_2 \sin\theta} - p_1 \sqrt{\cos\theta + s_1 \sin\theta} \right\} \right] \quad (4.74)$$

$$u_2^{\text{aux}} = K_{\text{I}}^{\text{aux}} \sqrt{\frac{2r}{\pi}} \, \text{Re} \left[ \frac{1}{s_1 - s_2} \left\{ s_1 q_2 \sqrt{\cos\theta + s_2 \sin\theta} - s_2 q_1 \sqrt{\cos\theta + s_1 \sin\theta} \right\} \right]$$

$$+ K_{\text{II}}^{\text{aux}} \sqrt{\frac{2r}{\pi}} \, \text{Re} \left[ \frac{1}{s_1 - s_2} \left\{ q_2 \sqrt{\cos\theta + s_2 \sin\theta} - q_1 \sqrt{\cos\theta + s_1 \sin\theta} \right\} \right] \quad (4.75)$$

and

$$\sigma_{11}^{\text{aux}} = \frac{K_{\text{I}}^{\text{aux}}}{\sqrt{2\pi r}} \, \text{Re} \left[ \frac{s_1 s_2}{s_1 - s_2} \left\{ \frac{s_2}{\sqrt{\cos\theta + s_2 \sin\theta}} - \frac{s_1}{\sqrt{\cos\theta + s_1 \sin\theta}} \right\} \right]$$

$$+ \frac{K_{\text{II}}^{\text{aux}}}{\sqrt{2\pi r}} \, \text{Re} \left[ \frac{1}{s_1 - s_2} \left\{ \frac{s_2^2}{\sqrt{\cos\theta + s_2 \sin\theta}} - \frac{s_1^2}{\sqrt{\cos\theta + s_1 \sin\theta}} \right\} \right] \quad (4.76)$$

$$\sigma_{22}^{\text{aux}} = \frac{K_{\text{I}}^{\text{aux}}}{\sqrt{2\pi r}} \, \text{Re} \left[ \frac{1}{s_1 - s_2} \left\{ \frac{s_1}{\sqrt{\cos\theta + s_2 \sin\theta}} - \frac{s_2}{\sqrt{\cos\theta + s_1 \sin\theta}} \right\} \right]$$

$$+ \frac{K_{\text{II}}^{\text{aux}}}{\sqrt{2\pi r}} \, \text{Re} \left[ \frac{1}{s_1 - s_2} \left\{ \frac{1}{\sqrt{\cos\theta + s_2 \sin\theta}} - \frac{1}{\sqrt{\cos\theta + s_1 \sin\theta}} \right\} \right] \quad (4.77)$$

$$\sigma_{12}^{\text{aux}} = \frac{K_{\text{I}}^{\text{aux}}}{\sqrt{2\pi r}} \, \text{Re} \left[ \frac{s_1 s_2}{s_1 - s_2} \left\{ \frac{1}{\sqrt{\cos\theta + s_1 \sin\theta}} - \frac{1}{\sqrt{\cos\theta + s_2 \sin\theta}} \right\} \right]$$

$$+ \frac{K_{\text{II}}^{\text{aux}}}{\sqrt{2\pi r}} \, \text{Re} \left[ \frac{1}{s_1 - s_2} \left\{ \frac{s_1}{\sqrt{\cos\theta + s_1 \sin\theta}} - \frac{s_2}{\sqrt{\cos\theta + s_2 \sin\theta}} \right\} \right] \quad (4.78)$$

After some manipulations (Asadpoure, Mohammadi and Vafai, 2007; Asadpoure and Mohammadi, 2007; Bayesteh and Mohammadi, 2011):

$$J^s = t_{11} \left( K_{\text{I}}^{\text{aux}} + K_{\text{I}} \right)^2 + t_{12} \left( K_{\text{I}}^{\text{aux}} + K_{\text{I}} \right) \left( K_{\text{II}}^{\text{aux}} + K_{\text{II}} \right) + t_{22} \left( K_{\text{II}}^{\text{aux}} + K_{\text{II}} \right)^2 \quad (4.79)$$

$$J^{\text{act}} = t_{11} (K_{\text{I}})^2 + t_{12} (K_{\text{I}} K_{\text{II}}) + t_{22} (K_{\text{II}})^2 \quad (4.80)$$

$$J^{\text{aux}} = t_{11} \left( K_{\text{I}}^{\text{aux}} \right)^2 + t_{12} \left( K_{\text{I}}^{\text{aux}} K_{\text{II}}^{\text{aux}} \right) + t_{22} \left( K_{\text{II}}^{\text{aux}} \right)^2 \quad (4.81)$$

and

$$M = 2t_{11} K_{\text{I}} K_{\text{I}}^{\text{aux}} + t_{12} \left( K_{\text{I}} K_{\text{II}}^{\text{aux}} + K_{\text{I}}^{\text{aux}} K_{\text{II}} \right) + 2t_{22} K_{\text{II}} K_{\text{II}}^{\text{aux}} \quad (4.82)$$

where

$$t_{11} = -\frac{c_{22}}{2} \text{Im} \left( \frac{s_1 + s_2}{s_1 s_2} \right) \quad (4.83)$$

$$t_{12} = -\frac{c_{22}}{2} \text{Im} \left( \frac{1}{s_1 s_2} \right) + \frac{c_{11}}{2} \text{Im} (s_1 s_2) \quad (4.84)$$

$$t_{22} = \frac{c_{11}}{2} \text{Im} \, (s_1 + s_2) \tag{4.85}$$

The stress intensity factor can then be obtained by considering the two states and solving a system of linear algebraic equations. These two states are state 1: $K_I^{\text{aux}} = 1$; $K_{II}^{\text{aux}} = 0$ and state 2: $K_I^{\text{aux}} = 0$; $K_{II}^{\text{aux}} = 1$. By calculating $M$ from both Eqs. (4.72) and (4.83) and solving a system of linear algebraic equations, the actual mixed mode stress intensity factors associated with state 1 and state 2 are obtained (Asadpoure, Mohammadi and Vafai, 2007; Asadpoure and Mohammadi, 2007):

$$M^{(1)} = 2t_{11}K_I + t_{12}K_{II} \tag{4.86}$$

$$M^{(2)} = t_{12}K_I + 2t_{22}K_{II} \tag{4.87}$$

### 4.4.4   Orthotropic Crack Propagation Criteria

In this section, the material behaviour is ideally assumed as a homogeneous orthotropic continuum to simplify the process of crack modelling. As a result, details of local failures of composites, such as cracked matrix, broken fibre, local delamination, and so on are not considered.

#### 4.4.4.1   Maximum Circumferential Tensile Stress Criterion

Saouma, Ayari and Leavell (1987) extended the original isotropic maximum circumferential tensile stress theory to anisotropic solids.

In this case, the fracture toughness is no longer uniquely defined. Instead, two values of $K_{Ic}^1$ and $K_{Ic}^2$ along the principal planes of elastic symmetry are required to characterize the brittle behaviour of the crack in a homogenous transversely isotropic solid with elastic constants $E_1$, $E_2$, and $\mu_{12}$ (Figure 4.4). According to Saouma, Ayari and Leavell (1987), even for a symmetrically aligned loading and material properties with respect to the crack, a parasitic crack sliding displacement occurs in addition to a pure mode I for anisotropic materials.

For a crack arbitrarily oriented with respect to direction 1, the toughness $K_{Ic}^\beta$ can be assumed as a polar function of $K_{Ic}^1$ and $K_{Ic}^2$ (Saouma, Ayari and Leavell, 1987; Ye and Ayari, 1994; Ayari

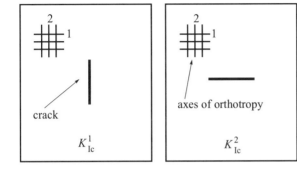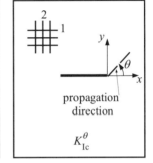

**Figure 4.4**   Fracture toughness for a homogeneous anisotropic solid (Saouma, Ayari and Leavell, 1987).

and Ye, 1995):

$$K_{\text{Ic}}^{\theta} = K_{\text{Ic}}^{1} \cos^2 \theta + K_{\text{Ic}}^{2} \sin^2 \theta \tag{4.88}$$

where $\theta$ is the crack propagation angle with respect to the $x_1 = x$ axis (See Figure 4.4).

In order to avoid performing two separate fracture toughness tests, it is sometimes assumed that the ratio of the fracture toughness in both directions is equal to the ratio of the elastic modulus (Saouma, Ayari and Leavell, 1987):

$$K_{\text{Ic}}^{2} = K_{\text{Ic}}^{1} \frac{E_1}{E_2} \tag{4.89}$$

The classical maximum circumferential/hoop tensile stress mode assumes that the crack propagation is along the direction of the maximum tangential stress $\sigma_{\theta\theta}$, while the shear stress is zero. The hoop stress component $\sigma_{\theta\theta}$ for an anisotropic body can be obtained from,

$$\sigma_{\theta\theta} = \frac{K_{\text{I}}}{\sqrt{2\pi r}} \operatorname{Re}\left[\frac{s_1 t_1 - s_2 t_2}{s_1 - s_2}\right] + \frac{K_{\text{II}}}{\sqrt{2\pi r}} \operatorname{Re}\left[\frac{t_1 - t_2}{s_1 - s_2}\right] \tag{4.90}$$

where

$$t_1^2 = (s_2 \sin\theta + \cos\theta)^3 \tag{4.91}$$

$$t_2^2 = (s_1 \sin\theta + \cos\theta)^3 \tag{4.92}$$

Since the angular variation of material toughness and the critical tensile strength is not constant for anisotropic materials, instead of maximizing the hoop stress component $\sigma_{\theta\theta}$, the ratio of $K_{\theta\theta}/K_{Ic}^{\theta}$ is maximized to determine the angle of crack propagation $\theta$,

$$\max\left\{\frac{K_{\theta\theta}}{K_{\text{Ic}}^{\theta}}\right\} = \max\left\{\frac{\sqrt{2\pi r}\sigma_{\theta\theta}}{K_{\text{Ic}}^{\theta}}\right\} = \max\left\{\frac{\left[K_{\text{I}}\operatorname{Re}\left[\frac{s_1 t_1 - s_2 t_2}{s_1 - s_2}\right] + K_{\text{II}}\operatorname{Re}\left[\frac{t_1 - t_2}{s_1 - s_2}\right]\right]}{\left[K_{\text{Ic}}^{1}\cos^2\theta + K_{\text{Ic}}^{2}\sin^2\theta\right]}\right\} \tag{4.93}$$

### 4.4.4.2 Minimum Strain Energy Density Criterion

It is also possible to extend the minimum strain energy density criterion to anisotropic problems. However, it is important to note that such a criterion is expressed in terms of a scalar quantity (strain energy density), which cannot be directly related to the tensorial (angular dependent) nature of toughness in anisotropic materials (Saouma, Ayari and Leavell, 1987).

Carloni and Nobile (2002) discussed the extension of the classical minimum strain energy density criterion to orthotropic materials. According to this model, propagation occurs at an angle where the ratio of strain energy density to the corresponding material toughness is minimized:

$$\min\left\{\frac{\dfrac{dw}{dV}}{\left(\dfrac{dw}{dV}\right)_c^{\theta}}\right\} \tag{4.94}$$

where

$$\frac{dw}{dV} = \frac{S}{r} = \frac{1}{r}\left\{c_{11}K_I^2 + 2c_{12}K_IK_{II} + c_{22}K_{II}^2\right\} \tag{4.95}$$

$$\left(\frac{dw}{dV}\right)_c^\theta = \left(\frac{dw}{dV}\right)_c^x \sin^2\theta + \left(\frac{dw}{dV}\right)_c^y \cos^2\theta \tag{4.96}$$

### 4.4.4.3 Maximum Circumferential Strain Criterion

Ayari and Ye (1995) have proposed a criterion based on maximization of the circumferential strain. The method accounts for angular variation of the toughness in orthotropic materials and can, reportedly, anticipate general mixed mode crack propagations in orthotropic problems. The method leads to rather implicit formulations such as Section 4.4.4.2 rather than explicit relations of (4.93). For details see Ayari and Ye (1995).

### 4.4.4.4 Simplified Models

Several simplified models are available for specific mixed mode propagation problems. Unfortunately, many of them cannot be used for general orthotropic problems, or at least need further experimental and numerical verifications. Among them, substantial work has been performed on wood and timber materials, which behave in specific orthotropic patterns. For a review of available techniques refer to Jernkevist and Thuvander (2001), and Jernkvist (2001a, 2001b).

## 4.5 Anisotropic XFEM

The general methodology of isotropic extended finite element can be similarly extended to include anisotropic problems, if anisotropic enrichment functions are embedded into an anisotropic finite element procedure. Generalized Heaviside and near crack-tip anisotropic enrichment functions are included as extra degrees of freedom in selected nodes near the discontinuities.

### 4.5.1 Governing Equation

Consider a body in the state of equilibrium with the boundary conditions in the form of traction and displacements that also include a crack (Figure 4.1). The strong form of the equilibrium equation can be written as:

$$\nabla.\sigma + \mathbf{f}^b = 0 \quad \text{in } \Omega \tag{4.97}$$

with the following boundary conditions:

$$\sigma \cdot \mathbf{n} = \mathbf{f}^t \quad \text{on } \Gamma_t : \quad \text{external traction} \tag{4.98}$$

$$\mathbf{u} = \bar{\mathbf{u}} \quad \text{on } \Gamma_u: \quad \text{prescribed displacement} \tag{4.99}$$

$$\sigma \cdot \mathbf{n} = 0 \quad \text{on } \Gamma_c : \quad \text{traction free crack} \tag{4.100}$$

where $\Gamma_t$, $\Gamma_u$ and $\Gamma_c$ are traction, displacement and crack boundaries, respectively, $\sigma$ is the stress tensor and $\mathbf{f}^b$ and $\mathbf{f}^t$ are the body force and external traction vectors, respectively (Figure 3.4).

The variational formulation of the boundary value problem can be defined as:

$$W^{\text{int}} = W^{\text{ext}} \tag{4.101}$$

or

$$\int_\Omega \sigma \cdot \delta\boldsymbol{\varepsilon} \, d\Omega = \int_\Omega \mathbf{f}^b \cdot \delta\mathbf{u} \, d\Omega + \int_\Gamma \mathbf{f}^t \cdot \delta\mathbf{u} \, d\Gamma \tag{4.102}$$

## 4.5.2 XFEM Discretization

In the extended finite element method, in order to include the effects of crack surfaces and crack tips, the approximation (3.28) is utilized (ignoring the weak discontinuity term) to calculate the displacement for point $\mathbf{x}$ locating within the domain

$$
\begin{aligned}
\mathbf{u}^h(\mathbf{x}) = {} & \left[ \sum_{j=1}^{n} N_j(\mathbf{x})\mathbf{u}_j \right] + \left[ \sum_{h=1}^{mh} N_h(\mathbf{x}) \left[ H(\xi(\mathbf{x})) - H(\xi(\mathbf{x}_h)) \right] \mathbf{a}_h \right] \\
& + \left[ \sum_{k=1}^{mt} N_k(\mathbf{x}) \left( \sum_{l=1}^{mf} [F_l(\mathbf{x}) - F_l(\mathbf{x}_k)] \, \mathbf{b}_k^l \right) \right]
\end{aligned} \tag{4.103}
$$

where $mh$ is the set of nodes that have crack face (but not crack tip) in their support domain, while $mt$ is the set of nodes associated with the crack tip in their influence domain; $\mathbf{u}_j$ are the nodal displacements (standard degrees of freedom); $\mathbf{a}_h$, $\mathbf{b}_k^l$ are vectors of additional degrees of nodal freedom for modelling crack faces and the crack tip, respectively. $F_l(\mathbf{x})$ represents the crack tip enrichment functions and $H(\mathbf{x})$ is the generalized Heaviside function.

Discretization of Eq. (4.102) using the XFEM procedure (4.103) results in a discrete system of linear equilibrium equations:

$$\mathbf{K}\mathbf{u}^h = \mathbf{f} \tag{4.104}$$

where $\mathbf{u}^h$ is the vector of nodal degrees of freedom (for both classical and enriched ones),

$$\mathbf{u}^h = \left\{ \mathbf{u} \quad \mathbf{a} \quad \mathbf{b}_1 \quad \mathbf{b}_2 \quad \mathbf{b}_3 \quad \mathbf{b}_4 \right\}^T \tag{4.105}$$

The global stiffness matrix $\mathbf{K}$ is assembled from the stiffness of each element $\Omega^e$,

$$\mathbf{K}_{ij}^e = \begin{bmatrix} \mathbf{K}_{ij}^{uu} & \mathbf{K}_{ij}^{ua} & \mathbf{K}_{ij}^{ub} \\ \mathbf{K}_{ij}^{au} & \mathbf{K}_{ij}^{aa} & \mathbf{K}_{ij}^{ab} \\ \mathbf{K}_{ij}^{bu} & \mathbf{K}_{ij}^{ba} & \mathbf{K}_{ij}^{bb} \end{bmatrix} \tag{4.106}$$

with

$$\mathbf{K}_{ij}^{rs} = \int_{\Omega^e} \left( \mathbf{B}_i^r \right)^T \mathbf{D} \mathbf{B}_j^s \, d\Omega \quad (r, s = \mathbf{u}, \mathbf{a}, \mathbf{b}) \tag{4.107}$$

where $\mathbf{B}_i^u$, $\mathbf{B}_i^a$ and $\mathbf{B}_i^b$ are derivatives of shape functions defined in Eqs. (3.49)–(3.55). $\mathbf{f}$ is the vector of external forces,

$$\mathbf{f}_i^e = \{\mathbf{f}_i^u \quad \mathbf{f}_i^a \quad \mathbf{f}_i^{b1} \quad \mathbf{f}_i^{b2} \quad \mathbf{f}_i^{b3} \quad \mathbf{f}_i^{b4}\}^T \tag{4.108}$$

and

$$\mathbf{f}_i^u = \int_{\Gamma_t} N_i \mathbf{f}^t \, d\Gamma + \int_{\Omega^e} N_i \mathbf{f}^b \, d\Omega \tag{4.109}$$

$$\mathbf{f}_i^a = \int_{\Gamma_t} N_i \left[H(\xi) - H(\xi_i)\right] \mathbf{f}^t \, d\Gamma + \int_{\Omega^e} N_i \left[H(\xi) - H(\xi_i)\right] \mathbf{f}^b \, d\Omega \tag{4.110}$$

$$\mathbf{f}_i^{b\alpha} = \int_{\Gamma_t} N_i \left(F_\alpha - F_{\alpha i}\right) \mathbf{f}^t \, d\Gamma + \int_{\Omega^e} N_i \left(F_\alpha - F_{\alpha i}\right) \mathbf{f}^b \, d\Omega \quad (\alpha = 1, 2, 3 \text{ and } 4) \tag{4.111}$$

### 4.5.3 Orthotropic Enrichment Functions

Crack-tip enrichment functions are obtained from the analytical solution for displacement in the vicinity of the crack tip. The general form of these functions can be defined as,

$$\{F_l(r, \theta)\}_{l=1}^4 = \left\{ \sqrt{r} \cos\frac{\theta_1}{2} \sqrt{g_1(\theta)}, \; \sqrt{r} \cos\frac{\theta_2}{2} \sqrt{g_2(\theta)}, \; \sqrt{r} \sin\frac{\theta_1}{2} \sqrt{g_1(\theta)}, \right.$$
$$\left. \sqrt{r} \sin\frac{\theta_2}{2} \sqrt{g_2(\theta)} \right\} \tag{4.112}$$

where functions $g_k(\theta)$ and $\theta_k$, in their general form for all orthotropic composites, are defined as:

$$g_k(\theta) = \sqrt{(\cos\theta + s_{kx}\sin\theta)^2 + (s_{ky}\sin\theta)^2} \tag{4.113}$$

$$\theta_k = \arctan\left(\frac{s_{ky}\sin\theta}{\cos\theta + s_{kx}\sin\theta}\right) \tag{4.114}$$

These functions for the type I composites $(a_1 > \sqrt{a_2})$ are

$$g_j(\theta) = \left(\cos^2\theta + \frac{\sin^2\theta}{e_j^2}\right)^{1/2} \quad j = 1, 2 \tag{4.115}$$

$$\theta_j = \tan^{-1}\left(\frac{x_2}{e_j x_1}\right) = \tan^{-1}\left(\frac{\tan\theta}{e_j}\right) \quad j = 1, 2 \tag{4.116}$$

where $e_1$ and $e_2$ are defined in Eq. (4.51)–(4.52).

Finally, in case of $|a_1| < \sqrt{a_2}$, (type II) the main contributing terms are.

$$g_j(\theta) - (\cos^2\theta + l^2\sin^2\theta \mid (\;1)^j l^2 \sin 2\theta)^{1/2} \quad j - 1, 2 \tag{4.117}$$

$$\theta_j = \arctan\left(\frac{\gamma_2 l^2 \sin\theta}{\cos\theta + (-1)^j \gamma_1 l^2 \sin\theta}\right) \quad j = 1, 2 \tag{4.118}$$

where $\gamma_1$ and $\gamma_2$ and $l$ are defined in Eqs. (4.55)–(4.57).

According to Eq. (4.112), the first two terms are continuous across the crack faces while the remaining ones are discontinuous.

Eq. (4.112) cannot be directly used in isotropic media because it may lead to indefinite (0/0) expressions (Mohammadi and Asadpoure, 2006). A straightforward remedy is to apply the original isotropic enrichment functions (3.31)

$$\{F_l(r, \theta)\}_{l=1}^4 = \left\{ \sqrt{r} \cos \frac{\theta}{2}, \ \sqrt{r} \sin \frac{\theta}{2}, \ \sqrt{r} \sin \theta \cos \frac{\theta}{2}, \ \sqrt{r} \sin \theta \sin \frac{\theta}{2} \right\} \quad (4.119)$$

## 4.6 Numerical Simulations

In order to examine the performance of the anisotropic XFEM approach, several examples including crack stability and propagation problems, are considered.

Stress intensity factors and $J$ integrals are calculated for all examples based on the orthotropic enriched XFEM and they are compared with other available analytical and numerical solutions. Results are presented in terms of the stress intensity factors $K_I$ and $K_{II}$ and/or normalized stress intensity factors $\bar{K}_I = K_I/\sigma_0\sqrt{\pi a}$ and $\bar{K}_{II} = K_{II}/\sigma_0\sqrt{\pi a}$ for the applied uniform stress $\sigma_0$ and $\bar{K}_I = K_I\sqrt{\delta}/\varepsilon_0 E\sqrt{\pi a}$ and $\bar{K}_{II} = K_{II}\sqrt{\delta}/\varepsilon_0 E\sqrt{\pi a}$ for fixed-grip loading $\varepsilon_0$.

A quadrature partitioning approach, as defined in Section 3.48, is used for integration over all cracked elements, and the Gauss quadrature rule is applied to each partition. A simple $2 \times 2$ Gauss rule is applied for any other uncracked finite elements.

### 4.6.1   *Plate with a Crack Parallel to the Material Axis of Orthotropy*

An orthotropic plate with a central crack aligned along one of the axes of orthotropy is simulated by XFEM. Figure 4.5a illustrates the geometry and boundary conditions of the

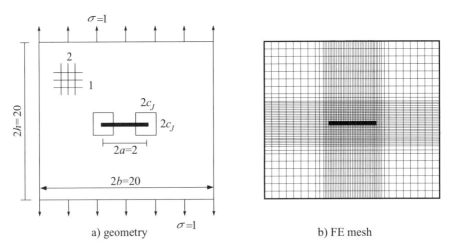

a) geometry                                         b) FE mesh

**Figure 4.5**   Geometry and boundary conditions for a plate with a crack parallel to material axis of orthotropy and the typical FE model (Mohammadi, 2008). (Reproduced by permission of John Wiley & Sons, Ltd.)

**Table 4.1**  Values of the stress intensity factor $K_I$ for a plate with a crack parallel to material orthotropic axes.

| Method | $n_{elem.}$ | $n_{nodes}$ | $n_{cells}$ | $n_{DOFs}$ | $c_J/a$ | $K_I$ |
|---|---|---|---|---|---|---|
| FEM (Kim and Paulino, 2002c) | 2001 | 5851 | — | 11702 | — | 1.767 |
| Meshless EFG (Ghorashi, Mohammadi and Sabbagh-Yazdi, 2011) | — | 2116 | 1849 | 3875 | — | 1.710 |
| Meshless XEFG (Ghorashi, Mohammadi and Sabbagh-Yazdi, 2011) | — | 2116 | 1849 | 4035 | — | 1.780 |
| XFEM (Asadpour and Mohammadi, 2007) | 400 | 441 | — | 904 | 1.5 | 1.781 |
|  | 784 | 841 | — | 1712 | 0.5 | 1.800 |
|  | 2025 | 2116 | — | 4278 | 0.5 | 1.807 |
| Bimaterial XFEM (Esna Ashari and Mohammadi, 2011a) | 2025 | 2116 | — | 4278 | — | 1.777 |
| XIGA (Ghorashi, Valizadeh and Mohammadi, 2012) | 1369 | — | — | 3507 | — | 1.744 |

plane stress cracked plate with a constant tensile traction ($\sigma = 1$). The plate is composed of a graphite-epoxy material with the following orthotropic properties:

$$E_1 = 114.8\,\text{GPa}, \quad E_2 = 11.7\,\text{GPa}, \quad G_{12} = 9.66\,\text{GPa}, \quad v_{12} = 0.21 \qquad (4.120)$$

The domain is discretized by adaptive finite element meshes, typically shown in Figure 4.5b. For instance, a fine mesh of 2025 quadrilateral finite elements and 2116 nodes is constructed from 45 rows of elements in each $x_1$ and $x_2$ directions, with the size of the finite elements around the crack tip of about one-sixteenth of the crack length.

Table 4.1 compares the stress intensity factors calculated by XFEM (Asadpoure and Mohammadi, 2007) with the results provided by Kim and Paulino (2002c) using a total of 2001 elements and 5851 nodes, which shows only a small difference of about 2.3% while the reference results were obtained with almost three times degrees of freedom (DOF).

Also, the results of two coarser finite element meshes with 784 and 400 elements illustrate that even with four times reduction in the number of degrees of freedom, the computed SIFs have changed only 1.5%; indicating the efficiency of the proposed enrichment approach in crack modelling.

Table 4.2 compares the computed stress intensity factors for various enrichment strategies in terms of different $J$ integral domain sizes $c_J$. Three cases are assumed: orthotropic enrichments (4.112), isotropic enrichments (4.119) and without crack-tip functions. It is observed that while for small $J$ integral domain sizes the analysis must include crack-tip enrichment functions to obtain accurate results and higher rates of convergence, there is only a small difference between the orthotropic and isotropic enrichments in this case. On the other hand, larger domains can be selected to avoid the local effects of the crack tip. Asadpoure, Mohammadi and Vafai (2006) have shown that when $c_J = 0.5a$, the values of the SIFs become nearly independent of the domain size.

**Table 4.2** Comparison of stress intensity factors for various enrichment strategies and different $J$ integral domain size.

| | | | | | $K_{\mathrm{I}}$ | | |
|---|---|---|---|---|---|---|---|
| Mesh | $n_{\mathrm{elem.}}$ | $n_{\mathrm{nodes}}$ | $n_{\mathrm{DOFs}}$ | $c_J/a$ | Ort. Enr. | Iso. Enr. | No Enr. |
| 1 | 2025 | 2116 | 4278 | 0.5 | 1.802 | | 1.796 |
| | | | | 1 | 1.802 | 1.810 | 1.798 |
| | | | | 2 | 1.804 | | 1.800 |
| 2 | 784 | 841 | 1712 | 0.5 | 1.799 | | — |
| | | | | 1 | 1.800 | 1.806 | — |
| | | | | 2 | 1.801 | | — |
| 3 | 400 | 441 | 904 | 0.5 | 1.769 | | — |
| | | | | 1 | 1.786 | 1.801 | — |
| | | | | 2 | 1.783 | | — |

Figure 4.6 illustrates the way crack-tip enrichments can reproduce the singular stress field around a crack. Clearly, inclusion of orthotropic enrichments closely resembles the analytical singular distribution of stress near a crack tip, whereas a finite and far lower value of stress is obtained if no enrichment is adopted.

The displacement and stress contours are illustrated in Figure 4.7, which clearly shows the effects of orthotropic behaviour and the existence of discontinuity (crack) on the displacement and stress fields.

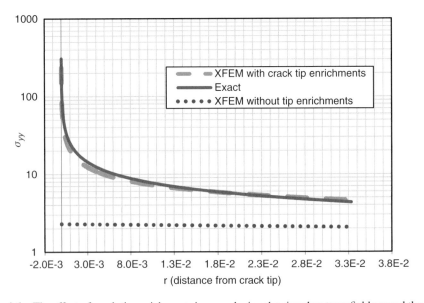

**Figure 4.6** The effect of crack tip enrichments in reproducing the singular stress field around the crack tip, compared with the finite stress in a non-enrichment solution.

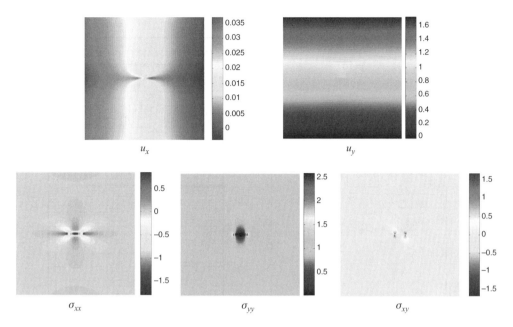

**Figure 4.7**   $x$ and $y$ displacement and $\sigma_{xx}$, $\sigma_{yy}$ and $\sigma_{xy}$ stress contours.

## 4.6.2   Edge Crack with Several Orientations of the Axes of Orthotropy

Consider a rectangular plane stress plate with an edge horizontal crack subjected to a tensile distributed load, as depicted in Figure 4.8a. The plate is composed of a graphite-epoxy material with the following orthotropic properties: $E_1$ =114.8 Gpa, $E_2$ = 11.7 GPa, $G_{12}$ = 9.66 GPa, $v_{12}$ = 0.21.

The effects of changing the orthotropy angle on mixed mode stress intensity factors are investigated by examining several orientations of material elastic axes. A fixed finite element model, composed of 1836 four-node elements with 1925 nodes (Figure 4.8b) is used to model all material angles. The model includes 3850 regular degrees of freedom and 42 additional degrees of freedom, associated with the enrichment part of the approximation.

A partitioning technique with the $5 \times 5$ Gauss quadrature rule in each partition is utilized for integration in elements containing the crack-tip and Heaviside enriched nodes, while a simple $2 \times 2$ Gauss integration is adopted for regular finite elements. Also, a study on variations of stress intensity factors for different domain sizes has shown that the domain size does not substantially affect the value of SIF in this example (Asadpoure and Mohammadi, 2007), so a fixed integration domain size $c_J$ of about $0.12a$ is considered.

Figure 4.9 compares the results of XFEM (Asadpoure, Mohammadi and Vafai, 2007; Asadpoure and Mohammadi, 2007) with the reference boundary element results by Aliabadi and Sollero (1998) and the meshless enriched EFG (XEFG) by Ghorashi, Mohammadi and Sabbagh-Yazdi (2011). The results indicate that the trends of mode I and II stress intensity factors change around $\alpha = 45°$ and $\alpha = 30°$, respectively. They both show an increasing trend before the turning point and then decrease afterwards.

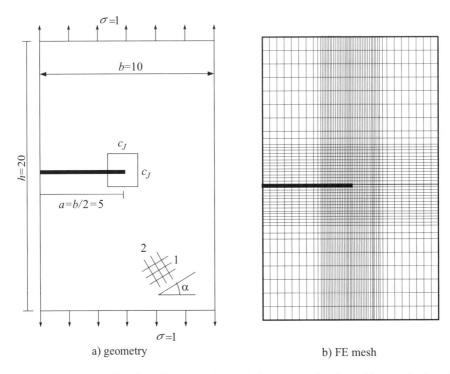

a) geometry                           b) FE mesh

**Figure 4.8**  Geometry and loading of a single edge crack in a rectangular plate with several orientations of the axes of orthotropy and the finite element mesh (Mohammadi, 2008). (Reproduced by permission of John Wiley & Sons, Ltd.)

### 4.6.3  Inclined Edge Notched Tensile Specimen

A plane strain tensile plate composed of orthotropic materials with an inclined edge crack, previously studied by Jernkvist (2001a, 2001b), is considered. The plate is composed of Norway spruce (Picea abies) with the material properties defined in Table 4.3. For this material, cracks usually propagate either along the wood fibres or perpendicular to them, so different inclined edge cracks, all along the material elastic axes 1, are analysed.

A fixed finite element model, depicted in Figure 4.10b, is utilized for all crack inclinations, indicating the efficiency and flexibility of XFEM in handling different crack patterns with the same finite element mesh. The sub-quad partitioning technique with 10 sections in both directions is adopted for enriched elements, and a $2 \times 2$ Gauss quadrature is used in each section. The results are calculated on the basis of converged values for various sizes of the $J$ integration domain size $c_J$ of about $0.1a$ to $0.5a$, which correspond to 2 to 10 elements far from the crack-tip position, respectively.

Mixed mode stress intensity factors are compared in Table 4.4 with the reference results by Jernkvist (2001b), based on a radial mesh of 36 rows of elements in the circumferential and 10 rows in the radial direction around the crack tip. The maximum differences between the XFEM and reference results of modes I and II for all crack angles are about 1.8 and

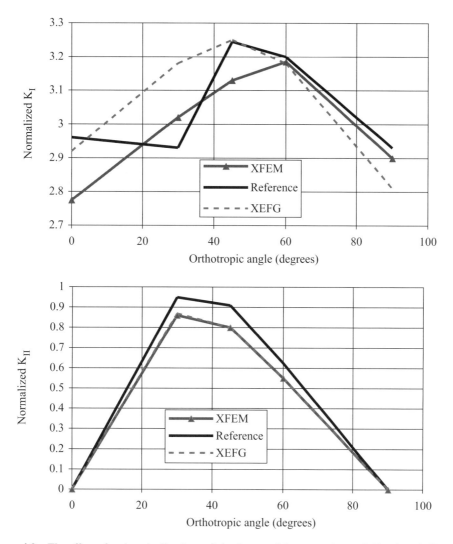

**Figure 4.9** The effect of various inclinations of elastic material axes on the mode I and mode II stress intensity factors (Ghorashi, Mohammadi and Sabbagh-Yazdi, 2011). (Reproduced by permissions of Elsevier.)

**Table 4.3** Orthotropic material properties of Norway spruce (Picea abies) (Jernkvist, 2001b).

| | | |
|---|---|---|
| $E_1 = 0.81$ GPa | $E_2 = 11.84$ GPa | $E_3 = 0.64$ GPa |
| $\nu_{12} = 0.38$ | $\nu_{13} = 0.56$ | $\nu_{23} = 0.4$ |
| $G_{12} = 0.63$ GPa | | |

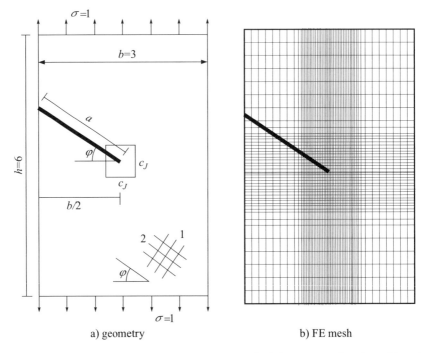

a) geometry                                                          b) FE mesh

**Figure 4.10**   Specimen geometry used for mixed mode analysis in a rectangular plate with an inclined edge crack and the finite element mesh (Mohammadi, 2008).

3.2%, respectively. Also, the maximum difference for both stress intensity factors occurs at the inclination angle $\varphi = 45°$.

Now, a number of simulations are performed to assess the effect of different numerical parameters on the quality of the solutions. First, the rate of convergence of the stress intensity factors for various relative integration domain sizes $c_J/a$ is investigated for crack angles $\varphi = 0°$ and $\varphi = 30°$, as depicted in Figure 4.11. Clearly, the results are not sensitive at all to the size of the $J$ integration domain.

**Table 4.4**   The effect of crack angle on the stress intensity factors in a rectangular plate with single notched cracked (Asadpoure and Mohammadi, 2007). (Reproduced by permission of John Wiley & Sons, Ltd.)

| $\varphi°$ | | 0 | 15 | 30 | 45 |
|---|---|---|---|---|---|
| FEM (Jernkvist, 2001b) | $\bar{K}_I$ | 3.028 | 3.033 | 3.020 | 2.806 |
| | $\bar{K}_{II}$ | 0 | 0.359 | 0.685 | 0.864 |
| XFEM (Asadpoure and Mohammadi, 2007) | $\bar{K}_I$ | 2.973 | 2.997 | 3.023 | 2.858 |
| | $\bar{K}_{II}$ | 0 | 0.361 | 0.691 | 0.892 |
| Differences | % | 1.9 | 1.2 | 0.01 | 1.8 |
| | % | 0 | 0.6 | 0.9 | 3.2 |

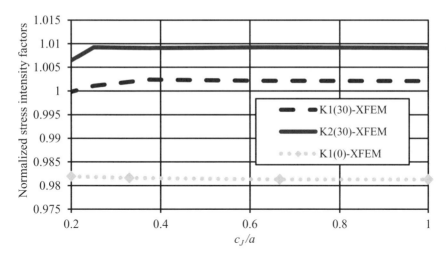

**Figure 4.11** Variations of the normalized mode I and II stress intensity factors with respect to $c_J/a$ for a horizontal and an inclined $\varphi = 30°$ crack.

In addition, the effect of the number of sub-quad partitioning in each element for the inclined crack $\varphi = 30°$ is depicted in Figure 4.12, which shows that an optimal value of 6 sub-quad partitioning can be used to determine both stress intensity factors.

Finally, variations of the SIFs with respect to the order of Gauss quadrature rule without any partitioning are examined in Figure 4.13. It is observed that a converged value for the normalized mode II stress intensity factor may not be achieved, even by applying a $10 \times 10$ Gauss quadrature rule; an important indication of the fact that a higher-order Gauss quadrature rule may not always replace the partitioning technique.

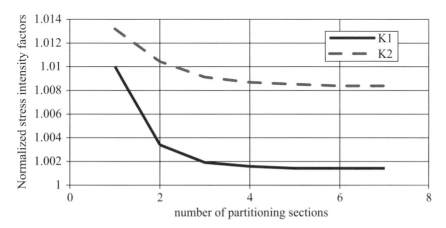

**Figure 4.12** The effect of the number of partitioning sections in each element on normalized mode I and II stress intensity factors for the inclined crack $\varphi = 30°$.

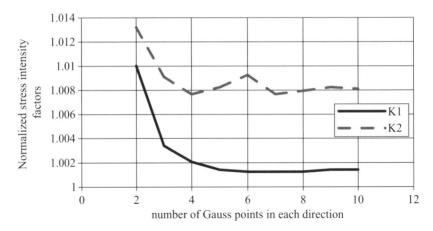

**Figure 4.13** The influence of the order of Gauss quadrature rule on stress intensity factors for the inclined crack $\varphi = 30°$.

### 4.6.4   Central Slanted Crack

A rectangular tensile orthotropic plate with a central slanted crack, as depicted in Figure 4.14a, is considered as a classic benchmark problem of a mixed mode stress intensity factor. Elastic properties of the plate are assumed as $E_1 = 3.5\,\mathrm{GPa}$, $E_2 = 12\,\mathrm{GPa}$, $G_{12} = 3\,\mathrm{GPa}$, $\nu_{21} = 0.7$.

The same problem has been investigated by several others; Sih, Paris and Irwin (1965) utilized a complex variable method, Atluri, Kobayashi and Nakagaki (1975b) used a hybrid-displacement finite element method, Wang, Yau and Corten (1980) adopted a conservation law of elasticity, Kim and Paulino (2002c) employed two methodologies of the modified crack closure (MCC) and the displacement correlation technique (DCT), Asadpoure, Mohammadi and Vafai (2006) and Asadpoure and Mohammadi (2007) implemented orthotropic XFEM, Ghorashi, Mohammadi and Sabbagh-Yazdi (2011) adopted conventional and enriched mesh-less EFG methods, and Ghorashi, Valizadeh and Mohammadi (2012) used the XIGA technique.

An adaptive fixed finite element model is constructed by 2400 quadrilateral elements and 2501 nodes, where element sizes are smaller in the vicinity of the crack than the other parts of the model (Figure 4.14b). The same mesh is used for all crack angles. In total, 5064 degrees of freedom are employed, which include 5002 normal DOF and 62 enrichment DOF. The size of the $J$ integration domain $c_J$ is about $0.85a$.

Table 4.5 compares the results of XFEM simulations with several available reference results. Again, in comparison to the analytical solution obtained by Sih, Paris and Irwin (1965), the maximum difference in mode II is about 3.9%, which is slightly higher than the mode I.

Figure 4.15 compares variations of the normalized SIFs with respect to different crack angles $\varphi = 0° - 90°$ for XFEM (Asadpoure, Mohammadi, Vafai, 2006) with XEFG (Ghorashi, Mohammadi and Sabbagh-Yazdi, 2011) and XIGA (Ghorashi, Valizadeh and Mohammadi, 2012) methods, which produce very close results. In all cases, the mode I stress intensity factor reduces steadily with an increase in crack angle, whereas the mode II stress intensity factor increases and reaches its maximum value at $\varphi = 45°$, and then decreases.

Similar comparisons can be made for variations of the value of $J$ for various relative integration domain sizes for the crack angle $45°$, as depicted in Figure 4.16. Clearly, XFEM

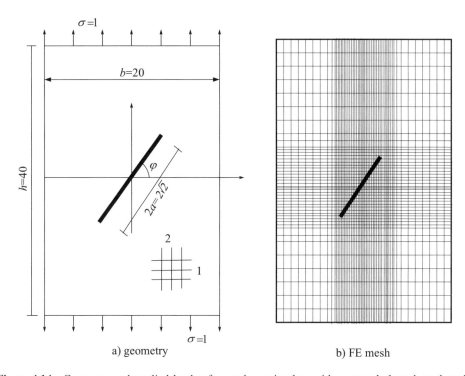

a) geometry                                          b) FE mesh

**Figure 4.14** Geometry and applied loads of an orthotropic plate with a central slanted crack under remote tension and the finite element mesh (Mohammadi, 2008). (Reproduced by permission of John Wiley & Sons, Ltd.)

**Table 4.5** Comparison of mixed mode stress intensity factors for an orthotropic plate with a central $\varphi = 45°$ slanted crack under remote tension.

| Method | $K_\mathrm{I}$ | $\bar{K}_\mathrm{I}$ | $K_\mathrm{II}$ | $\bar{K}_\mathrm{II}$ |
|---|---|---|---|---|
| Complex variable method (Sih, Paris and Irwin, 1965) | 1.054 | 0.500 | 1.054 | 0.500 |
| Hybrid-displacement FEM (Atluri, Kobayashi and Nakagaki, 1975b) | 1.020 | 0.484 | 1.080 | 0.512 |
| Conservation law of elasticity (Wang, Yau and Corten, 1980) | 1.023 | 0.485 | 1.049 | 0.498 |
| Displacement correlation technique (Kim and Paulino, 2002c) | 1.077 | 0.511 | 1.035 | 0.491 |
| Modified crack closure (Kim and Paulino, 2002c) | 1.067 | 0.506 | 1.044 | 0.495 |
| Meshless enriched XEFG (Ghorashi, Mohammadi and Sabbagh-Yazdi, 2011) | 1.079 | 0.512 | 1.117 | 0.530 |
| XFEM (Asadpoure and Mohammadi, 2007) | 1.084 | 0.514 | 1.095 | 0.519 |
| XIGA (Ghorashi, Valizadeh and Mohammadi, 2012) | 1.060 | 0.498 | 1.054 | 0.500 |

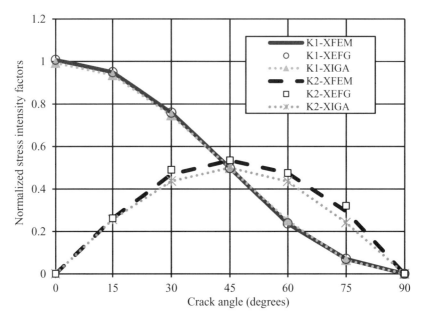

**Figure 4.15**  Variations of normalized mode I and II stress intensity factors with respect to different crack angles for XFEM, XEFG and XIGA methods.

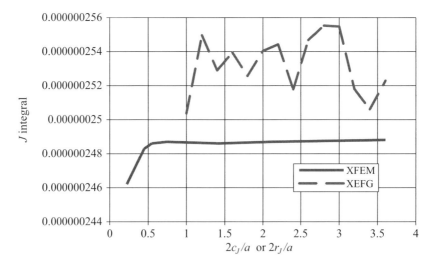

**Figure 4.16**  The value of the $J$ integral for various relative integration domains for the crack angle $\varphi = 45°$ (Asadpoure and Mohammadi, 2007) in comparison with the meshless enriched XEFG (Ghorashi, Mohammadi and Sabbagh-Yazdi, 2011).

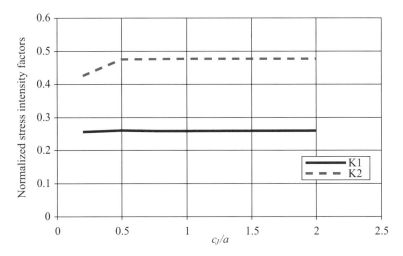

**Figure 4.17** Rate of convergence of normalized mode I and II stress intensity factors for $\varphi = 60°$ with respect to relative domain size in a plate with a slanted crack under remote tension.

is less sensitive to the size of the $J$ integral domain than XEFG, which shows an oscillating response. A similar insensitive trend is observed for XFEM in Figure 4.17 which shows values of SIFs corresponding to different sizes of the $J$ integration domain.

To further investigate the difference between orthotropic and isotropic enrichment functions, the rate of convergence of SIFs for both cases is studied for a crack angle $\varphi = 30°$ with different relative sizes of the $J$ integral domain (Figure 4.18). While almost similar variations are obtained for both enrichment functions, the differences in computed mode II stress intensity factors may not be negligible.

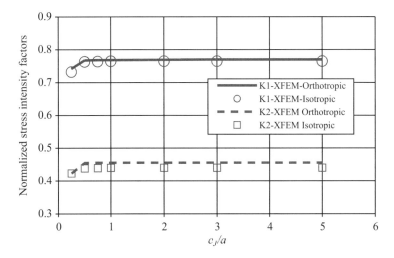

**Figure 4.18** Rate of convergence of normalized mode I and II stress intensity factors with $\varphi = 30°$ with respect to relative radius of $J$ integration domain for isotropic and orthotropic enrichment functions (Asadpoure, Mohammadi and Vafai, 2006).

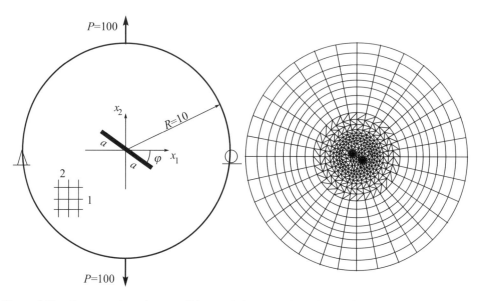

**Figure 4.19**   Geometry, boundary conditions and the FEM discretisation for an inclined centre crack in a disk subjected to point loads (Mohammadi, 2008). (Reproduced by permission of John Wiley & Sons, Ltd.)

### 4.6.5   An Inclined Centre Crack in a Disk Subjected to Point Loads

A circular disk subjected to double point loads with an inclined central crack is considered (Figure 4.19). This example somehow resembles the Brazilian test in orthotropic materials. The material elastic axes are assumed to be coincident with $x_1$ and $x_2$ axes with the following normalized material properties: $E_{11} = 0.1$, $E_{22} = 1.0$, $G_{12} = 0.5$, $v_{12} = 0.03$.

Kim and Paulino (2002c) used FEM to study this problem, while Asadpoure, Mohammadi and Vafai (2006) and Asadpoure and Mohammadi (2007) adopted XFEM for their simulation and Ghorashi, Mohammadi and Sabbagh-Yazdi (2011) employed the meshless XEFG method to analyse the cracked disk.

A fixed finite element model is generated by 920 finite elements and 1960 classical DOFs to simulate the model by XFEM for all crack inclinations, as depicted in Figure 4.19. According to Figure 4.20 the computed values of mode I and II stress intensity factors for crack angle $\varphi = 45°$ are insensitive to the size of the $J$ integral domain. More or less similar results are anticipated for other crack angles.

Table 4.6 compares the values of mixed mode stress intensity factors for crack angle $\varphi = 30°$, evaluated by XFEM and in comparison with the results reported by Kim and Paulino (2002c) based on using 999 elements and 2712 nodes for their classic finite element crack simulation, and the meshless XEFG with 877 nodes (Ghorashi, Mohammadi and Sabbagh-Yazdi, 2011).

Finally, Figure 4.21 compares the pseudo-sinusoidal variations of mixed mode stress intensity factors predicted by XFEM and XEFG methods for the range of $\varphi = 0°$ to $\varphi = 45°$. Very close results are achieved.

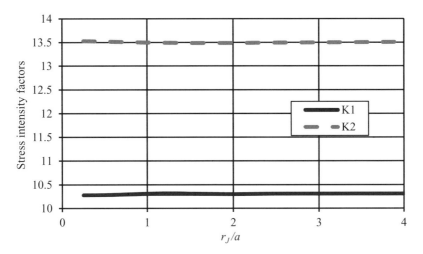

**Figure 4.20**  The rate of convergence of mode I and II stress intensity factors with respect to the relative integration domain size ($\varphi = 45°$).

**Table 4.6**  Values of stress intensity factors for an inclined centre crack in a disk subjected to point loads when angle of crack alignment with respect to $x_1$ axis is $\varphi = 30°$.

| Method | $n_{\text{nodes}}$ | $n_{\text{elem.}}$ | $n_{\text{cells}}$ | $n_{\text{DOFs}}$ | $K_{\text{I}}$ | $K_{\text{II}}$ |
|---|---|---|---|---|---|---|
| MCC (Kim and Paulino, 2002c) | 2712 | 999 | – | 5424 | 16.73 | 11.33 |
| M-integral (Kim and Paulino, 2002c) | 2712 | 999 | – | 5424 | 16.75 | 11.38 |
| XEFG (Ghorashi, Mohammadi and Sabbagh-Yazdi, 2011) | 877 | – | 641 | 1507 | 16.80 | 11.79 |
| XFEM (Asadpoure and Mohammadi, 2007) | 877 | 920 | – | 1960 | 17.12 | 11.72 |

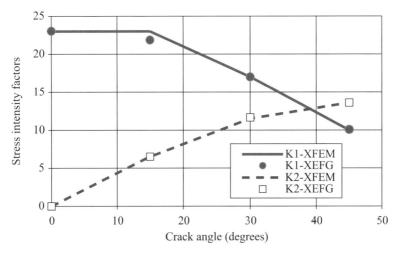

**Figure 4.21**  Effects of various crack inclinations on the mode I and II stress intensity factors.

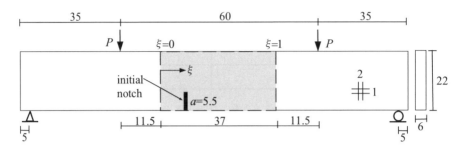

**Figure 4.22**    Geometry and boundary conditions of the 4-point bending beam.

## 4.6.6    Crack Propagation in an Orthotropic Beam

In the last example of this chapter, crack propagation in an orthotropic beam is investigated for various orthotropic material properties. The simply supported beam is subjected to two concentrated point loads, as depicted in Figure 4.22. Rousseau and Tippur (2000) investigated mixed-mode crack propagation of the same problem but with functionally graded material in the mid-part of the beam and different notch positions $\xi$, which will be comprehensively discussed in Section 6.6.6. Here, only a homogeneous orthotropic case with the notch position at $\xi = 0.17$ is studied.

The orthotropic material properties are defined as:

$$\bar{E} = \sqrt{E_1 E_2} = 5000$$

$$\bar{\nu} = \sqrt{\nu_1 \nu_2} = 0.3 \tag{4.121}$$

$$2G_{12} = \bar{E}/(\kappa_0 + \bar{\nu}) = 3125$$

$$K_{Ic}^x = K_{Ic}^y E_2/E_1 = 2$$

two different material stiffness ratios $\lambda = E_2/E_1 = 0.1, 10$, corresponding to major stiffness axes of 1 and 2, respectively, are assumed to further investigate the crack propagation patterns.

An unstructured finite element mesh of 2547 nodes and 4848 three-node elements is used to model this problem, as depicted in Figure 4.23. A fixed enriched area with radius $r = 0.2a$ and a constant $J$ contour radius of $r_J = 0.7a$ have been used for crack-tip enrichment in all propagation steps.

Crack trajectories for $\lambda = 0.1$ and $10$ are depicted in Figure 4.24. It is clearly observed that for $\lambda = 10$ the crack propagates almost vertically along the original notch direction, whereas for $\lambda = 0.1$(i.e. strong $x$-direction) the crack has a tendency to propagate towards the $x$-direction.

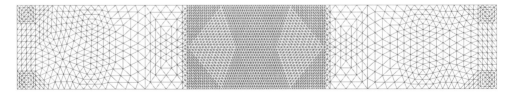

**Figure 4.23**    The finite element mesh of the composite beam.

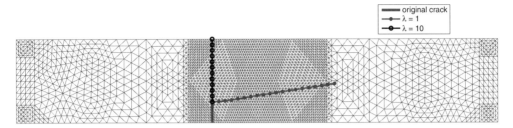

**Figure 4.24** Initial notch and crack trajectories for λ = 0.1 and 10.

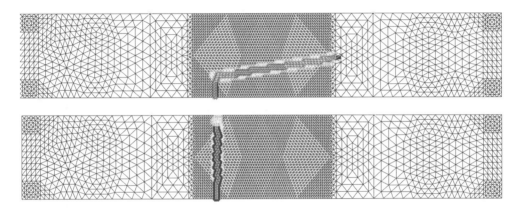

**Figure 4.25** Illustration of enriched nodes for the crack propagation for the two cases of λ = 0.1 and λ = 10.

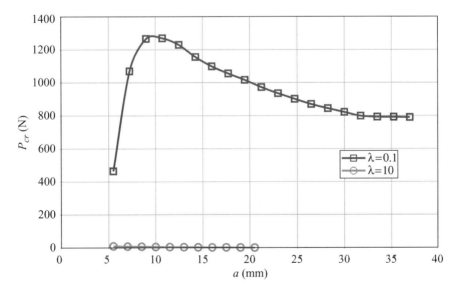

**Figure 4.26** Variations of load $P_{cr}$ versus crack length for different values of λ.

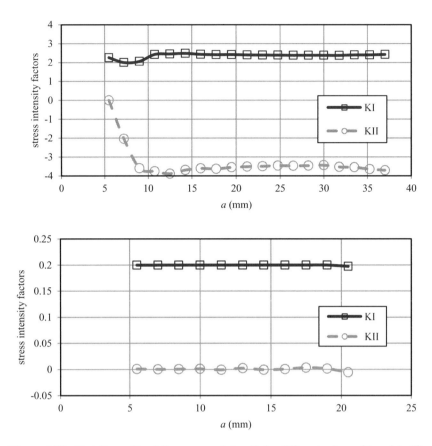

**Figure 4.27**   Variations of load $P$ versus crack length for different values of $\lambda = 0.1, 10$.

During the propagation process, crack enrichments (both the Heaviside and crack-tip enrichment functions) have to be updated according to the position of the propagated crack. This is clearly observed from Figure 4.25, where the crack-tip enriched nodes during the propagation process for the two cases of $\lambda = 0.1$ and 10 are presented.

For the case of $\lambda = 0.1$, the crack starts to deviate from the vertical direction and quickly propagates at an initial angle of $4°$. In contrast, for the case of $\lambda = 10$, the crack propagates in an almost self-similar pattern from its initial notch. Variations of the critical load $P_{cr}$ associated with each propagation step are illustrated in Figure 4.26. It is also observed that the critical load capacity of $\lambda = 0.1$ is far higher than $\lambda = 10$, which is a direct result of the existence of higher strength fibres causing the bending action for the $\lambda = 0.1$ case. A very low critical load of about 24 is anticipated for $\lambda = 10$.

Figure 4.27 shows variations of mode I and II stress intensity factors with respect to the total crack length $a$. For $\lambda = 10$, the stress intensity factors remain almost unchanged during the whole propagation process with a vanishing $K_{II}$. In contrast, for $\lambda = 0.1$, the initial crack propagation is in a mixed mode of fracture with varying $K_{II}$, which quickly changes to a mode with almost constant $K_I$ and $K_{II}$.

# 5

# Dynamic Fracture Analysis of Composites

## 5.1  Introduction

### 5.1.1  Dynamic Fracture Mechanics

The earliest empirical studies of dynamic fracture were probably related to the bursting of military cannon or the impact loading of industrial machines in the nineteenth century (Kirkaldy, 1863; Freund, 1990).

In the early days of fracture mechanics, a structure was assumed to completely fail once a crack became unstable and began to grow (Tipper, 1962). Accordingly, the rapid crack growth phase was only studied by some researchers out of curiosity, without any particular practical importance. Gradually in the 1970s, however, the importance of understanding crack propagation, crack arrest and dynamic effects was recognized in engineering applications and earth sciences, leading to significant continuous progress thereafter (Freund, 1990).

Dynamic fracture mechanics, broadly classified as stationary and propagating cracks, is one of the important subjects of classical fracture mechanics in problems where the role of material inertia and strain rate dependent material properties are significant in studying the stability and propagation of existing cracks (Freund, 1990).

Inertia effects can arise either from dynamic loadings on a cracked solid or from rapid crack propagation. Once a dynamic loading is applied on a cracked solid, the generated stress waves influence all parts of the body including the existing crack. The transient driving force acting on the crack determines whether the crack remains stationary or propagates. The interaction between the rapid crack motion and the stress wave field radiated from a moving crack is important in some material testing techniques and a number of important engineering applications, such as seismology (Freund, 1990).

In the brittle fracture of elastic solids, cracks can propagate at velocities large enough for the stress and displacement fields to be significantly influenced by elastodynamic effects (Achenbach and Bazant, 1975). Nevertheless, the dynamic effects are not always significant in any given fracture problem. For example, once the characteristic loading time is small compared with the time required for a stress wave to travel over the crack length or the

*XFEM Fracture Analysis of Composites*, First Edition. Soheil Mohammadi.
© 2012 John Wiley & Sons, Ltd. Published 2012 by John Wiley & Sons, Ltd.

distance from the crack edge to the loaded boundary, it can be expected that the inertia effects will become insignificant. In contrast, when the speed of the crack tip or edge is a significant fraction of the lowest characteristic wave speed of the material, the dynamic effects must be considered (Freund, 1990).

Despite extensive research, there are several disputed issues, such as terminal crack speed or dynamic crack bifurcation under symmetric tensile loading, or the fact that running cracks seem to have a maximum speed that depends on the material but not on the loading or geometrical configuration (Edgerton and Barstow, 1941), indicating the need for further in-depth research.

Experimental, analytical and numerical studies of dynamic fracture phenomena have all been largely complicated, in comparison with the static ones, by the time dependence of all contributing parameters.

Special data loggers for several accurate sequential measurements in a very short time are required in dynamic fracture experiments. The tests become more complicated in the case of crack growth, where the main place of data gathering varies in the very same short period of time. Optical methods of observation and special techniques of high-speed photography for study of the behavior of growing cracks have been developed since the late 1940s (Schardin, 1959). They include, for instance, the optical shadow spot method, introduced by Manogg (1966) to infer directly the instantaneous crack-tip stress intensity factor and the crack-tip position, the multiple spark arrangement by Wells and Post (1958), and the superimposed ultrasonic technique by Kerkhof (1970, 1973) to illustrate that the crack tip itself does not have an effective inertia, and a brittle crack subjected to oblique incident stress waves could gradually curve towards the local direction of maximum tensile stress. Reviews of earlier experimental works on the effects of stress waves on fractures in brittle materials can be found in Kolsky (1953), Dally (1987), Kalthoff (1987) and Freund (1990).

On the other hand, the time dependency of dynamic fracture processes requires complex analytical approaches based on complicated mathematical formulations. Yoffe (1951), Craggs (1960) and Baker (1962) illustrated that the maximum values of the stress field near a propagating crack tip move out of the plane of crack propagation when the speed of the crack tip exceeds a certain critical value. Rice (1986a, 1986b) analysed the steady-state elastodynamic problems to investigate the general nature of near-tip stress fields in isotropic materials by means of complex-variable techniques. Achenbach (1974), Freund and Clifton (1974) and Achenbach and Bazant (1975) developed analytical methods for the analysis of elastodynamic stress fields near propagating cracks, based on the original work of Cotterell (1964) using the mathematical models of Knein (1972) and Williams (1952), and showed that the steady state solution is also valid for transient cracks in the near tip region. Similar efforts were made by Freund and Clifton (1974) and Freund (1976) for non-uniform crack propagation. In addition, Nilsson (1974) investigated the angle dependence of near-tip fields, Chen (1978) studied the stress and displacement fields of dynamic stationary crack problems and Nishioka and Atluri (1983, 1984) and Freund (1990) examined crack propagations with high speeds.

## 5.1.2   Dynamic Fracture Mechanics of Composites

The fracture mechanics of composite structures has been studied by many researchers in the static and quasi-static states (Atluri, Kobayashi and Nakagaki, 1975a; Forschi and Barret, 1976; Boone, Wawrzynek and Ingraffea, 1987, among others) either through analytical solutions or numerical methodologies. Nevertheless, many of them do not represent real conditions because of the dynamic nature of applied loading or the fast moving state of propagating cracks. They are, in fact, only used as efficient simplified models for highly complex dynamic phenomena, which unfortunately may not even lead to conservative solutions. Therefore, it is of paramount importance to study the dynamic fracture mechanics of composites in their real dynamic states.

Several researchers have studied the problem of finding the elastodynamic fields around a propagating crack within an anisotropic medium, including Achenbach and Bazant (1975), Arcisz and Sih (1984), Piva and Viola (1988), Viola, Piva and Radi (1989), Shindo and Hiroaki (1990), De and Patra (1992), Gentilini, Piva and Viola (2004), Lee, Hawong, Choi (1996), Rubio-Gonzales and Mason (1998), Broberg (1999), Lim, Choi and Sankar (2001) and Nobile and Carloni (2005), among others.

Gentilini, Piva and Viola (2004) studied the case of a degenerated state of a steadily propagating shear crack in anisotropic media which is associated with repeated eigenvalues of the matrix of elastodynamic coefficients. They proposed an analytical approach based on the complex variable formulation of fundamental equations for dealing with this case which is a critical point in the topic of crack dynamics.

Later, Lee, Hawong and Choi (1996) derived the dynamic stress and displacement components around the crack tip of a steady state propagating crack in an orthotropic material under the dynamic loading. They discussed the crack propagation characteristics in terms of the mechanical properties of the orthotropic material and the crack speed, and concluded that the stress values in the vicinity of the crack tip increase by the increased crack velocity.

Federici et al. (2001) revisited the elastodynamic response of an orthotropic material with a semi-infinite line crack, which propagates intersonically for Mode I and Mode II of steady state crack propagation. They used a strain energy release rate analysis to discuss the physical admissibility of Mode I and II in terms of the order of stress singularity, which is larger and smaller than one half for Modes I and II, respectively. They also concluded that a steady state intersonic propagation is allowed only for a particular crack-tip velocity which is a function of the material orthotropy.

Sethi et al. (2011) used the method of variable separation to study the propagation of Love waves in a non-homogeneous orthotropic layer under variable compression overlying a semi-infinite non-homogeneous medium. They illustrated that the velocity of Love waves lies between two quantities which are dependent on the non-homogeneities of the two media. In a related study, the propagation of shear waves in a non-homogeneous anisotropic incompressible gravity field and initially stressed medium was recently investigated by Abd-Alla et al. (2011).

The limitations and inflexible nature of analytical methods in handling arbitrary complex problems, especially in anisotropic dynamic problems, have resulted in extensive use of several numerical techniques to solve composite fracture mechanics problems in the past decades. The main classes of numerical methods for solving general crack stability and propagation problems are the boundary element method (BEM), meshless methods, the finite element method (FEM), the extended finite element method (XFEM) and, more recently, various multiscale techniques.

For the first class, Aliabadi and Sollero (1998) and García-Sánchez, Zhang and Sáez (2008) developed boundary element solutions for quasi-static crack propagation and dynamic analysis of cracks in orthotropic media. Unfortunately, the boundary element method cannot be readily extended to non-linear and dynamic propagation systems.

Meanwhile, meshless methods such as the element-free Galerkin method (EFG) have been adopted by Belytschko and Tabbara (1996) and Belytschko, Organ and Gerlach (2000) to investigate a number of dynamic fracture problems. Despite higher accuracy and flexible adaptive schemes, the majority of meshless methods are yet to be user friendly because of high numerical expense and the need for difficult stabilization schemes in many of them (Motamedi and Mohammadi, 2012).

In contrast, the finite element method is well developed in nonlinear, inhomogeneous, anisotropic, multilayer, large deformation and dynamic problems, and can be easily adapted to many types of boundary conditions and geometries. It is important to note that the classical continuum-based formulation of FEM is not capable of reproducing the singular stress field at a crack tip, unless widely used quarter-point singular elements are adopted for stationary cracks. Unfortunately, these elements cannot be easily used in general propagation problems.

One of the powerful FEM procedures for simulation of progressive crack propagation under quasi-static and dynamic loadings is the adaptive remeshing technique. This technique, employed by a number of commercial finite element softwares for several years, is based on the discrete crack model and combines the simplicity of the classical finite element method with the sophistication of error estimation and adaptive remeshing techniques in each timestep or load increment when the cracks propagate (Moosavi and Mohammadi, 2007). Both regular and singular finite elements can be employed in an adaptive procedure. The method, however, is numerically expensive and the nodal alignments may cause numerical difficulties and mesh dependency to some extent in propagation problems.

The use of XFEM for modelling dynamic crack propagation problems was begun by Belytschko *et al.* (2003) and continued by several others for the dynamic analysis of crack propagation in isotropic media using the static isotropic enrichment functions, and by Motamedi (2008) and Motamedi and Mohammadi (2010a, 2010b, 2012) to study the dynamic crack stability and propagation in composites, based on dynamic orthotropic enrichment functions.

## 5.1.3 Dynamic Fracture by XFEM

The basis of XFEM was extended to dynamic applications by Belytschko *et al.* (2003) and Belytschko and Chen (2004), followed by several others, such as Ventura, Budyn and Belytschko (2003), Zi *et al.* (2005), Rethore, Gravouil and Combescure (2005a), Rethore *et al.* (2005), Menouillard *et al.* (2006), Gregoire *et al.* (2007), Nistor, Pantale and Caperaa (2008), Prabel, Marie and Combescure (2008), Combescure *et al.* (2008), Gregoire, Maigre and Combescure (2008), Motamedi (2008), Kabiri (2009), Rezaei (2010) and Motamedi and Mohammadi (2010a, 2010b, 2012), among others. Motamedi and Mohammadi (2012) developed new dynamic orthotropic enrichment functions and comprehensively discussed the dynamic fracture of composite structures.

Compared with a static crack analysis by XFEM, three extra major parts are involved in a dynamic crack propagation analysis by XFEM. First, a crack tracking procedure, such as the well-developed level set method and the fast marching approach, is required to determine the crack path for quasi-static or dynamic crack evolution problems.

The second part is related to the available time integration schemes and the potential for a time-discontinuous XFEM. Chessa and Belytschko (2004, 2006) presented a locally enriched space–time extended finite element method (TXFEM) for solving hyperbolic problems with discontinuities, through a weak enforcement of the flux continuity between the space–time and semi-discrete domains. They successfully applied TXFEM to the Rankine–Hugoniot jump conditions for solving linear first order wave and nonlinear Burgers equations. The idea of TXFEM was further developed by Rethore, Gravouil and Combescure (2005b) in the form of a combined space–time extended finite element method, which satisfied the stability and energy conservation criteria. The same approach was successfully implemented by Rezaei (2010) to solve several dynamic fracture problems. Also, Menouillard *et al.* (2006) and Gravouil, Elguedj and Maigre (2009a, 2009b) developed a pure explicit XFEM formulation based on a lumped mass matrix for enriched elements.

The last and most challenging part is the way dynamic fracture properties are determined. This part includes definitions of basic concepts of energy release rate, dynamic toughness and stress intensity factors, reformulation of the contour integral, the equivalent domain integral and the interaction integral to account for the effects of velocity and acceleration terms, and the way the dynamic crack propagation criterion is expressed. For instance, Peerlings *et al.* (2002), Belytschko *et al.* (2003) and Oliver *et al.* (2003) used the original idea of Gao and Klein (1998) and developed a methodology for switching from a continuum to a discrete discontinuity based on the loss of hyperbolicity for rate-independent materials.

## 5.2   Analytical Solutions for Near Crack Tips in Dynamic States

Consider a crack in an anisotropic body with general boundary conditions which is subjected to arbitrary forces, as depicted in Figure 5.1. In addition to global Cartesian co-ordinates $(X_1, X_2)$, local Cartesian co-ordinates $(x, y)$ and local polar co-ordinates $(r, \theta)$ are defined on the crack tip.

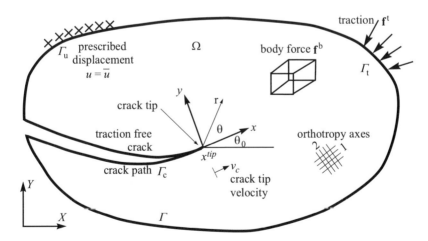

**Figure 5.1**   An arbitrary orthotropic body subjected to traction **t** and arbitrary boundary conditions in a state of elastodynamic equilibrium.

### 5.2.1 Analytical Solution for a Propagating Crack in Isotropic Material

Sih (1968), Sih and Irwin (1970) and Freund (1990) developed the main asymptotic solution for dynamic analysis of propagating cracks in isotropic materials. Later, Nishioka and Atluri (1983, 1984) and Swenson (1986) used the same formulation to present a unified fashion for all fracture modes. Accordingly, the asymptotic displacement field in the local crack tip coordinate system $(x, y)$ in terms of the dynamic stress intensity factors $K_{\mathrm{I}}$ and $K_{\mathrm{II}}$ can be written as:

$$
u_x = \frac{K_{\mathrm{I}}\left(1 + \beta_s^2\right)}{\mu D(v_c)} \sqrt{\frac{2}{\pi}} \left\{ \sqrt{r_d} \cos \frac{\theta_d}{2} - \frac{2\beta_d \beta_s}{\left(1 + \beta_s^2\right)} \sqrt{r_s} \cos \frac{\theta_s}{2} \right\}
$$
$$
+ \frac{2K_{\mathrm{II}}\beta_s}{\mu D(v_c)} \sqrt{\frac{2}{\pi}} \left\{ \sqrt{r_d} \sin \frac{\theta_d}{2} - \frac{\left(1 + \beta_s^2\right)}{2} \sqrt{r_s} \sin \frac{\theta_s}{2} \right\} \tag{5.1}
$$

$$
u_y = \frac{K_{\mathrm{I}}\left(1 + \beta_s^2\right)}{\mu D(v_c)} \sqrt{\frac{2}{\pi}} \left\{ -\beta_d \sqrt{r_d} \sin \frac{\theta_d}{2} + \frac{2\beta_d}{\left(1 + \beta_s^2\right)} \sqrt{r_s} \sin \frac{\theta_s}{2} \right\}
$$
$$
+ \frac{2K_{\mathrm{II}}\beta_s}{\mu D(v_c)} \sqrt{\frac{2}{\pi}} \left\{ \beta_d \sqrt{r_d} \cos \frac{\theta_d}{2} - \frac{\left(1 + \beta_s^2\right)}{2\beta_s} \sqrt{r_s} \cos \frac{\theta_s}{2} \right\} \tag{5.2}
$$

and the stress componenets in the same local crack-tip coordinates $(x, y)$ are:

$$
\sigma_{xx} = \frac{K_{\mathrm{I}}\left(1 + \beta_s^2\right)}{\sqrt{2\pi}D(v_c)} \left\{ \left(1 + 2\beta_d^2 - \beta_s^2\right) r_d^{-\frac{1}{2}} \cos \frac{\theta_d}{2} - \frac{4\beta_d \beta_s}{\left(1 + \beta_s^2\right)} r_s^{-\frac{1}{2}} \cos \frac{\theta_s}{2} \right\}
$$
$$
+ \frac{2K_{\mathrm{II}}\beta_s}{\sqrt{2\pi}D(v_c)} \left\{ -\left(1 + 2\beta_d^2 - \beta_s^2\right) r_d^{-\frac{1}{2}} \sin \frac{\theta_d}{2} + \left(1 + \beta_s^2\right) r_s^{-\frac{1}{2}} \sin \frac{\theta_s}{2} \right\} \tag{5.3}
$$

$$
\sigma_{yy} = \frac{K_{\mathrm{I}}\left(1 + \beta_s^2\right)}{\sqrt{2\pi}D(v_c)} \left\{ -\left(1 + \beta_s^2\right) r_d^{-\frac{1}{2}} \cos \frac{\theta_d}{2} + \frac{4\beta_d \beta_s}{\left(1 + \beta_s^2\right)} r_s^{-\frac{1}{2}} \cos \frac{\theta_s}{2} \right\}
$$
$$
+ \frac{2K_{\mathrm{II}}\beta_s}{\sqrt{2\pi}D(v_c)} \left\{ \left(1 + \beta_s^2\right) r_d^{-\frac{1}{2}} \sin \frac{\theta_d}{2} - \left(1 + \beta_s^2\right) r_s^{-\frac{1}{2}} \sin \frac{\theta_s}{2} \right\} \tag{5.4}
$$

$$
\sigma_{xy} = \frac{K_{\mathrm{I}}\left(1 + \beta_s^2\right)}{\sqrt{2\pi}D(v_c)} \left\{ 2\beta_d r_d^{-\frac{1}{2}} \sin \frac{\theta_d}{2} - 2\beta_d \frac{4\beta_d \beta_s}{\left(1 + \beta_s^2\right)} r_s^{-\frac{1}{2}} \sin \frac{\theta_s}{2} \right\}
$$
$$
+ \frac{2K_{\mathrm{II}}\beta_s}{\sqrt{2\pi}D(v_c)} \left\{ 2\beta_d r_d^{-\frac{1}{2}} \cos \frac{\theta_d}{2} - \frac{\left(1 + \beta_s^2\right)^2}{2\beta_s} r_s^{-\frac{1}{2}} \cos \frac{\theta_s}{2} \right\} \tag{5.5}
$$

$\sigma_{zz}$ is zero for plane stress problems, whereas it takes the following form for plane strains,

$$
\sigma_{zz} = \frac{K_{\mathrm{I}}\left(1 + \beta_s^2\right)}{\sqrt{2\pi}D(v_c)} \left\{ 2v \left(\beta_d^2 - \beta_s^2\right) r_d^{-\frac{1}{2}} \cos \frac{\theta_d}{2} \right\}
$$
$$
+ \frac{2K_{\mathrm{II}}\beta_s}{\sqrt{2\pi}D(v_c)} \left\{ -2v \left(\beta_d^2 - \beta_s^2\right) r_d^{-\frac{1}{2}} \sin \frac{\theta_d}{2} \right\} \tag{5.6}
$$

$v_c$ is the crack-tip speed, and

$$\beta_d^2 = 1 - \frac{v_c^2}{c_d^2} \tag{5.7}$$

$$\beta_s^2 = 1 - \frac{v_c^2}{c_s^2} \tag{5.8}$$

$$D(v_c) = 4\beta_d\beta_s - \left(1 + \beta_s^2\right)^2 \tag{5.9}$$

where $c_d$ and $c_s$ are the dilatational and shear wave speeds, respectively,

$$c_d = \sqrt{\frac{\lambda + 2\mu}{\rho}} \tag{5.10}$$

$$c_s = \sqrt{\frac{\mu}{\rho}} \tag{5.11}$$

and $\mu$ and $\lambda$ are the Lame coefficients and $\rho$ is the material density. $(r_j, \theta_j)$ are defined from the following equation ($i^2 = -1$),

$$r_j e^{i\theta_j} = x + i\beta_j y, \quad j = d, s \tag{5.12}$$

or more explicitly,

$$r_d^2 = x^2 + \beta_d^2 y^2 \tag{5.13}$$

$$r_s^2 = x^2 + \beta_s^2 y^2 \tag{5.14}$$

$$\tan \theta_d = \beta_d \tan \theta \tag{5.15}$$

$$\tan \theta_s = \beta_s \tan \theta \tag{5.16}$$

### 5.2.2  *Asymptotic Solution for a Stationary Crack in Orthotropic Media*

The contracted form of Hooke's law between linear strain $\varepsilon_\alpha$ and stress $\sigma_\beta$ in an arbitrary linear elastic homogeneous material can be written as

$$\varepsilon_\alpha = c_{\alpha\beta}\sigma_\beta \, (\alpha, \beta = 1 - 6) \tag{5.17}$$

where coefficients $c_{\alpha\beta}$ are defined in (4.8) and (4.12) (see Section 4.2.1).

The displacements and stress fields in the vicinity of the crack tip, which were elicited by Sih, Paris and Irwin (1965) in terms of mode I and mode II stress intensity factors, $K_I$ and $K_{II}$, respectively, have been defined in Eqs. (4.36)–(4.45) in terms of the roots ($\bar{s}_k = \bar{s}_{kx} + i\bar{s}_{ky}$, $k = 1, 2$) of the characteristic equation (4.22)

$$c_{11}\bar{s}^4 - 2c_{16}\bar{s}^3 + (2c_{12} + c_{66})\bar{s}^2 - 2c_{26}\bar{s} + c_{22} = 0 \tag{5.18}$$

and variables $\bar{p}_k$ and $\bar{q}_k$, as defined in (4.29) and (4.30), respectively,

$$\bar{p}_k = c_{11}\bar{s}_k^2 + c_{12} - c_{16}\bar{s}_k \tag{5.19}$$

$$\bar{q}_k = c_{12}\bar{s}_k + \frac{c_{22}}{\bar{s}_k} - c_{26} \tag{5.20}$$

### 5.2.3 Analytical Solution for Near Crack Tip of a Propagating Crack in Orthotropic Material

The elastodynamic equilibrium equation in the absence of body forces for an orthotropic material, where the material axes are assumed coincident with the local $(x, y)$ coordinate system, can be obtained from the extension of in-plane elastostatic equations (4.34) and (4.35): (Piva and Viola, 1988; Brock, Georgiadis and Hanson, 2001; Federici *et al.*, 2001)

$$d_{11}\frac{\partial^2 u_x}{\partial x^2} + d_{66}\frac{\partial^2 u_x}{\partial y^2} + (d_{12} + d_{66})\frac{\partial^2 u_y}{\partial x \partial y} = \rho\frac{\partial^2 u_x}{\partial t^2} \tag{5.21}$$

$$d_{66}\frac{\partial^2 u_y}{\partial x^2} + d_{22}\frac{\partial^2 u_y}{\partial y^2} + (d_{12} + d_{66})\frac{\partial^2 u_x}{\partial x \partial y} = \rho\frac{\partial^2 u_y}{\partial t^2} \tag{5.22}$$

Now consider that the crack moves with constant velocity $v_c$ along the local $x$ axis ($x = X - v_c t$ and $y = Y$). The equilibrium equations can be transformed into the moving coordinate system in terms of the new complex variable $z = x + my$(see Figure 5.2) using the linear elastic orthotropic constitutive equations (4.31)–(4–33): (Lee, Hawong and Choi, 1996)

$$\left(1 - \rho v_c^2 c_{11}\right)\frac{\partial \sigma_x}{\partial z} - \rho v_c^2 c_{12}\frac{\partial \sigma_y}{\partial z} + m\frac{\partial \tau_{xy}}{\partial z} = 0 \tag{5.23}$$

$$\left(-\rho v_c^2 c_{12}\right)\frac{\partial \sigma_x}{\partial z} + \left(m^2 - \rho v_c^2 c_{22}\right)\frac{\partial \sigma_y}{\partial z} + m\frac{\partial \tau_{xy}}{\partial z} = 0 \tag{5.24}$$

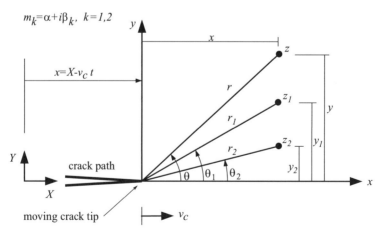

**Figure 5.2** Definitions of the dynamic crack propagation under mode I and II loading conditions (Lee, Hawong and Choi, 1996).

Lee, Hawong and Choi (1996) proposed a method to obtain the solution for a crack which is propagating with constant velocity. The near crack tip displacement fields for a mixed mode dynamic propagating crack can then be expressed as (for details see Lee, Hawong and Choi, 1996):

$$
u_x = K_I \sqrt{\frac{2r}{\pi}} \operatorname{Re} \left\{ \frac{\left[ p_2 s_1 \sqrt{\cos\theta + m_2 \sin\theta} - p_1 s_2 \sqrt{\cos\theta + m_1 \sin\theta} \right]}{\left[ 1 + \rho v_c^2 (c_{12} - c_{11}) \right] (s_1 - s_2)} \right\}
$$

$$
+ K_{II} \sqrt{\frac{2r}{\pi}} \operatorname{Re} \left\{ \frac{\left[ p_2 \sqrt{\cos\theta + m_2 \sin\theta} - p_1 \sqrt{\cos\theta + m_1 \sin\theta} \right]}{(s_1 - s_2)} \right\} \tag{5.25}
$$

$$
u_y = K_I \sqrt{\frac{2r}{\pi}} \operatorname{Re} \left\{ \frac{\left[ q_2 s_1 \sqrt{\cos\theta + m_2 \sin\theta} - q_1 s_2 \sqrt{\cos\theta + m_1 \sin\theta} \right]}{\left[ 1 + \rho v_c^2 (c_{12} - c_{11}) \right] (s_1 - s_2)} \right\}
$$

$$
+ K_{II} \sqrt{\frac{2r}{\pi}} \operatorname{Re} \left\{ \frac{\left[ q_2 \sqrt{\cos\theta + m_2 \sin\theta} - q_1 \sqrt{\cos\theta + m_1 \sin\theta} \right]}{(s_1 - s_2)} \right\} \tag{5.26}
$$

and the near crack tip stress components are described as:

$$
\sigma_x = \frac{K_I}{\sqrt{2\pi r}} \operatorname{Re} \left\{ \frac{\dfrac{\left[ 1 + m_2^2 + \rho v_c^2 (c_{12} - c_{22}) \right] s_1}{\sqrt{\cos\theta + m_2 \sin\theta}} - \dfrac{\left[ 1 + m_1^2 + \rho v_c^2 (c_{12} - c_{22}) \right] s_2}{\sqrt{\cos\theta + m_1 \sin\theta}}}{\left[ 1 + \rho v_c^2 (c_{12} - c_{11}) \right] (s_1 - s_2)} \right\}
$$

$$
+ \frac{K_{II}}{\sqrt{2\pi r}} \operatorname{Re} \left\{ \frac{\left[ \dfrac{\left[ 1 + m_2^2 + \rho v_c^2 (c_{12} - c_{22}) \right]}{\sqrt{\cos\theta + m_2 \sin\theta}} - \dfrac{\left[ 1 + m_1^2 + \rho v_c^2 (c_{12} - c_{22}) \right]}{\sqrt{\cos\theta + m_1 \sin\theta}} \right]}{(s_1 - s_2)} \right\} \tag{5.27}
$$

$$
\sigma_y = \frac{K_I}{\sqrt{2\pi r}} \operatorname{Re} \left\{ \frac{1}{(s_1 - s_2)} \left[ \frac{s_1}{\sqrt{\cos\theta + m_2 \sin\theta}} - \frac{s_2}{\sqrt{\cos\theta + m_1 \sin\theta}} \right] \right\}
$$

$$
+ \frac{K_{II}}{\sqrt{2\pi r}} \operatorname{Re} \left\{ \frac{\left[ 1 + \rho v_c^2 (c_{12} - c_{22}) \right]}{(s_1 - s_2)} \left[ \frac{1}{\sqrt{\cos\theta + m_2 \sin\theta}} - \frac{1}{\sqrt{\cos\theta + m_1 \sin\theta}} \right] \right\} \tag{5.28}
$$

$$
\tau_{xy} = \frac{K_I}{\sqrt{2\pi r}} \operatorname{Re} \left\{ \frac{s_1 s_2 \left[ \dfrac{s_1}{\sqrt{\cos\theta + m_1 \sin\theta}} - \dfrac{s_2}{\sqrt{\cos\theta + m_2 \sin\theta}} \right]}{\left[ 1 + \rho v_c^2 (c_{12} - c_{22}) \right] (s_1 - s_2)} \right\}
$$

$$
+ \frac{K_{II}}{\sqrt{2\pi r}} \operatorname{Re} \left\{ \frac{1}{(s_1 - s_2)} \left[ \frac{s_1}{\sqrt{\cos\theta + m_1 \sin\theta}} - \frac{s_2}{\sqrt{\cos\theta + m_2 \sin\theta}} \right] \right\} \tag{5.29}
$$

where $p_i$, $q_i$ and $s_i$ are defined as:

$$p_1 = c_{11}\left[1 + m_1^2 + \rho v_c^2 (c_{12} - c_{22})\right] + c_{12}\left[1 + \rho v_c^2 (c_{12} - c_{11})\right] \tag{5.30}$$

$$p_2 = c_{11}\left[1 + m_2^2 + \rho v_c^2 (c_{12} - c_{22})\right] + c_{12}\left[1 + \rho v_c^2 (c_{12} - c_{11})\right] \tag{5.31}$$

$$q_1 = \frac{c_{12}\left[1 + m_1^2 + \rho v_c^2 (c_{12} - c_{22})\right] + c_{22}\left[1 + \rho v_c^2 (c_{12} - c_{11})\right]}{m_1} \tag{5.32}$$

$$q_2 = \frac{c_{12}\left[1 + m_2^2 + \rho v_c^2 (c_{12} - c_{22})\right] + c_{22}\left[1 + \rho v_c^2 (c_{12} - c_{11})\right]}{m_2} \tag{5.33}$$

$$s_1 = m_1 - \rho v_c^2 c_{22}/m_1 - m_1 \rho v_c^2 c_{11} + \rho^2 v_c^4 \left(c_{11} c_{22} - c_{12}^2\right)/m_1 \tag{5.34}$$

$$s_2 = m_2 - \rho v_c^2 c_{22}/m_2 - m_2 \rho v_c^2 c_{11} + \rho^2 v_c^4 \left(c_{11} c_{22} - c_{12}^2\right)/m_2 \tag{5.35}$$

Coefficients $m_j (j = 1.2)$ are calculated from the following characteristic equation:

$$(m^4 + 2B_{12}m^2 + B_{66}) = 0 \tag{5.36}$$

where $B_{12}$ and $B_{66}$ are defined as:

$$2B_{12} = \frac{1}{c_{11}}\left[2c_{12} + c_{66} + \rho v_c^2 \left(c_{12}^2 - c_{11}c_{66} - c_{11}c_{22}\right)\right] \tag{5.37}$$

$$B_{66} = \frac{1}{c_{11}}\left\{c_{22} + \rho v_c^2 \left[c_{12}^2 - c_{22}c_{66} - c_{11}c_{22} + \rho v_c^2 c_{66}\left(c_{11}c_{22} - c_{12}^2\right)\right]\right\} \tag{5.38}$$

## 5.3  Dynamic Stress Intensity Factors

Several methods have been proposed for computation of stress intensity factors. They include formulations proposed by Nishioka nad Atluri (1984), Freund (1990), Dongye and Ting (1989), Aliabadi and Sollero (1998), Wu (2000) and Kim and Paulino (2002c), and so on.

### 5.3.1  Stationary and Moving Crack Dynamic Stress Intensity Factors

Dynamic stress intensity factors, which play an important role in dynamic fracture problems, are used to calculate the maximum hoop stress intensity factor and to evaluate dynamic crack propagation properties.

Freund (1972) obtained the ideal mode I and II dynamic stress intensity factors for a stationary crack subjected to normal and shear incident plane stress pulses with a jump in stress of magnitude $\sigma_0$ and $\tau_0$, respectively, (Freund, 1990)

$$K_I(0, t) = \frac{2\sigma_0}{1 - \nu}\sqrt{\frac{c_d t(1 - 2\nu)}{\pi}} \tag{5.39}$$

$$K_{II}(0, t) = 2\tau_0 \sqrt{\frac{2c_s t}{\pi(1 - \nu)}} \tag{5.40}$$

Equations (5.39) and (5.40) can be extended to moving crack-tip problems by (Freund, 1972)

$$K_{\mathrm{I}}(v_{\mathrm{c}}, t) = f_{\mathrm{I}}^{\mathrm{d}}(v_{\mathrm{c}})K_{\mathrm{I}}(0, t) \tag{5.41}$$

$$K_{\mathrm{II}}(v_{\mathrm{c}}, t) = f_{\mathrm{II}}^{\mathrm{d}}(v_{\mathrm{c}})K_{\mathrm{II}}(0, t) \tag{5.42}$$

where $f_{\mathrm{I}}^{\mathrm{d}}$ and $f_{\mathrm{II}}^{\mathrm{d}}$ are universal functions of the crack tip speed $v_{\mathrm{c}}$, which can be approximated for isotropic problems by (Freund, 1972, 1990)

$$f_{\mathrm{I}}^{\mathrm{d}}(v_{\mathrm{c}}) = \frac{1 - \dfrac{v_c}{c_R}}{\sqrt{1 - \dfrac{v_c}{c_{\mathrm{d}}}}} \tag{5.43}$$

$$f_{\mathrm{II}}^{\mathrm{d}}(v_{\mathrm{c}}) = \frac{1 - \dfrac{v_{\mathrm{c}}}{c_R}}{\sqrt{1 - \dfrac{v_{\mathrm{c}}}{c_{\mathrm{s}}}}} \tag{5.44}$$

where $c_R$ is the Rayleigh wave speed, which is the positive root of the following equation (Destrade, 2007),

$$c_{\mathrm{R}}^4\left(d_{11} - c_{\mathrm{R}}^2\right) = \frac{\rho^2 d_{22}}{d_{66}}\left(d_0 - c_{\mathrm{R}}^2\right)^2\left(d_{66} - c_{\mathrm{R}}^2\right) \tag{5.45}$$

where

$$d_0 = d_{11} - \frac{d_{12}^2}{d_{22}} \tag{5.46}$$

Equation (5.45) for isotropic problems is reduced to (Pichugin, Askes and Tyas, 2008)

$$c_{\mathrm{Rs}}^6 - 8c_{\mathrm{Rs}}^4 + \frac{8(2 - v)}{1 - v}c_{\mathrm{Rs}}^2 - \frac{8}{1 - v} = 0, \quad c_{Rs} = \frac{c_R}{c_s} \tag{5.47}$$

with a final approximate solution for $c_R$ by Bower (2012),

$$c_{\mathrm{R}} = c_{\mathrm{s}}\left[0.875 - 0.2v - 0.05(v + 0.25)^3\right] \tag{5.48}$$

### 5.3.2   Dynamic Fracture Criteria

The dynamic crack propagation law should realistically be chosen based upon the type of simulated or experimental material. Despite the fact that simulation of dynamic brittle crack propagation remains a difficult challenge, numerical modelling of fracture phenomenon can be performed by relatively simple concepts. First, a criterion is required to describe the state of stability of an existing crack. Then, a rule is needed to anticipate the propagation direction, if the crack becomes unstable. Finally, another equation is necessary to specify the crack propagation speed (Gregoire et al., 2007).

The main idea is to compare the stress intensity factor with the dynamic crack initiation toughness to evaluate the crack stability/instability. The dynamic crack initiation toughness is a material property and can be obtained by experiment. If the crack violates such a stability criterion, it starts to propagate and the direction of propagation is evaluated from a propagation criterion such as the maximum hoop stress (Maigre and Rittel, 1993).

During the crack propagation, the crack speed adapts its value in such a way as to make the maximum hoop stress intensity factor become equal to the dynamic crack growth toughness. To evaluate the dynamic crack growth toughness, Kanninen and Popelar (1985) proposed to replace the quasi-static toughness by the dynamic crack initiation toughness. In this method, the dynamic crack growth toughness is considered to be (Gregoire *et al.*, 2007):

$$
\begin{cases}
K_{tt} < K_{\mathrm{Ic}}^{\mathrm{d}} & : \text{stable crack} \\
K_{tt}(t,0) = K_{\mathrm{Ic}}^{\mathrm{d}} & : \text{initiation at } \theta = \theta_c \\
K_{tt}(t,v_c) \geq K_{\mathrm{Ic}}^{\mathrm{d}} \Rightarrow K_{tt}(t,v_c) = K_{\mathrm{Ic}}^{\mathrm{D}}(v_c) & : \text{propagation}
\end{cases} \tag{5.49}
$$

where $K_{tt}$ is the dynamic maximum hoop stress intensity factor, $K_{\mathrm{Ic}}^{\mathrm{d}}$ is the dynamic crack initiation toughness and $K_{\mathrm{Ic}}^{\mathrm{D}}$ is the dynamic crack growth (propagation) toughness (Gregoire *et al.*, 2007; Elguedj, Gravouil and Maigre, 2009)

$$
K_{\mathrm{Ic}}^{\mathrm{D}}(v_c) = \frac{K_{\mathrm{Ic}}^{\mathrm{d}}}{\left(1 - \dfrac{v_c}{c_{\mathrm{R}}}\right)} \tag{5.50}
$$

$v_c$ can be elicited from (5.49) and (5.50) on propagation (Freund (1972,1990), Freund and Douglass (1982) and Rosakis and Freund (1982)):

$$
v_c = c_{\mathrm{R}} \left(1 - \frac{K_{\mathrm{Ic}}^{\mathrm{d}}}{K_{tt}}\right) \tag{5.51}
$$

Alternatively, Menouillard *et al.* (2010) have used the following approximation

$$
v_c = c_{\mathrm{R}} \left(1 - \left(\frac{K_{\mathrm{Ic}}^{\mathrm{d}}}{K_{tt}}\right)^2\right) \tag{5.52}
$$

The maximum hoop/circumferential stress criterion (Section 4.4.4.1) or the minimum strain energy density criterion (Section 4.4.4.2) can be used to determine the propagation angle.

### 5.3.3   J Integral for Dynamic Problems

The analytical form of the dynamic $J$ integral ($J_k^{\mathrm{d}}$) in the global coordinate system $k = X, Y$, developed by Nishioka and Atluri (1984), for an infinitesimally small internal contour $\Gamma_1$ can be written as:

$$
J_k^{\mathrm{d}} = \int_{\Gamma + \Gamma_c} \left((w_s + w_{\mathrm{d}})n_k - t_i u_{i,k}\right) \mathrm{d}\Gamma + \int_{A - A_1} \left((\rho \ddot{u}_i - f_i^{\mathrm{b}}) u_{i,k} - \rho \dot{u}_{i,k} \dot{u}_i\right) \mathrm{d}A \tag{5.53}
$$

where $u_i$, $t_i$ and $f_i^{\mathrm{b}}$ denote the displacement, traction and body force, respectively, $n_k$ is the component of the normal outward vector $\mathbf{n} = (n_X, n_Y)$ and the integral paths $\Gamma_1$, $\Gamma$, and $\Gamma_c$ denote near-field, far-field and crack surface paths, respectively. $A$ and $A_1$ are the regions

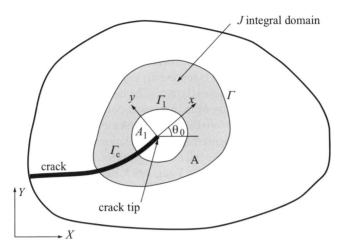

**Figure 5.3** Definitions of crack path contours and the local crack tip coordinates for evaluation of the $J$ integral.

surrounded by $\Gamma$ and $\Gamma_1$, respectively (Figure 5.3). The strain energy density $w_s$ and the kinetic energy density $w_d$ are defined as,

$$w_s = \frac{1}{2}\sigma_{ij}\varepsilon_{ij} \tag{5.54}$$

$$w_d = \frac{1}{2}\rho\dot{u}_i\dot{u}_i \tag{5.55}$$

### 5.3.4  Domain Integral for Orthotropic Media

Kim and Paulino (2003d) proposed an equivalent form of Eq. (5.53), using the divergence theorem and some additional assumptions for homogeneous materials, which is better suited for the finite element method (in the global system $k = X, Y$),

$$J_k^d = \int_A \left( \sigma_{ij}u_{i,k} - (w_s + w_d) \right) q_{,k}\mathrm{d}A + \int_A \left( (\rho\ddot{u}_i - f_i^b)u_{i,k} - \rho\dot{u}_{i,k}\dot{u}_i \right) q\mathrm{d}A \tag{5.56}$$

where $q$ is a smooth function, assumed constant $q = 1$ inside $A_1$ and varying linearly from 1 at $\Gamma_1$ near the crack tip to $q = 0$ at the exterior boundary $\Gamma$, as discussed in Section 2.7.4. Note that the value of $q$ is constant near the crack-tip area, and the gradient of $q$ vanishes inside $A_1$.

It should be noted that in a general transient analysis, potential oscillations of $\ddot{u}_i$ may affect the quality of $J$ integral computations through Eq. (5.53) or (5.54). Fortunately, this is not the case for dynamic analysis of stationary cracks. Among several investigations on this issue, Gregoire *et al.* (2007) have performed an extensive study on the stability of dynamic fracture mechanics and conservation of energy. They showed that utilizing the Newmark's implicit mean acceleration approach combined with enriching the cracked elements will provide sufficient accuracy for finding $u_i$, $\dot{u}_i$, $\ddot{u}_i$ and hence the $J$ integral.

### 5.3.5   Interaction Integral

Evaluation of the mixed mode $J$ integral and associated stress intensity factors in XFEM is usually performed by means of interaction integral methods. Since the interaction integral method requires several auxiliary field components, equations of dynamic near tip displacement and stress fields in orthotropic materials have to be derived in order to extract the required auxiliary displacement and stress fields. Moreover, more complex computations are required to calculate the 'auxiliary velocity' and 'auxiliary strain velocity' fields, especially in orthotropic media. These require a substantial theoretical and simulation experience and cost.

The general form of the dynamic interaction integral for a moving crack can be defined as (Suo and Combescure, 1992; Attigui and Petit, 1997; Rethore, Gravouil and Combescure, 2004)

$$
M^{\mathrm{d}}(t) = \frac{1}{2} \int_{\Gamma} \left[ \left( \sigma_{ij,1}^{\mathrm{aux}} u_i - \sigma_{ij} u_{i,1}^{\mathrm{aux}} \right) n_j - \rho \dot{u}_i \dot{u}_i^{\mathrm{aux}} n_i - v_{\mathrm{c}} \rho \left( \dot{u}_i \dot{u}_{i,1}^{\mathrm{aux}} + \dot{u}_i^{\mathrm{aux}} \dot{u}_{i,1} \right) n_1 \right] \mathrm{d}\Gamma
$$
$$
+ \frac{1}{2} \frac{\mathrm{d}}{\mathrm{d}t} \int_{A} \left[ \rho \left( \dot{u}_i \dot{u}_{i,1}^{\mathrm{aux}} + \dot{u}_i^{\mathrm{aux}} \dot{u}_{i,1} \right) \right] \mathrm{d}A
\tag{5.57}
$$

Neglecting the effect of crack-tip velocity, the equivalent domain form of the dynamic interaction integral can be written as (Rethore *et al.*, 2005; Gregoire *et al.*, 2007; Combescure *et al.*, 2008; Elguedj, Gravouil and Maigre, 2009)

$$
M^{\mathrm{d}}(t) = - \int_{A} \left[ \left( \sigma_{ml}^{\mathrm{aux}} u_{m,l} - \rho \dot{u}_l \dot{u}_l^{\mathrm{aux}} \right) \delta_{kj} - \left( \sigma_{ij}^{\mathrm{aux}} u_{i,k} + \sigma_{ij} u_{i,k}^{\mathrm{aux}} \right) \right] q_{k,j} \mathrm{d}A
$$
$$
+ 2 \int_{A} \left[ \sigma_{ij,j}^{\mathrm{aux}} u_{i,k} + \rho \ddot{u}_i u_{i,k}^{\mathrm{aux}} + \rho \dot{u}_{i,k}^{\mathrm{aux}} \dot{u}_{i,k} + \rho \dot{u}_i \dot{u}_{i,k}^{\mathrm{aux}} \right] q_k \mathrm{d}A
\tag{5.58}
$$

Evaluation of the dynamic stress intensity factors $K_{\mathrm{I}}$ and $K_{\mathrm{II}}$ from the $M^{\mathrm{d}}$ integral for isotropic problems can be performed from (Menouillard *et al.*, 2010).

$$
M^{\mathrm{d}}(t) = \frac{2}{E'} \left[ f_{\mathrm{I}}(v_{\mathrm{c}}) K_{\mathrm{I}} K_{\mathrm{I}}^{\mathrm{aux}} + f_{\mathrm{II}}(v_{\mathrm{c}}) K_{\mathrm{II}} K_{\mathrm{II}}^{\mathrm{aux}} \right]
\tag{5.59}
$$

by setting $K_{\mathrm{I}}^{\mathrm{aux}} = 1, K_{\mathrm{II}}^{\mathrm{aux}} = 0$ and $K_{\mathrm{I}}^{\mathrm{aux}} = 0, K_{\mathrm{II}}^{\mathrm{aux}} = 1$, respectively. Functions $f_{\mathrm{I}}(v_{\mathrm{c}})$ and $f_{\mathrm{II}}(v_{\mathrm{c}})$ are defined as

$$
f_{\mathrm{I}}(v_{\mathrm{c}}) = \frac{4\beta_{\mathrm{d}} \left( 1 - \beta_{\mathrm{s}}^2 \right)}{(\kappa + 1) D(v_{\mathrm{c}})}
\tag{5.60}
$$

$$
f_{\mathrm{II}}(v_{\mathrm{c}}) = \frac{4\beta_{\mathrm{s}} \left( 1 - \beta_{\mathrm{s}}^2 \right)}{(\kappa + 1) D(v_{\mathrm{c}})}
\tag{5.61}
$$

It should be noted that evaluation of the interaction integral requires careful attention as the main fields are usually obtained from the finite element solution in a crack-independent global or local coordinate system, while the auxiliary fields are defined in the local crack-tip polar coordinate system. Therefore, necessary transformations are required to use a unified coordinate system.

### 5.3.6   Crack-Axis Component of the Dynamic J Integral

Computation of $J_X^d$ from (5.56) is usually less affected by the way the crack tip is modelled, whereas $J_Y^d$ is sensitive to modelling of the crack tip duo to the singularity of $w_s + w_d$ in evaluation of $J_Y^d$. To overcome this difficulty in the direct method, Nishioka, Tokudome and Kinoshita (2001), adopted the component separation method.

The local crack-axis (tangential) components $(J_l^d)$ of the dynamic $J$ integral, which corresponds to the rate of change in the potential energy per unit crack extension $(G)$, can be evaluated by the following coordinate transformation (Kim and Paulino, 2003d; Nishioka, Tokudome and Kinoshita, 2001):

$$J_l^d = \alpha_{lk}(\theta_0) J_k^d, \quad k = X, Y \quad \text{and} \quad l = x, y \tag{5.62}$$

where $\alpha_{lk}$ is the coordinate transformation tensor and $\theta_0$ is the crack angle with horizontal axis. Wu (2000) showed that the dynamic energy release rate $G$ can be related to the instantaneous stress intensity factors for an elastodynamic self-similar propagating crack with the velocity $v_c$,

$$G = J_x^d = J_X^d \cos \theta_0 + J_Y^d \sin \theta_0 \tag{5.63}$$

#### 5.3.6.1   Isotropic Formulation

The displacements of crack edges $(\mathbf{u}^+, \mathbf{u}^-)$ for each loading mode can be calculated by transforming the numerically computed displacement field into the local polar crack-tip coordinates $(x, y)$,

$$\delta_I = u_y^+ - u_y^- \tag{5.64}$$

$$\delta_{II} = u_x^+ - u_x^- \tag{5.65}$$

Nishioka and Atluri (1980a, 1980b, 1984) illustrated that while $G$ can be computed from (5.63) for linear and nonlinear elastic homogeneous or inhomogeneous materials, the stress intensity factors for homogenous linear elastic materials can be obtained from the following equations,

$$K_I = \delta_I \left\{ \frac{(2\mu \beta_s D(v_c)) J_x^d}{\beta_d \left(1 - \beta_s^2\right) \left[\delta_I^2 \beta_s + \delta_{II}^2 \beta_d\right]} \right\}^{\frac{1}{2}} \tag{5.66}$$

$$K_{II} = \delta_{II} \left\{ \frac{(2\mu \beta_d D(v_c)) J_x^d}{\beta_s \left(1 - \beta_s^2\right) \left[\delta_I^2 \beta_s + \delta_{II}^2 \beta_d\right]} \right\}^{\frac{1}{2}} \tag{5.67}$$

### 5.3.6.2 Orthotropic Formulation

Now, the extension of the method to orthotropic media is explained. Beginning with the definition of $G$ (Kim and Paulino, 2003d)

$$G = \frac{1}{2}K^T L^{-1}(v_c)K \tag{5.68}$$

where $L(v_c)$ has been presented by Dongye and Ting (1989) for orthotropic materials with the symmetry planes coinciding with the coordinate planes, and $K$ includes the stress intensity factors.

The nonzero components of $L(v_c)$ are:

$$\sqrt{d_{66}^d d_{22}} L_{11}(v_c) = \sqrt{d_{11}^d d_{66}} L_{22}(v_c) = \Omega \Psi^{-\frac{1}{2}} \tag{5.69}$$

where $d_{ij}^d = d_{ij} - \rho v_c^2 \delta_{ij}$, $d_{ij}$ are the constitutive coefficients defined in (4.10) and

$$\Omega = \left(d_{11}^d d_{22} - d_{12}^2\right)\sqrt{d_{66}^d d_{66}} - \rho v_c^2 d_{66}\sqrt{d_{11}^d d_{22}} \tag{5.70}$$

$$\Psi = \left(\sqrt{d_{66}^d d_{66}} + \sqrt{d_{11}^d d_{22}}\right)^2 - (d_{12} + d_{66})^2 \tag{5.71}$$

To accurately evaluate the in-plane mixed-mode stress intensity factors from the dynamic $J$ integral, the component separation method, proposed by Aliabadi and Sollero (1998), is implemented. The method is based on the following relationship between the dynamic stress intensity factors and the relative sliding and opening displacements of the crack face:

$$\begin{Bmatrix} \delta_I \\ \delta_{II} \end{Bmatrix} = \sqrt{\left(\frac{8r}{\pi}\right)} \begin{bmatrix} A_{11} & A_{12} \\ A_{21} & A_{22} \end{bmatrix} \begin{Bmatrix} K_I \\ K_{II} \end{Bmatrix} \tag{5.72}$$

with

$$A_{11} = \mathrm{Im}\left(\frac{\bar{s}_2 \bar{p}_1 - \bar{s}_1 \bar{p}_2}{\bar{s}_1 - \bar{s}_2}\right), \quad A_{12} = \mathrm{Im}\left(\frac{\bar{p}_1 - \bar{p}_2}{\bar{s}_1 - \bar{s}_2}\right)$$

$$A_{21} = \mathrm{Im}\left(\frac{\bar{s}_2 \bar{q}_1 - \bar{s}_1 \bar{q}_2}{\bar{s}_1 - \bar{s}_2}\right), \quad A_{22} = \mathrm{Im}\left(\frac{\bar{q}_1 - \bar{q}_2}{\bar{s}_1 - \bar{s}_2}\right) \tag{5.73}$$

where $\bar{s}_k$, $\bar{p}_k$ and $\bar{q}_k$ are defined in Eqs. (5.18)–(5–20). The ratio of opening to sliding displacements, $R_\delta$, can be written as:

$$R_\delta = \frac{\delta_{II}}{\delta_I} = \frac{A_{21}K_I + A_{22}K_{II}}{A_{11}K_I + A_{12}K_{II}} \tag{5.74}$$

Therefore, the ratio of dynamic stress intensity factors, $R_K$, is obtained as:

$$R_K = \frac{K_I}{K_{II}} = \frac{R_\delta A_{21} - A_{22}}{A_{21} + R_\delta A_{11}} \tag{5.75}$$

Substitution for $K_I$ from Eq. (5.75) into Eq. (5.74) leads to the following relationship for $K_{II}$:

$$K_{II} = \sqrt{\frac{2G}{L_{11}(v_c)R_K^2 + L_{22}(v_c)}} \qquad (5.76)$$

### 5.3.7 Field Decomposition Technique

Decomposition of displacement and stress fields into two independent symmetric and anti-symmetric modes, as described in Section 2.5.5, can also be used in combination with the analytical solutions of Section 5.2.3 to compute the stress intensity factors. This is considered as a direct approach to determining the stress intensity factors, either on a locally symmetric mesh around the crack, or based on symmetric points with respect to the crack direction on any arbitrarily generated mesh. For further details see Nishioka and Atluri, 1984.

## 5.4 Dynamic XFEM

### 5.4.1 Dynamic Equations of Motion

Consider a body $\Omega$ with an initial traction-free crack in the state of dynamic equilibrium, as depicted in Figure 5.1. The fundamental elastodynamic equation can be expressed as:

$$\nabla \cdot \boldsymbol{\sigma} + \mathbf{f}^b = \rho \ddot{\mathbf{u}} \qquad (5.77)$$

with the following boundary conditions:

$$\mathbf{u}(\mathbf{x}, t) = \bar{\mathbf{u}}(\mathbf{x}, t) \text{ on } \Gamma_u \qquad (5.78)$$

$$\boldsymbol{\sigma} \cdot \mathbf{n} = \mathbf{f}^t \text{ on } \Gamma_t \qquad (5.79)$$

$$\boldsymbol{\sigma} \cdot \mathbf{n} = 0 \text{ on } \Gamma_c \qquad (5.80)$$

and initial conditions:

$$\mathbf{u}(\mathbf{x}, t = 0) = \bar{\mathbf{u}}(0) \qquad (5.81)$$

$$\dot{\mathbf{u}}(\mathbf{x}, t = 0) = \bar{\dot{\mathbf{u}}}(0) \qquad (5.82)$$

where $\Gamma_t$, $\Gamma_u$ and $\Gamma_c$ are traction, displacement and crack boundaries, respectively; $\boldsymbol{\sigma}$ is the stress tensor and $\mathbf{f}^b$ and $\mathbf{f}^t$ are the body force and external traction vectors, respectively.

The variational formulation of the initial/boundary value problem of Eq. (5.77) can be written as:

$$\int_\Omega \rho \ddot{\mathbf{u}} \cdot \delta \mathbf{u} \, d\Omega + \int_\Omega \boldsymbol{\sigma} \cdot \delta \boldsymbol{\varepsilon} \, d\Omega = \int_\Omega \mathbf{f}^b \cdot \delta \mathbf{u} \, d\Omega + \int_\Gamma \mathbf{f}^t \cdot \delta \mathbf{u} \, d\Gamma \qquad (5.83)$$

### 5.4.2 XFEM Discretization

Assuming a discontinuity (a crack) within an independent finite element mesh, the displacement field for any typical point $\mathbf{x}$ inside the domain can be written in terms of the classical finite element approximation and the XFEM enriched fields (see Section 3.4.3):

$$\mathbf{u}^h(\mathbf{x}) = \mathbf{u}(\mathbf{x}) + \mathbf{u}^H(\mathbf{x}) + \mathbf{u}^{tip}(\mathbf{x})$$

$$
= \left[ \sum_{j=1}^{n} N_j(\mathbf{x}) \mathbf{u}_j \right] + \left[ \sum_{h=1}^{mh} N_h(\mathbf{x}) \left( H(\mathbf{x}) - H(\mathbf{x}_h) \right) \mathbf{a}_h \right]
$$

$$
+ \left[ \sum_{k=1}^{mt} N_k(\mathbf{x}) \left( \sum_{l=1}^{mf} [F_l(\mathbf{x}) - F_l(\mathbf{x}_k)] \mathbf{b}_k^l \right) \right]
\tag{5.84}
$$

where $N_i$ is the shape function associated with the node $i$, $H(\mathbf{x})$ is the Heaviside enrichment function and $F_l(\mathbf{x})$ are crack tip enrichment functions. $\mathbf{u}_j$ are the displacement degrees of freedom, and $\mathbf{a}_h$ and $\mathbf{b}_k^l$ are additional Heaviside and tip enrichment degrees of freedom, respectively.

The discretized form of Eq. (5.83) using the XFEM approximation (5.84) can be written as:

$$\mathbf{M}\ddot{\mathbf{u}}^h + \mathbf{K}\mathbf{u}^h = \mathbf{f} \tag{5.85}$$

where $\mathbf{u}^h$ and $\ddot{\mathbf{u}}^h$ denote the vector of nodal parameters (displacements $\mathbf{u}$, Heaviside and crack-tip enrichment degrees of freedom $\mathbf{a}$ and $\mathbf{b}$, respectively) and its second time derivative, respectively:

$$\mathbf{u}^h = \{\mathbf{u}, \mathbf{a}, \mathbf{b}\}^{\mathrm{T}} \tag{5.86}$$

$$\ddot{\mathbf{u}}^h = \{\ddot{\mathbf{u}}, \ddot{\mathbf{a}}, \ddot{\mathbf{b}}\}^{\mathrm{T}} \tag{5.87}$$

The stiffness matrix $\mathbf{K}$, mass matrix $\mathbf{M}$ and the external load vector $\mathbf{f}$ are defined as:

$$
\mathbf{K}_{ij}^e = 
\begin{bmatrix}
\mathbf{K}_{ij}^{\mathbf{uu}} & \mathbf{K}_{ij}^{\mathbf{ua}} & \mathbf{K}_{ij}^{\mathbf{ub}} \\
\mathbf{K}_{ij}^{\mathbf{au}} & \mathbf{K}_{ij}^{\mathbf{aa}} & \mathbf{K}_{ij}^{\mathbf{ab}} \\
\mathbf{K}_{ij}^{\mathbf{bu}} & \mathbf{K}_{ij}^{\mathbf{ba}} & \mathbf{K}_{ij}^{\mathbf{bb}}
\end{bmatrix}
\tag{5.88}
$$

$$
\mathbf{M}_{ij}^e = 
\begin{bmatrix}
\mathbf{M}_{ij}^{\mathbf{uu}} & \mathbf{M}_{ij}^{\mathbf{ua}} & \mathbf{M}_{ij}^{\mathbf{ub}} \\
\mathbf{M}_{ij}^{\mathbf{au}} & \mathbf{M}_{ij}^{\mathbf{aa}} & \mathbf{M}_{ij}^{\mathbf{ab}} \\
\mathbf{M}_{ij}^{\mathbf{bu}} & \mathbf{M}_{ij}^{\mathbf{ba}} & \mathbf{M}_{ij}^{\mathbf{bb}}
\end{bmatrix}
\tag{5.89}
$$

$$\mathbf{f}_i = \{\mathbf{f}_i^{\mathbf{u}}, \mathbf{f}_i^{\mathbf{a}}, \mathbf{f}_i^{\mathbf{b}}\}^{\mathrm{T}} \tag{5.90}$$

The stiffness components $\mathbf{K}_{ij}^{rs}$ $(r, s = \mathbf{u}, \mathbf{a}, \mathbf{b})$ include the classical FEM ($\mathbf{uu}$), Heaviside enrichment ($\mathbf{aa}$), orthotropic crack-tip enrichment ($\mathbf{bb}$) and the coupled parts of the XFEM approximation:

$$\mathbf{K}_{ij}^{rs} = \int_{\Omega^e} \left( \mathbf{B}_i^r \right)^{\mathrm{T}} \mathbf{D} \mathbf{B}_j^s \, \mathrm{d}\Omega \quad (r, s = \mathbf{u}, \mathbf{a}, \mathbf{b}) \tag{5.91}$$

where $\mathbf{B}_i^{\mathbf{u}}$, $\mathbf{B}_i^{\mathbf{a}}$ and $\mathbf{B}_i^{\mathbf{b}}$ are matrices of derivatives of shape functions, defined in equations (3.49), (3.50) and (3.51), respectively.

Classical and enrichment components of the consistent mass matrix can be expressed as:

$$\mathbf{M}_{ij}^{\mathbf{uu}} = \int_{\Omega} \rho N_i N_j \, d\Omega \tag{5.92}$$

$$\mathbf{M}_{ij}^{\mathbf{aa}} = \int_{\Omega} \rho \left[ N_i \left( H(\xi) - H(\xi_i) \right) \right] \left[ N_j \left( H(\xi) - H(\xi_j) \right) \right] d\Omega \tag{5.93}$$

$$\mathbf{M}_{ij}^{\mathbf{bb}} = \int_{\Omega} \rho \left[ N_i \left( F_\alpha - F_{\alpha i} \right) \right] \left[ N_j \left( F_\alpha - F_{\alpha j} \right) \right] d\Omega \quad (\alpha = 1, 2, 3 \text{ and } 4) \tag{5.94}$$

$$\mathbf{M}_{ij}^{\mathbf{ua}} = \mathbf{M}_{ij}^{\mathbf{au}} = \int_{\Omega} \rho N_i \left[ N_j \left( H(\xi) - H(\xi_j) \right) \right] d\Omega \tag{5.95}$$

$$\mathbf{M}_{ij}^{\mathbf{ub}} = \mathbf{M}_{ij}^{\mathbf{bu}} = \int_{\Omega} \rho N_i \left[ N_j \left( F_\alpha - F_{\alpha j} \right) \right] d\Omega \quad (\alpha = 1, 2, 3 \text{ and } 4) \tag{5.96}$$

$$\mathbf{M}_{ij}^{\mathbf{ab}} = \int_{\Omega} \rho \left[ N_i \left( H(\xi) - H(\xi_i) \right) \right] \left[ N_j \left( F_\alpha - F_{\alpha j} \right) \right] d\Omega \quad (\alpha = 1, 2, 3 \text{ and } 4) \tag{5.97}$$

The diagonal mass matrix can be computed by the lumping technique proposed by Menouillard *et al.* (2008).

Finally, the force vectors associated with the classical and enrichment degrees of freedom are defined in (3.45)–(3.47):

$$\mathbf{f}_i^{\mathbf{u}} = \int_{\Gamma_t} N_i \mathbf{f}^{\mathbf{t}} \, d\Gamma + \int_{\Omega^e} N_i \mathbf{f}^{\mathbf{b}} \, d\Omega \tag{5.98}$$

$$\mathbf{f}_i^{\mathbf{a}} = \int_{\Gamma_t} N_i \left[ H(\xi) - H(\xi_i) \right] \mathbf{f}^{\mathbf{t}} d\Gamma + \int_{\Omega^e} N_i \left[ H(\xi) - H(\xi_i) \right] \mathbf{f}^{\mathbf{b}} \, d\Omega \tag{5.99}$$

$$\mathbf{f}_i^{\mathbf{b}\alpha} = \int_{\Gamma_t} N_i \left( F_\alpha - F_{\alpha i} \right) \mathbf{f}^{\mathbf{t}} \, d\Gamma + \int_{\Omega^e} N_i \left( F_\alpha - F_{\alpha i} \right) \mathbf{f}^{\mathbf{b}} \, d\Omega \quad (\alpha = 1, 2, 3 \text{ and } 4) \tag{5.100}$$

### 5.4.3 XFEM Enrichment Functions

#### 5.4.3.1 Orthotropic Crack-Tip Functions for Stationary Cracks

The following orthotropic crack-tip enrichment functions, derived by Asadpoure and Moham-madi (2007), can be adopted for stationary cracks:

$$\{F_l(r, \theta)\}_{l=1}^4 = \left\{ \sqrt{r} \cos \frac{\bar{\theta}_1}{2} \sqrt{\bar{g}_1(\theta)}, \sqrt{r} \cos \frac{\bar{\theta}_2}{2} \sqrt{\bar{g}_2(\theta)}, \right.$$
$$\left. \sqrt{r} \sin \frac{\bar{\theta}_1}{2} \sqrt{\bar{g}_1(\theta)}, \sqrt{r} \sin \frac{\bar{\theta}_2}{2} \sqrt{\bar{g}_2(\theta)} \right\} \tag{5.101}$$

where $r$ and $\theta$ define the crack-tip local polar coordinates, and $\bar{g}_k(\theta)$ and $\bar{\theta}_k$, $(k = 1, 2)$ are defined as:

$$\bar{g}_k(\theta) = \sqrt{(\cos \theta + \bar{s}_{kx} \sin \theta)^2 + \left( \bar{s}_{ky} \sin \theta \right)^2} \tag{5.102}$$

$$\bar{\theta}_k = \text{arctg} \left( \frac{\bar{s}_{ky} \sin \theta}{\cos \theta + \bar{s}_{kx} \sin \theta} \right) \tag{5.103}$$

and $\bar{s}_k = \bar{s}_{kx} + i \bar{s}_{ky}$ are the roots of Eq. (5.18).

In addition to static and quasi-static problems, the static orthotropic enrichments (5.101) have already been used for analysis of stationary and moving cracks subjected to dynamic loadings by Motamedi and Mohammadi (2010a, 2010b).

### 5.4.3.2  Orthotropic Crack-Tip Functions for Moving Cracks

In order to derive crack-tip enrichment functions from the analytical solutions (5.25) and (5.26), the following polar transformation is adopted:

$$Z_k^{\text{aux}} = r_k e^{i\theta_k} = r\left(\cos\theta + m_k \sin\theta\right), \quad k = 1, 2 \tag{5.104}$$

where $r_k$ and $g_k(\theta)$ are defined as:

$$r_k = r g_k(\theta), \quad k = 1, 2 \tag{5.105}$$

$$g_k(\theta) = \sqrt{\left(\cos\theta + m_{kx}\sin\theta\right)^2 + \left(m_{ky}\sin\theta\right)^2}, \quad k = 1, 2 \tag{5.106}$$

$$\theta_k = \text{arctg}\left(\frac{m_{ky}\sin\theta}{\cos\theta + m_{kx}\sin\theta}\right), \quad k = 1, 2 \tag{5.107}$$

and $(r, \theta)$ are considered from the crack-tip position, as depicted in Figure 5.1. $m_{kx}$ and $m_{ky}$ are the real and imaginary parts of $m_k$ (roots of Eq. (5.36)):

$$m_{kx} = \text{Re}(m_k), \quad k = 1, 2 \tag{5.108}$$

$$m_{ky} = \text{Im}(m_k), \quad k = 1, 2 \tag{5.109}$$

Thereafter, the imaginary and real parts of the main displacement fields can be rewritten as (Figure 5.2):

$$\sqrt{\left(\cos\theta + m_k \sin\theta\right)}, \, (k = 1, 2) \tag{5.110}$$

$$\text{Im}\left(\sqrt{Z_k^{\text{aux}}}\right) = r^{1/2}\sqrt{g_k(\theta)}\sin\frac{\theta_k}{2}, \quad k = 1, 2 \tag{5.111}$$

$$\text{Re}\left(\sqrt{Z_k^{\text{aux}}}\right) = r^{1/2}\sqrt{g_k(\theta)}\cos\frac{\theta_k}{2}, \quad k = 1, 2 \tag{5.112}$$

Finally, the following dynamic orthotropic enrichment functions can be derived for moving crack-tip problems:

$$\{F_l(r, \theta)\}_{l=1}^4 = \left\{\sqrt{r}\cos\frac{\theta_1}{2}\sqrt{g_1(\theta)}, \sqrt{r}\cos\frac{\theta_2}{2}\sqrt{g_2(\theta)}, \right.$$
$$\left. \sqrt{r}\sin\frac{\theta_1}{2}\sqrt{g_1(\theta)}, \sqrt{r}\sin\frac{\theta_2}{2}\sqrt{g_2(\theta)}\right\} \tag{5.113}$$

where $g_k(\theta)$ and $\theta_k$, $(k = 1, 2)$ are defined in equations (5.106) and (5.107).

The present formulation can be used for modelling all groups of composites and, interestingly, is in a similar form to those presented in Asadpoure and Mohammadi (2007) but with different definitions for the parameters. These functions, in fact, can be degenerated to the static orthotropic enrichment functions developed by Asadpoure and Mohammadi (2007) if dynamic effects are neglected. Similarly, these functions can be used for the special case of isotropic enrichment functions of moving cracks, as proposed by Belytschko and Chen (2004) and discussed in Section 5.4.3.3.

### 5.4.3.3   Isotropic Crack-Tip Functions for Moving Cracks

Belytschko and Chen (2004) used the analytical asymptotic solutions (5.1) and (5.2) near a propagating crack in an isotropic medium to develop the dynamic enrichment functions. A simplified form of these functions can be written as,

$$\{F_l(r, \theta)\}_{l=1}^4 = \left\{ r_d \cos \frac{\theta_d}{2}, r_s \cos \frac{\theta_s}{2}, r_d \sin \frac{\theta_d}{2}, r_d \sin \frac{\theta_s}{2} \right\} \tag{5.114}$$

where $r_d$, $r_s$, $\theta_d$ and $\theta_s$ are defined in equation (5.13)–(5.16).

Alternatively, Kabiri (2009) proposed the following enrichment functions for dynamic problems based on a simple idea of adding the influence of the crack-tip velocity on classical isotropic enrichment functions,

$$\{F_l(r, \theta)\}_{l=1}^4 = \left\{ \sqrt{r} \cos \frac{\theta}{2}, \sqrt{r} \sin \frac{\theta}{2}, \sqrt{r} \sin \theta \sin \frac{\theta}{2}, \sqrt{r} \cos \theta \sin \frac{\theta}{2} \right\} \sqrt{\frac{\gamma_d^2 + \gamma_s^2}{2}} \tag{5.115}$$

with

$$\gamma_d^2 = 1 - \frac{v_c^2}{c_d^2} \sin^2 \theta \tag{5.116}$$

$$\gamma_s^2 = 1 - \frac{v_c^2}{c_s^2} \sin^2 \theta \tag{5.117}$$

Initial results have shown improved dynamic responses, but further investigation is required to assess the performance of (5.115) for general dynamic problems. For further details see Kabiri (2009), Rezaei (2010) and Parchei (2012).

### 5.4.3.4   Propagating Enrichments

It is important to change the additional degrees of freedom according to crack propagation. As proposed by Combescure *et al.* (2008) and Rethore, Gravouil and Combescure (2005a) and adopted by Motamedi and Mohammadi (2010a, 2010b, 2012) for dynamic analysis of stationary and moving cracks in orthotropic media, once the crack tip enters an element, new tip enrichment DOFs are assumed for that element. The previous crack-tip element contains sets of both Heaviside and crack-tip DOFs from a previous step, but evaluated from the present crack-tip position (Figure 5.4). As a result, the singularity effect is expected to significantly reduce in the previous crack-tip element because of its larger distance from the present crack-tip position. Several numerical simulations have proved that this method substantially reduces the usual oscillations in the dynamic response which are due to a sudden change in DOFs. The method preserves the energy conservation, allowing the acceleration to be reliable for evaluation of the *J* integral and for post-processing purposes.

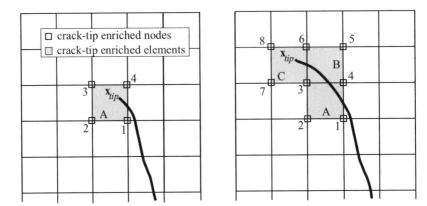

**Figure 5.4** Crack-tip enriched degrees of freedom for a propagating crack. Enrichments of all tip enriched elements in successive time-steps are based on the current position of $\mathbf{x}_{\text{tip}}$.

### 5.4.3.5  Time Variations of XFEM Enrichment Functions

Crack-tip enrichment functions for propagating cracks (5.113)–(5.114) are functions of time. As a result, time variations of these functions should be considered in discretizing the dynamic equilibrium equation (5.83).

Nevertheless, while many numerical solutions have considered the changing nature of these functions in different time steps, several other simulations have been performed without considering the time derivatives of crack-tip enrichment functions, and good results have been reported for most of the cases (Belytschko *et al.*, 2003; Belytschko and Chen, 2004; Rethore, Gravouil and Combescure, 2005a; Motamedi and Mohammadi, 2010a, 2010b, 2012). In general, however, it is necessary to account for time derivatives of these functions for problems with fast propagation patterns, and probably for high strain rates.

The crack-tip enrichment part of the XFEM approximation (5.84), $\mathbf{u}_{\text{tip}}^{\text{Enr}}(\mathbf{x})$, is in fact a function of time $\mathbf{u}_{\text{tip}}^{\text{Enr}}(\mathbf{x}, t)$ as it is a function of the current position of the crack tip $\mathbf{x}_{\text{tip}}$. For a typical crack-tip enrichment part of the approximation (without shifting),

$$\mathbf{u}_{\text{tip}}^{\text{Enr}}(\mathbf{x}, t) = \sum_k \phi_k(\mathbf{x}) F(\mathbf{x}, t) \mathbf{b}_k \tag{5.118}$$

the first and second time derivatives of $\mathbf{u}_{\text{tip}}^{\text{Enr}}(\mathbf{x}, t)$ can be computed from:

$$\dot{\mathbf{u}}_{\text{tip}}^{\text{Enr}}(\mathbf{x}, t) = \sum_k \phi_k(\mathbf{x}) \dot{F}(\mathbf{x}, t) \mathbf{b}_k(t) + \sum_k \phi_k(\mathbf{x}) F(\mathbf{x}, t) \dot{\mathbf{b}}_k(t) \tag{5.119}$$

$$\ddot{\mathbf{u}}_{\text{tip}}^{\text{Enr}}(\mathbf{x}, t) = \sum_k \phi_k(\mathbf{x}) \ddot{F}(\mathbf{x}, t) \mathbf{b}_k(t) + \sum_k \phi_k(\mathbf{x}) F(\mathbf{x}, t) \ddot{\mathbf{b}}_k(t)$$

$$+ 2 \sum_k \phi_k(\mathbf{x}) \dot{F}(\mathbf{x}, t) \dot{\mathbf{b}}_k(t) \tag{5.120}$$

In order to demonstrate further steps of computation, assume one of the terms of isotropic crack-tip enrichment functions (3.31) in crack tip local coordinates, $F(r, \theta) = \sqrt{r} \sin \theta / 2$.

Therefore, for a typical point $\mathbf{x}$ with respect to the time-dependent crack-tip coordinate $\mathbf{x}_{\text{tip}}(t)$ (in terms of global coordinates)

$$F(r, \theta) = F(\mathbf{x}, t) = \left\| \mathbf{x} - \mathbf{x}_{\text{tip}}(t) \right\|^{\frac{1}{2}} \sin \frac{\theta}{2} \tag{5.121}$$

and its time derivatives become (Menouillard *et al.*, 2010),

$$\dot{F}(\mathbf{x}, t) = -\dot{\mathbf{x}}_{\text{tip}}(t) \nabla F(r, \theta) \tag{5.122}$$

$$\ddot{F}(\mathbf{x}, t) = -\ddot{\mathbf{x}}_{\text{tip}}(t) \nabla F(r, \theta) - \left\| \dot{\mathbf{x}}_{\text{tip}}(t) \right\|^2 \nabla^2 F(r, \theta) \tag{5.123}$$

As a result, Eq. (5.85) is modified to: (Menouillard *et al.*, 2010),

$$\mathbf{M}\ddot{\mathbf{u}}^h + \mathbf{K}\mathbf{u}^h = \mathbf{f} - \mathbf{f}_1^b - \mathbf{f}_2^b - \mathbf{f}_3^b \tag{5.124}$$

where new $\mathbf{f}_1^b$, $\mathbf{f}_2^b$ and $\mathbf{f}_3^b$ vectors appear in addition to the original tip enrichment force vector $\mathbf{f}_i^b$, as defined in (5.100),

$$\mathbf{f}_1^b = \left[ \int_{\Omega^e} N_i N_j \ddot{F}_\alpha \, d\Omega \right] \mathbf{b} + \left[ \int_{\Omega^e} 2 N_i N_j \dot{F}_\alpha \, d\Omega \right] \dot{\mathbf{b}}, \quad \alpha = 1 - 4 \tag{5.125}$$

$$\mathbf{f}_2^b = \left[ \int_{\Omega} (N_i H_i) \left( N_j \ddot{F}_{\alpha j} \right) d\Omega \right] \mathbf{b} + \left[ \int_{\Omega} 2 (N_i H_i) \left( N_j \dot{F}_{\alpha j} \right) d\Omega \right] \dot{\mathbf{b}}, \quad \alpha = 1 - 4 \tag{5.126}$$

$$\mathbf{f}_3^b = \left[ \int_{\Omega} (N_i F_{\alpha i}) \left( N_j \ddot{F}_{\alpha j} \right) d\Omega \right] \mathbf{b} + \left[ \int_{\Omega} 2 (N_i F_{\alpha i}) \left( N_j \dot{F}_{\alpha j} \right) d\Omega \right] \dot{\mathbf{b}}, \quad \alpha = 1 - 4 \tag{5.127}$$

Equations (5.125)–(5.127) can be extended to their shifted forms, similar to the procedure used in the original definitions (5.99) and (5.100). It is also noted that the derivatives $\dot{F}(\mathbf{x}, t)$ and $\ddot{F}(\mathbf{x}, t)$ vanish when there is no propagation, $\dot{\mathbf{x}}_{\text{tip}} = 0$. As a result, Eq. (5.124) reduces to the original form of (5.85).

### 5.4.4  Time Integration Schemes

There are a number of time integration techniques that are frequently used in various dynamic problems. They include the explicit and implicit approaches and span a wide range of solutions associated with different complexity, speed, efficiency and accuracy. While the explicit approach is simple and fast, it is less accurate and suffers from stability conditions. On the other hand, implicit approaches are generally more accurate and stable, but they are computationally more expensive. General single time step algorithms are usually based on residual formulations (the time finite element method) or the truncated Taylor series (Rezaei, 2010).

SS$pj$ (single step with approximation of degree $p$ for equations of order $j$) methods and GN$pj$ (generalized Newmark with degree $p$ and order $j$) methods generate a wide range of time integration schemes, such as the Euler method, the Crank–Nicolson approach, the Galerkin formulation, the Newmark method and the backward finite difference formulation (Zienkiewicz et al., 2005; Rezaei, 2010).

Here, only two methods are briefly reviewed. First, the classical Newmark approach is briefly presented and then the concept of time-discontinuity is embedded within an extended finite element method, resulting in a space-time XFEM (STXFEM).

### 5.4.4.1   Newmark Time Integration

It is possible to similarly extend the formulation (5.85) to include the effects of velocity-based global damping,

$$\mathbf{M\ddot{u}}^h + \mathbf{C\dot{u}}^h + \mathbf{Ku}^h = \mathbf{f} \tag{5.128}$$

The Newmark time integration scheme can then be adopted to solve for the extended finite element equation of motion at time step $n$:

$$\mathbf{M\ddot{u}}_n^h + \mathbf{C\dot{u}}_n^h + \mathbf{Ku}_n^h = \mathbf{f}_n \tag{5.129}$$

where $\mathbf{M}$, $\mathbf{K}$ and $\mathbf{C}$ are the mass, stiffness and damping matrices, respectively, and $\mathbf{f}_n$ is the load vector at the time step $n$. The final time-discretized simultaneous equations are expressed as:

$$\left( \mathbf{M} + \beta \Delta t^2 \mathbf{K} + \alpha \Delta t \mathbf{C} \right) \ddot{\mathbf{u}}_n^h = \mathbf{f_n} - \mathbf{K} \left( \mathbf{u}_{n-1}^h + \Delta t \dot{\mathbf{u}}_{n-1}^h + (1 - 2\beta) \frac{\Delta t^2}{2} \ddot{\mathbf{u}}_{n-1}^h \right)$$
$$- \mathbf{C} \left( \dot{\mathbf{u}}_{n-1}^h + (1 - \alpha) \Delta t \ddot{\mathbf{u}}_{n-1}^h \right) \tag{5.130}$$

$$\dot{\mathbf{u}}_n^h = \dot{\mathbf{u}}_{n-1}^h + (1 - \alpha) \Delta t \ddot{\mathbf{u}}_{n-1}^h + \alpha \Delta t \ddot{\mathbf{u}}_n^h \tag{5.131}$$

$$\mathbf{u}_n^h = \mathbf{u}_{n-1}^h + \Delta t \dot{\mathbf{u}}_{n-1}^h + (1 - 2\beta) \frac{\Delta t^2}{2} \ddot{\mathbf{u}}_{n-1}^h + \beta \Delta t^2 \ddot{\mathbf{u}}_n^h \tag{5.132}$$

where $\mathbf{u}_n$, $\dot{\mathbf{u}}_n$ and $\ddot{\mathbf{u}}_n$ are the global vectors of nodal displacements, nodal velocities and nodal accelerations, respectively. $\Delta t$ is the time increment at the present time step and the Newmark's parameters are chosen to be $\beta = 1/4$ and $\alpha = 1/2$ to fulfil the unconditionally stable criterion.

### 5.4.4.2   Space–Time XFEM (STXFEM)

The space–time extended finite element method provides a time integration approximation which guarantees continuity of displacements $\mathbf{u}$ in time, generates linear variations of velocity field $\mathbf{v}$ which can become discontinuous at each time step, and ensures satisfaction of $\dot{\mathbf{u}} = \mathbf{v}$ (Rethore, Gravouil and Combescure, 2005b).

In this section the method proposed by Rethore, Gravouil and Combescure (2005b) is briefly presented. Using the partition of unity concept to approximate the velocity field $\mathbf{v}(t)$ leads to (Rethore *et al.*, 2005; Chessa and Belytschko, 2004):

$$\mathbf{v}(t) = \sum_{i=0}^{N} \Lambda_i(t) \mathbf{v}_i^c + \sum_{j=0}^{M} \sum_{i \in T_j} \Lambda_i(t) H(t - t_j) \mathbf{v}_{i,j}^e \tag{5.133}$$

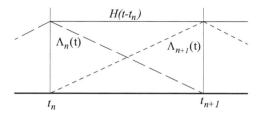

**Figure 5.5** Linear shape functions and the Heaviside function within the time interval $I_{n+1}$ (Rethore, Gravouil and Combescure, 2005b).

where $v_i^c$ are classical velocity degrees of freedom, $v_{i,j}^e$ are additional degrees of freedom, $\Lambda_j$ is the time interval shape function, $H(t - t_j)$ is the Heaviside enrichment function to resemble the potential discontinuity in time (see Figure 5.5) and $T_j$ is the time interval that is enriched by the enrichment function $H(t - t_j)$. $T_j$ at each time step is limited to only that step and $H(t - t_j)$ vanishes for times before $t_j$, so $\Lambda_j$ is defined in the $\lfloor t_{j-1}, t_{j+1} \rfloor$ time interval. Therefore, within the time interval $I_{n+1} = ]t_n; t_{n+1}[$, only the Heaviside enrichment $H(t - t_n)$ remains present.

Velocity and displacement fields can now be defined from their continuum and enriched parts,

$$\mathbf{v}(t) = \mathbf{v}^c(t) + \mathbf{v}^e(t) \tag{5.134}$$

$$\mathbf{u}(t) = \mathbf{u}^c(t) + \mathbf{u}^e(t) \tag{5.135}$$

where

$$\mathbf{v}^c(t) = \mathbf{v}_n^c \Lambda_n(t) + \mathbf{v}_{n+1}^c \Lambda_{n+1}(t) \tag{5.136}$$

$$\mathbf{v}^e(t) = \mathbf{v}_{n+1}^e \Lambda_n(t) H(t - t_n) \tag{5.137}$$

and

$$\Lambda_n(t) = \frac{t_{n+1} - t}{t_{n+1} - t_n} \tag{5.138}$$

$$\Lambda_{n+1}(t) = \frac{t - t_n}{t_{n+1} - t_n} \tag{5.139}$$

Intentionally, $\mathbf{v}_{n,n}^e$ has been replaced by $\mathbf{v}_{n+1}^e$ to insist that it has to be evaluated from the solution in $I_{n+1}$ time step.

In order to satisfy the conditions $\dot{\mathbf{u}} = \mathbf{v}$ and $\mathbf{u}(t_n^+) = \mathbf{u}(t_n^-)$ at each time step, the displacement field is obtained as,

$$\mathbf{u}^c(t) = \mathbf{u}_n^c + \int_{t_n}^{t} \mathbf{v}^c(\tau) d\tau \tag{5.140}$$

$$\mathbf{u}^e(t) = \mathbf{u}_n^e + \int_{t_n}^{t} \mathbf{v}^e(\tau) d\tau \tag{5.141}$$

and the weighted residual form of the equation of motion in time step $I_{n+1}$ can be written as,

$$\int_{t_n^+}^{t_{n+1}^-} w(t) \left[ \mathbf{m}\dot{\mathbf{v}}(t) + \mathbf{k}\mathbf{u}(t) \right] dt + w(t_n^+) \mathbf{m}\mathbf{v}_{n+1}^e = \int_{t_n^+}^{t_{n+1}^-} w(t)\mathbf{f}(t) dt \tag{5.142}$$

Two independent set of weightings have to be considered to account for additional degrees of freedom. The discrete form of (5.142) can be written as (Rethore, Gravouil and Combescure, 2005b),

$$\boldsymbol{\Pi}_1 \bar{\mathbf{u}}_{n+1} = \boldsymbol{\Pi}_0 \bar{\mathbf{u}}_n + \mathbf{F}_{n+1} \tag{5.143}$$

$$\bar{\mathbf{u}}_n = \begin{bmatrix} \mathbf{u}_n^c \\ \mathbf{u}_n^e \\ \mathbf{v}_n^c \\ \mathbf{v}_n^e \end{bmatrix} \tag{5.144}$$

$$\boldsymbol{\Pi}_0 = \begin{bmatrix} 1 & 0 & \dfrac{\Delta t}{2} & 0 \\ 0 & 1 & 0 & 0 \\ -\Delta t \mathbf{k} & -\Delta t \mathbf{k} & \mathbf{m} + (\beta_1 - \gamma_1)\Delta t^2 \mathbf{k} & 0 \\ -\Delta t \mathbf{k} & -\Delta t \mathbf{k} & \mathbf{m} + (\beta_2 - \gamma_2)\Delta t^2 \mathbf{k} & 0 \end{bmatrix} \tag{5.145}$$

$$\boldsymbol{\Pi}_1 = \begin{bmatrix} 1 & 0 & -\dfrac{\Delta t}{2} & 0 \\ 0 & 1 & 0 & -\dfrac{\Delta t}{2} \\ 0 & 0 & \mathbf{m} + \beta_1 \Delta t^2 \mathbf{k} & (\delta_1 - 1)\mathbf{m} + (\gamma_1 - \beta_1)\Delta t^2 \mathbf{k} \\ 0 & 0 & \mathbf{m} + \beta_2 \Delta t^2 \mathbf{k} & (\delta_2 - 1)\mathbf{m} + (\gamma_2 - \beta_2)\Delta t^2 \mathbf{k} \end{bmatrix} \tag{5.146}$$

$$\mathbf{F}_{n+1} = \begin{bmatrix} 0 \\ 0 \\ \Delta t \left[ (1 - \gamma_1)\mathbf{f}_n + \gamma_1 \mathbf{f}_{n+1} \right] \\ \Delta t \left[ (1 - \gamma_2)\mathbf{f}_n + \gamma_2 \mathbf{f}_{n+1} \right] \end{bmatrix} \tag{5.147}$$

and

$$2\beta_i \frac{1}{\Delta t} \int_{t_n^+}^{t_{n+1}^-} w_i(t)\mathrm{d}t = \frac{1}{\Delta t^3} \int_{t_n^+}^{t_{n+1}^-} w_i(t) \left( t - t_n^+ \right)^2 \mathrm{d}t \tag{5.148}$$

$$\gamma_i \frac{1}{\Delta t} \int_{t_n^+}^{t_{n+1}^-} w_i(t)\mathrm{d}t = \frac{1}{\Delta t^2} \int_{t_n^+}^{t_{n+1}^-} w_i(t) \left( t - t_n^+ \right) \mathrm{d}t \tag{5.149}$$

$$\delta_i \frac{1}{\Delta t} \int_{t_n^+}^{t_{n+1}^-} w_i(t)\mathrm{d}t = w_i \left( t_n^+ \right) \tag{5.150}$$

The solution procedure consists of evaluation of unknown variables $\mathbf{v}_{n+1}^e, \mathbf{v}_{n+1}^c, \mathbf{u}_{n+1}^e$ and $\mathbf{u}_{n+1}^c$ from the known values of $\mathbf{v}_n^e, \mathbf{v}_n^c, \mathbf{u}_n^e$ and $\mathbf{u}_n^c$ at each time step. Six parameters $\{\beta_i, \gamma_i, \delta_i, i = 1, 2\}$ are determined from Eqs. (5.148)–(5.150), which need proper selection of weight functions $w_i, i = 1, 2$ to achieve the optimal solution (such as (5.152) and (5.153)).

The constant piecewise acceleration within the time interval $I_{n+1}$ is obtained from the derivative of (5.140) and (5.141)

$$\mathbf{a}(t) = \mathbf{a}_{n+1} = \frac{1}{\Delta t} \left( \mathbf{v}_{n+1}^c - \mathbf{v}_{n+1}^e - \mathbf{v}_n^c \right) \tag{5.151}$$

STXFEM can be degenerated to the discontinuous Galerkin time integration scheme (Li *et al.*, 2003) by setting,

$$w_1 = \lambda_n \tag{5.152}$$

$$w_2 = \lambda_{n+1} \tag{5.153}$$

where $\lambda_n$ and $\lambda_{n+1}$ are determined from (5.138) and (5.139). The resulting $\{\beta_i, \gamma_i, \delta_i, i = 1, 2\}$ coefficients are:

$$\beta_1 = \frac{1}{12}, \beta_2 = \frac{1}{4}, \gamma_1 = \frac{1}{3}, \gamma_2 = \frac{2}{3}, \delta_1 = 2, \delta_2 = 2 \tag{5.154}$$

Rethore, Gravouil and Combescure (2005b) have computed the spectral radius of the STXFEM approach for an undamped SDOF system subjected to an initial velocity. They have illustrated that the spectral radius remains always less than 1, indicating that the method is unconditionally stable. Similar conclusions have been made by Sukumar and Prevost (2003) and Hughes (2000).

## 5.5   Numerical Simulations

### 5.5.1   Plate with a Stationary Central Crack

For the first example, a plane stress rectangular plate with a central horizontal crack subjected to a step tensile distributed load is considered (Figure 5.7), and several orientations of material elastic axes are studied. The size of the cracked plate is $b = 20$ mm, $h = 40$ mm and $2a = 4.8$ mm. Material properties are defined in Table 5.1. The time-step is selected from $\Delta t = h/50c_L$, where $c_L = \sqrt{d_{22}/\rho}$ is the wave velocity along the principal material axis 2 and $d_{22}$ is defined in (4.16). The relative integration domain radius, $r_J/a$, is set to be 0.4.

A structured uniform $50 \times 100$ finite element mesh is used with $2 \times 2$ and $6 \times 6$ Gauss quadrature rules for ordinary and enriched elements, respectively (Figure 5.6).

The time history results of stress intensity factors, calculated by the $J$ integral domain approach and presented in Figure 5.7, show that the maximum values of the mode I stress intensity factor decrease by increasing $\varphi$ from $0°$ to $60°$. Comparison of the results with the reference BEM approach, reported by Garcia-Sanchez, Zhang and Saez (2008), shows a well conforming trend, and the existing differences may be attributed to the fact that these methods have adopted different approaches to calculate dynamic stress intensity factors. In addition to the smoother and less-oscillatory results of XFEM, another advantage of the present results is that the predicted mode I dynamic stress intensity factors remain always positive.

Motamedi and Mohammadi (2010a) investigated the effects of different finite element meshes on dynamic stress intensity factors by using four different meshes $40 \times 80$, $50 \times 100$, $60 \times 120$ and $70 \times 140$ for the material angle zero and the time-step $\Delta t = h/40c_L$.

**Table 5.1**   Orthotropic material properties.

| $E_1$(GPa) | $E_2$(Gpa) | $G_{12}$(GPa) | $\nu_{12}$ | $\rho$(kg m$^{-3}$) |
|---|---|---|---|---|
| 118.3 | 54.8 | 8.79 | 0.083 | 1900 |

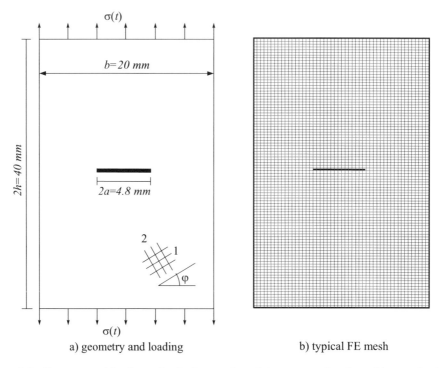

a) geometry and loading                                    b) typical FE mesh

**Figure 5.6**   Geometry and loadings of a single central crack in a rectangular plate with several orientations of the axes of orthotropy.

Performed numerical sensitivity analyses proved that the results for different meshes remain close together, illustrating that even with a coarse finite element mesh accurate solutions can be obtained from the XFEM modelling.

Figure 5.8 illustrates the $\sigma_{yy}$ stress contour with a clear indication of stress concentration around the crack tips. The maximum concentration factor is predicted to be around 6.6.

## 5.5.2   Mode I Plate with an Edge Crack

A $40 \times 52$mm plane stress tensile plate with a 12 mm edge crack and material properties of Table 5.2 is considered (Figure 5.9). The stationary crack is analysed in a dynamic step loading condition. Two different time-steps $\Delta t = a/3c_L$ and $\Delta t = a/10c_L$ are selected to evaluate the effect of different time-steps. $c_L = \sqrt{d_{22}/\rho}$ is the wave velocity along the principal axis 2 of the material, and $d_{22}$ is defined in (4.16) .

A $78 \times 60$ finite element mesh is used for discretization of the model (Figure 5.9) with $2 \times 2$ and $5 \times 5$ Gauss quadrature rules for ordinary and enriched elements, respectively. The $J$ integration domain radius is $r_J = 0.4a$.

This problem was simulated by Motamedi and Mohammadi (2010a) using an XFEM approach. The results of XFEM analysis for the mode I stress intensity factor are depicted in Figure 5.10 and compared with the reference BEM results, reported by Garcia-Sanchez, Zhang and Saez (2008). Very good agreement is clearly observed.

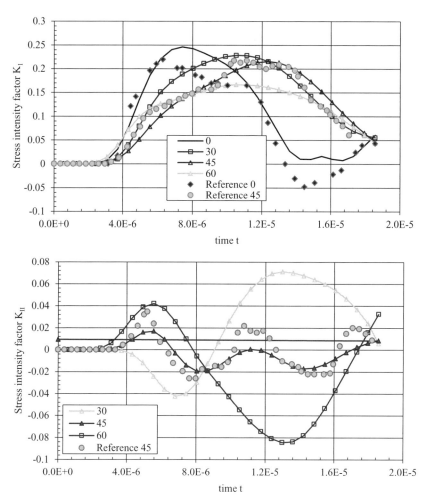

**Figure 5.7**   Time histories of mode I and II dynamic SIFs for a single central crack in a rectangular plate with several orientations of the axes of orthotropy, predicted by XFEM, and compared with the reference boundary element results of Garcia-Sanchez, Zhang and Saez (2008). (Motamedi and Mohammadi, 2010a).

Variations of the stress intensity factor for various relative $J$ integration domain sizes, $r_J/a$, are also depicted in Figure 5.10. For most of the cases, the results are less sensitive to this ratio when $r_J/a$ is between 0.2 and 0.6. For higher values of $r_J/a$, the results show a slightly different trend which can be interpreted as a sign of improper size of the $J$ integration domain with respect to the crack length.

A sensitivity analysis is performed to assess the effects of a different number of Gauss integration points using five different Gauss quadrature rules $(2 \times 2 - 5 \times 5)$, $(2 \times 2 - 6 \times 6)$, $(4 \times 4 - 4 \times 4)$, $(4 \times 4 - 7 \times 7)$ and $(5 \times 5 - 7 \times 7)$ for (ordinary–enriched) elements on the mode I stress intensity factor based on the $78 \times 60$ finite element mesh. Figure 5.11 illustrates that the results are virtually the same, indicating that an accurate result can be obtained even with the minimum order of the Gauss quadrature rule.

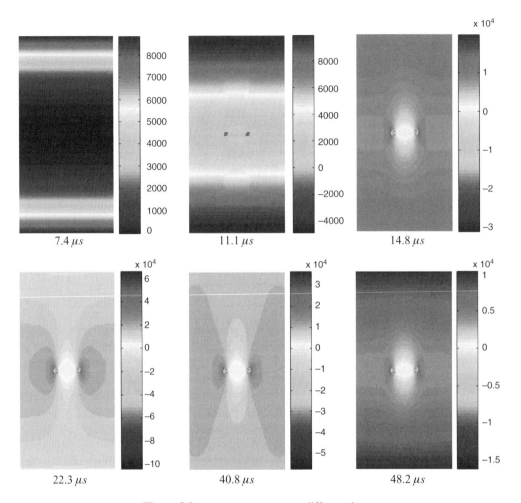

**Figure 5.8**   $\sigma_{yy}$ stress contours at different times.

Furthermore, the effects of the size of the finite elements or the number of finite elements on dynamic stress intensity factors are investigated in Figure 5.11 by examining three other meshes $50 \times 42$, $100 \times 84$ and $125 \times 105$. Again, similar results are obtained.

Figure 5.12 illustrates the $\sigma_{yy}$ stress contour at different stages of the loading. At first, the stress wave starts to propagate from the plate edges towards the crack. Then, a stress concentration is observed as the stress wave reaches the crack tip. It will remain concentrated, but with reduced intensity as the wave peak passes the crack.

**Table 5.2**   Orthotropic material properties.

| $E_1(\text{GPa})$ | $E_2(\text{GPa})$ | $G_{12}(\text{GPa})$ | $\nu_{12}$ | $\rho(\text{kg m}^{-3})$ |
|---|---|---|---|---|
| 118.3 | 54.8 | 8.79 | 0.083 | 1900 |

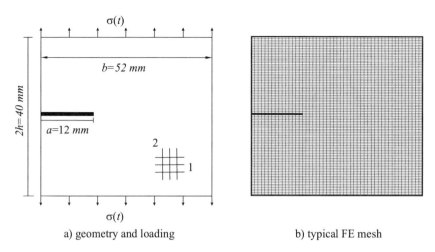

a) geometry and loading                                    b) typical FE mesh

**Figure 5.9** Geometry, loading and the finite element mesh of a single edge crack in a rectangular plate.

## 5.5.3   Mixed Mode Edge Crack in Composite Plates

### 5.5.3.1   Edge Crack in an Orthotropic Plate under Tensile Stress Loading

This example was studied by Aliabadi and Sollero (1998) using a quasi-static BEM, and Motamedi and Mohammadi (2010b) based on a dynamic XFEM. The problem consists of a plane stress rectangular graphite-epoxy plate with an edge horizontal crack subjected to

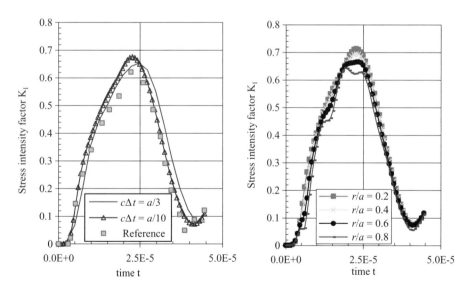

**Figure 5.10** The effect of various time-steps and relative integration domain sizes on the mode I stress intensity factor in a rectangular plate with a single edge crack and comparison with the reference boundary element method by Garcia-Sanchez, Zhang and Saez (2008). (Motamedi and Mohammadi, 2010a).

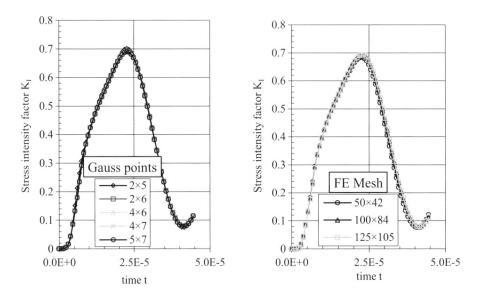

**Figure 5.11**    The effect of various Gauss quadrature rules and finite element meshes on the mode I stress intensity factor (Motamedi and Mohammadi, 2010a).

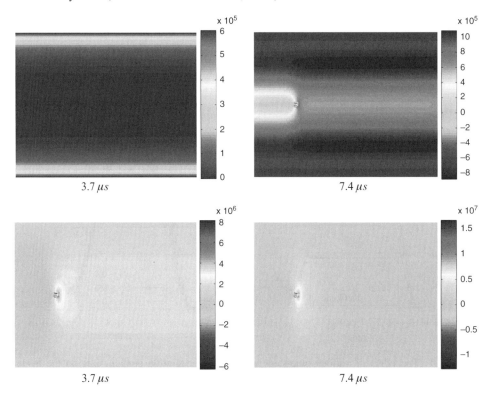

**Figure 5.12**    $\sigma_{yy}$ stress contours for a horizontal edge crack in a rectangular plate.

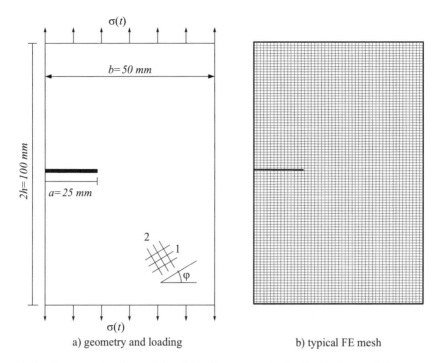

a) geometry and loading                                    b) typical FE mesh

**Figure 5.13**  Geometry, loading and the finite element mesh of a single edge crack in a rectangular plate with several orientations of the axes of orthotropy.

a tensile distributed step loading (Figure 5.13). Six different material orientation angles, including $0°$, $30°$, $45°$, $60°$ and $90°$ are studied. The material properties are defined in Table 5.3. The time-step is selected to be $\Delta t = 0.5\mu s$ and the radius of the $J$ integration domain is $r_J = 0.4a$. A structured $40 \times 80$ finite element mesh is adopted, and the integrations are performed by $4 \times 4$ and $7 \times 7$ Gauss quadrature rules for ordinary and enriched elements, respectively.

The time histories of mode I and mode II dynamic stress intensity factors for different orthotropic material angles are illustrated in Figure 5.14. The results include the maximum and minimum quasi-static boundary element solutions of Aliabadi and Sollero (1998) (dash-dot lines in Figure 5.14, with zero reference minimum solution for $K_{II}$). Although the present dynamic solutions are oscillatory, an acceptable trend is generally observed, with the maximum value of mode I dynamic stress intensity factor occurring for the case of $\varphi = 30°$, which is similar to the conclusion of Aliabadi and Sollero (1998).

The crack propagation is governed by the maximum hoop stress intensity factor $K_{tt}$. Figure 5.15 illustrates time history variations of $K_{tt}$ and its corresponding crack tip

**Table 5.3**  Orthotropic material properties of graphite-epoxy.

| $E_1(GPa)$ | $E_2(GPa)$ | $G_{12}(GPa)$ | $v_{12}$ | $\rho(kg\,m^{-3})$ |
|---|---|---|---|---|
| 114.8 | 11.7 | 9.66 | 0.21 | 1500 |

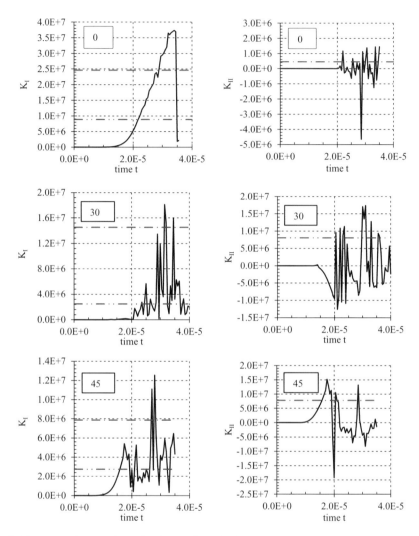

**Figure 5.14** Dynamic stress intensity factors for a single edge crack plate with $\varphi = 0°, 30°, 45°, 60°$ and $90°$ material angles. Dash-dot lines show the reference maximum and minimum results, obtained by Aliabadi and Sollero (1998). (Also see Motamedi and Mohammadi, 2010b). (Reproduced by permission of Springer.)

velocity $v_c$ for the special case of material orthotropic angle $60°$. Apparently, the crack starts to propagate at about $11 \mu s$.

Figure 5.16 illustrates the predicted crack propagation paths for three orthotropic material angles $\varphi = 30°, 45°, 60°$, and compares them with the reference results by Aliabadi and Sollero (1998). Also, Motamedi and Mohammadi (2010b) have reported that XFEM predicts a self-similar extension for $\varphi = 0°, 90°$ as the crack-tip sliding displacement remains zero. This is slightly different from the report by Aliabadi and Sollero (1998) which predicted an inclined propagation for $\varphi = 0°$. For a comprehensive discussion on potential reasons see Motamedi and Mohammadi (2010b).

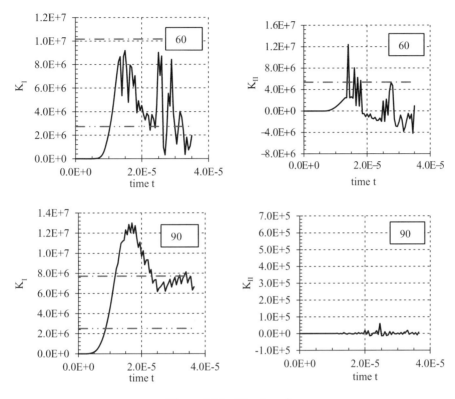

**Figure 5.14** (*Continued*)

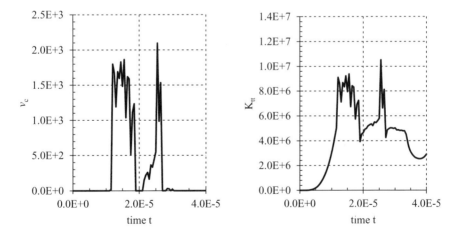

**Figure 5.15** $v_c$ and $K_{tt}$ for a single edge crack plate with $\varphi = 60°$ material angle.

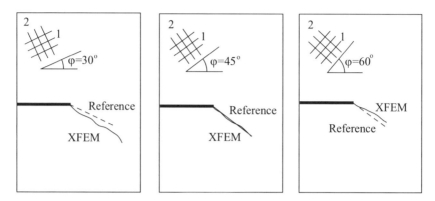

**Figure 5.16** Crack paths for a single edge crack plate for angles $\varphi = 0°$, $30°$, $45°$, $60°$ and $90°$ material angles. Solid lines represent XFEM solutions (Motamedi and Mohammadi, 2010b) and the reference Aliabadi and Sollero (1998) results are depicted by dashed lines. (Reproduced by permission of Springer.)

### 5.5.3.2  Semi-Infinite Crack in an Infinite Plate Subjected to a Tensile Stress Wave

A rectangular plane strain plate with an edge crack, subjected to an impact tensile step traction, is considered (Figure 5.17). This problem has been frequently adopted for dynamic fracture analysis by a number of researchers such as Lu, Belytschko and Tabbara (1995), Krysl and Belytschko (1999) and Belytschko *et al.* (2000) using EFGM, Duarte, Babuska and Oden (1998) using GFEM, Belytschko *et al.* (2003), Gravouil, Elguedj and Maigre (2009a, 2009b) and Menouillard *et al.* (2010) using isotropic XFEM and Motamedi and Mohammadi (2010a, 2010b, 2012) using orthotropic stationary and moving enrichments within XFEM.

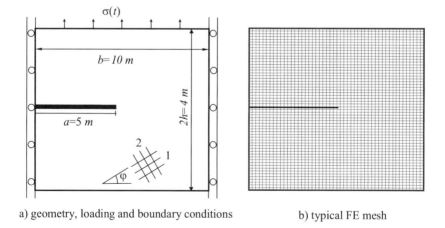

a) geometry, loading and boundary conditions          b) typical FE mesh

**Figure 5.17**  A tensile plate with an edge crack, and its finite element mesh.

The size of initial notch is $a = 5\,\text{m}$. The step tensile loading is defined as $\sigma_0 H(t)$, where $H$ is the Heaviside function and $\sigma_0$ is the intensity of the loading. The time-step is selected to be $\Delta t = 0.5\,\mu\text{s}$.

A $40 \times 100$ finite element mesh is used to define the model. The integrations are performed by $4 \times 4$ and $7 \times 7$ Gauss quadrature rules for ordinary and enriched elements, respectively. The size of the $J$ integration domain is set to $r_J = 0.4a$.

#### 5.5.3.2.1 Stationary and Propagating Cracks in an Isotropic Plate

In the first stage, a stationary crack in an isotropic plate is simulated to assess the accuracy of the orthotropic enrichments for a simplified isotropic case, with the material properties defined in Table 5.4. For this ideal case, the analytical solution of dynamic stress intensity factor $K_I$ was presented by Freund (1976) and defined in (5.39).

In order to show the robustness of the proposed method in modelling dynamic fracture problems three different analyses are performed. First, the crack is assumed to be stationary. The time $t_0$ that the wave takes to reach the crack tip (after imposition of the loading) can be computed from

$$t_0 = \frac{h}{c_d} \approx 3.31 \times 10^{-4}\,\text{s} \tag{5.155}$$

Therefore, in order to simulate the semi-infinite nature of the problem, the numerical simulation has to be terminated before or at about $3t_0$, which is associated with the time that the numerical reflected wave (from the bottom boundary) reaches the crack tip.

Variations of the normalized dynamic mode I stress intensity factor

$$\bar{K}_I = K_I / \sigma_0 \sqrt{h} \tag{5.156}$$

with respect to the normalized time $t/t_0$ has been computed and compared with the analytical solution in Figure 5.18a. A close agreement between the results obtained by the dynamic XFEM with orthotropic enrichments and the analytical solution is observed. It should be mentioned that the error near $t/t_0 = 1$ is due to the tensile stress wave entering the $J$ integral contour area for calculating the stress intensity factor, even before it actually reaches the crack tip (Menouillard et al., 2010). This type of error is observed in all $J$-based dynamic fracture analysis.

Then, the crack tip is instantly forced to propagate with a constant velocity $v_c = 1500\,\text{m s}^{-1}$, while in the third analysis, the crack is allowed to rest until $t = 1.5t_0$ and then the crack tip is forced to propagate with the same constant velocity $v_c = 1500\,\text{m s}^{-1}$. For all cases, the crack tip velocity is imposed as a boundary condition.

For a propagating crack, Eq. (5.39) is replaced by (5.41). In contrast to the smooth results for the stationary crack (Figure 5.18a), oscillating results around the analytical solution are obtained (Figures 5.18b,c). In addition to conventional numerical oscillations, the reason for

**Table 5.4** Isotropic material properties.

| $E$ (GPa) | $\nu$ | $\rho$ (kg m$^{-3}$) | $\sigma_0$ (MPa) | $c_d$ (dilational) (m s$^{-1}$) | $c_s$ (shear) (m s$^{-1}$) |
| --- | --- | --- | --- | --- | --- |
| 211 | 0.3 | 7800 | 500 | 6035 | 3226 |

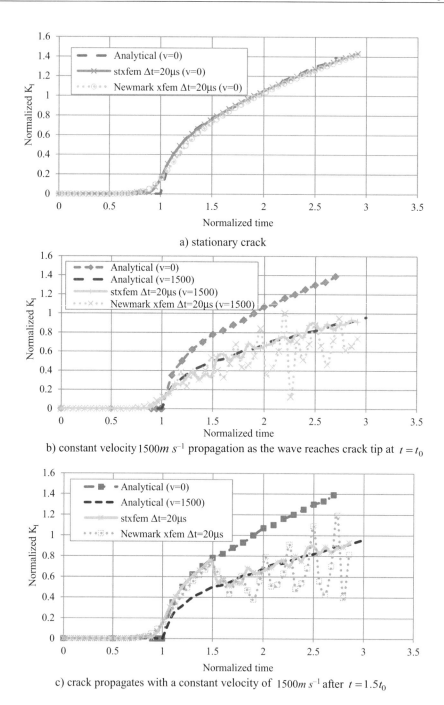

a) stationary crack

b) constant velocity $1500m\ s^{-1}$ propagation as the wave reaches crack tip at $t = t_0$

c) crack propagates with a constant velocity of $1500m\ s^{-1}$ after $t = 1.5t_0$

**Figure 5.18** Comparison of the normalized dynamic stress intensity factor predicted by STXFEM and Newmark XFEM for (a) stationary crack ($v_c = 0$), (b) crack propagates with a constant $v_c = 1500\,\mathrm{m\,s^{-1}}$ as soon as the wave reaches its tip at $t = t_0$, (c) crack propagates with a constant velocity of $v_c = 1500\,\mathrm{m\,s^{-1}}$ after $t = 1.5t_0$ (Kabiri, 2009; Rezaei, 2010).

the occurrence of such oscillation may be attributed to the generated numerical noise when some elements are crossed through during the crack tip advancing process, causing some sort of time discontinuity in their stiffness matrix.

Here, the efficiency of the STXFEM approach is compared with the conventional Newmark time integration scheme. Clearly, the extensive oscillating nature of the Newmark solution has been well smoothed by the proposed STXFEM approach. Numerical sensitivity analyses have shown that the Newmark time step should be reduced by a factor of at least 10 in order to reduce the oscillatory response to a level comparable with STXFEM (Rezaei, 2010). More or less similar results have been reported by Gravouil, Elguedj and Maigre (2009a, 2009b).

Another important aspect of the approach is to change the additional degrees of freedom according to the crack propagation stage. As proposed by Combescure *et al.* (2008) and Rethore, Gravouil and Combescure (2005a) and adopted by Motamedi and Mohammadi (2010a, 2010b), once the crack tip enters an element, while new crack tip DOFs are assumed for that element, the previous crack-tip element now contains both sets of Heaviside and crack-tip DOFs from a previous step (but evaluated from the present crack-tip position). As a result, the singularity effect is expected to significantly reduce in the previous crack-tip element because of its larger distance from the present crack-tip position. Figure 5.19 shows that this method, designated by enriched STXFEM, further reduces the existing oscillations. It also improves, to some extent, the oscillatory solution of Newmark-based XFEM.

### 5.5.3.2.2  *Stationary and Propagating Cracks in an Orthotropic Plate*
After examining the performance of various time integration schemes for simulating a benchmark dynamic isotropic plate, the same methodology is used to simulate stationary and propagating cracks in an orthotropic plate, with material properties defined in Table 5.5.

In the case of a propagating crack, the crack is allowed to propagate with a varying velocity, which can be calculated from equation (5.42). In orthotropic materials, the Rayleigh wave speed $c_R$ should be replaced with the shear wave speed $c_s$ (Motamedi and Mohammadi, 2012).

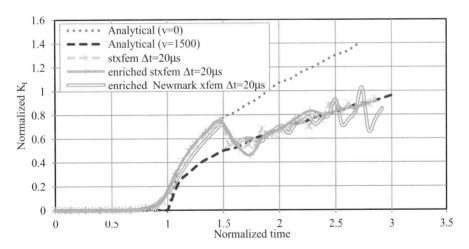

**Figure 5.19**  Comparison of the normalized dynamic stress intensity factor predicted by extended enrichment of STXFEM and Newmark XFEM for a crack that propagates with a constant velocity of 1500 m s$^{-1}$ after $t = 1.5t_0$ (Kabiri, 2009; Rezaei, 2010).

**Table 5.5**  Orthotropic material properties.

| $E_1$ (GPa) | $E_2$ (GPa) | $G_{12}$ (GPa) | $v_{12}$ | $\rho$ (kg m$^{-3}$) | $\sigma_0$ (GPa) | $K_{\mathrm{ID}} = K_{\mathrm{IC}}$ (MPa $\sqrt{\mathrm{m}}$) |
|---|---|---|---|---|---|---|
| 114.8 | 11.7 | 9.66 | 0.21 | 1500 | 1 | 2 |

The time histories of the mode I dynamic stress intensity factor ($K_{\mathrm{I}}$) for stationary and prop-agating cracks are compared in Figure 5.20. Again, an oscillating response is obtained for the propagation case. This figure also illustrates the time-history of the crack-tip velocity $v_{\mathrm{c}}$. It is observed that whenever the dynamic stress intensity factor $K_{\mathrm{I}}$ exceeds the dynamic toughness $K_{\mathrm{ID}}$, the crack starts to propagate and the extra energy is released, instantly decreasing the dy-namic stress intensity factor. At the same time, with an increased total crack length, the overall instability escalates and the dynamic stress intensity factor should increase. As a result, an os-cillating behaviour for the dynamic stress intensity factor in propagating cracks is predictable.

Motamedi and Mohammadi (2012) have investigated the sensitivity of dynamic stress intensity factors for different finite element meshes. Figure 5.21 compares the results for two different finite element meshes 30 × 80 and 40 × 100 for the 30° orthotropic angle. The results indicate that while the general trends remain similar, a sufficiently fine mesh is required to obtain accurate solutions. The coarser mesh has predicted an earlier propagation time. Further studies are required to obtain the optimum finite element mesh for a pre-set level of accuracy in a dynamic crack propagation problem.

### 5.5.3.2.3  Mixed Mode Crack Propagation in 45° and 60° Orthotropic Plates

In the last part of the present example, the mixed-mode crack propagation is simulated by implementing different 45° and 60° orthotropic orientation angles (with respect to the $x$-axis). The results for the dynamic stress intensity factors, the energy release rate and the crack-tip velocity for each orthotropy angle on a 40 × 100 mesh are depicted in Figures 5.22 and 5.23, respectively. For the same results for the 30° orthotropy angle refer to Motamedi and Mohammadi (2012).

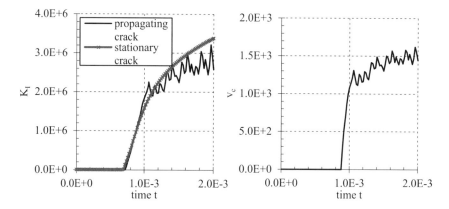

**Figure 5.20**  Variations of $K_{\mathrm{I}}$ for stationary and propagating cracks in the tensile orthotropic plate with 0° material angle. Variations of the crack tip velocity for the propagation case are also presented. (Motamedi and Mohammadi, 2012). (Reproduced by permission of Elsevier.)

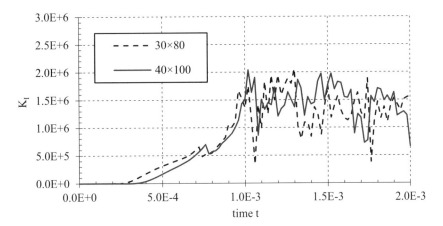

**Figure 5.21**  Comparison of the mode I stress intensity factor computed from different meshes for the 30° orthotropic problem (Motamedi and Mohammadi, 2012). (Reproduced by permission of Elsevier.)

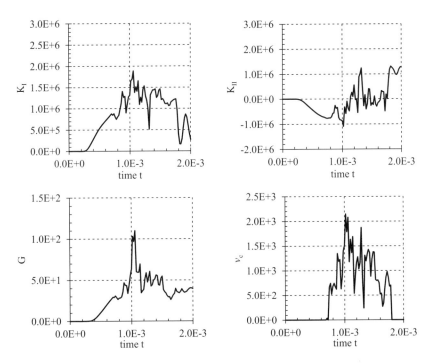

**Figure 5.22**  Variations of $K_I$, $K_{II}$, $G$ and $v_c$ for a propagating crack in the tensile 45° orthotropic plate.

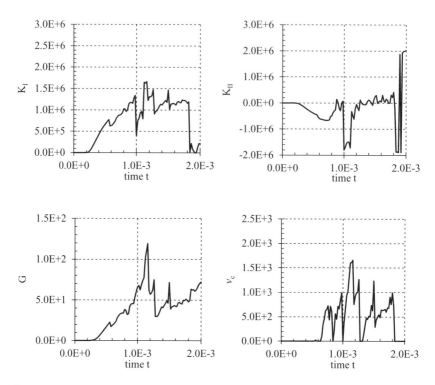

**Figure 5.23**  Variations of $K_I$, $K_{II}$, $G$ and $v_c$ for a propagating crack in the tensile $60°$ orthotropic plate.

## 5.5.4  A Composite Plate with Double Edge Cracks under Impulsive Loading

An impact by a projectile on a plate with two edge notches is now simulated by XFEM (Figure 5.24). This problem was investigated experimentally by Kalthoff and Winkler (1987) and Kalthoff (2000). Kalthoff and Winkler (1987) showed that in low speed impacts a relatively low strain rate is generated and a brittle fracture mode with a crack propagation angle of approximately $70°$ is anticipated. This benchmark test has been frequently studied by several numerical research groups, such as Belytschko and Chen (2004) and Motamedi and Mohammadi (2010a, 2012), among others.

### 5.5.4.1  Isotropic Analysis

In the first part of the study, the material is assumed to be an isotropic material, as it was in the original experiment by Kalthoff and Winkler (1987), with the material properties of Table 5.6. The results of XFEM analysis can then be compared with the results of Belytschko and Chen (2004), which were based on the loss of hyperbolicity criterion within an isotropic XFEM methodology.

A $50 \times 50$ finite element mesh is utilized to simulate one half of the plate due to the symmetry, and the radius of the $J$ integration domain is set to $r_J = 0.4a$. The timestep

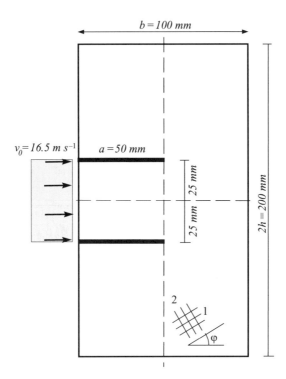

**Figure 5.24**   Geometry and boundary condition of the double edge crack in a rectangular plate.

$\Delta t = 1\,\mu s$ is considered and the simulation is performed up to $80\,\mu s$, which is similar to the time interval of the available reference results for the crack propagation speed (Figure 5.25). Belytschko and Chen (2004) reported an initiation time of $t = 26.17\,\mu s$ which is very close to the XFEM prediction at $t = 27\,\mu s$. The existing differences probably originate from the different definitions adopted for estimation of the crack speed in the two simulations.

Motamedi and Mohammadi (2010b) have reported that the crack propagation angle initiates at about $67.5°$ (very close to the results of Kalthoff and Winkler (1987) and Belytschko and Chen (2004)) and gradually decreases (with oscillations) to $59.4°$ during the crack growth process.

**Table 5.6**   Isotropic and orthotropic material properties.

|  | Isotropic beam | Orthotropic beam |
| --- | --- | --- |
| $E_1\,(\text{GPa})$ | 190 | 8.6 |
| $E_2\,(\text{GPa})$ |  | 39 |
| $G_{12}\,(\text{GPA})$ |  | 3.8 |
| $\nu_{12}$ | 0.3 | 0.061 |
| $\rho\,(\text{kgm}^{-3})$ | 8000 | 2100 |
| $K_{\text{IC}}\,(\text{MPa}\sqrt{m})$ |  | 1 |
| $c_d$ (dilational) $(\text{m s}^{-1})$ | 68 | 2071.3 |
| $c_s$(shear) $(\text{m s}^{-1})$ | 3022.4 | 1545.2 |
| $c_R$(Rayleigh) $(\text{m s}^{-1})$ | 2799.2 | 1345.6 |

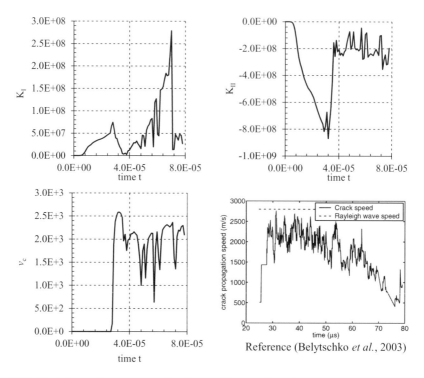

**Figure 5.25** Variations of stress intensity factors $K_I$ and $K_{II}$ and the crack propagation velocity $v_c$ (Motamedi and Mohammadi, 2010b) for the Kalthoff experiment in comparison with the reference velocity reported by Belytschko *et al.* (2003). (Reproduced by permission of Springer and John & Wiley Sons, Ltd.)

### 5.5.4.2 Orthotropic Analysis

The same problem is now considered as an orthotropic material with the dedicated material properties of Table 5.6. A similar finite element mesh and radius of the $J$ integral domain are used in this simulation, but the time-step is increased to $\Delta t = 1.5\,\mu\text{s}$. Also, the impact speed is reduced to $5\,\text{m s}^{-1}$ in order to ensure the low strain rate condition.

Figures 5.26 and 5.27 illustrate the time histories of different stress intensity factors and the crack propagation velocity for two orthotropic material angles of $\varphi = 45°$ and $90°$. According to the results, the initial crack starts to propagate just after $20$ and $23\,\mu\text{s}$ for material angles $\varphi = 45°$ and $\varphi = 90°$, respectively. The maximum crack tip speed reaches $1600\,\text{ms}^{-1}$ for $\varphi = 45°$, while it is limited to the much lower value of $1200\,\text{m s}^{-1}$ for $\varphi = 90°$.

The predicted propagation patterns for these two material angles are illustrated in Figure 5.28, which shows that the initial notch in material angle $\varphi = 45°$ propagates with an angle of $63.4°$ at the propagation initiation time and then decreases to $54.1°$ at the end of crack growth at $80\,\mu\text{s}$. In contrast, the crack propagation path remains similar for the material angle $\varphi = 90°$.

A sensitivity analysis is now performed on four different finite element meshes: $50 \times 50$, $60 \times 60$, $70 \times 70$ and $80 \times 80$. Figure 5.29 illustrates the time histories of mode I and II stress intensity factors for all meshes for the material angle of $45°$. Similar results have been reported

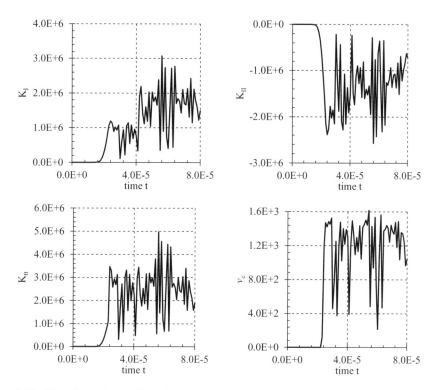

**Figure 5.26** Variations of stress intensity factors $K_I$, $K_{II}$, $K_{tt}$ and the crack propagation velocity $v_c$ for the Kalthoff experiment with $45°$ orthotropic angle.

by Motamedi and Mohammadi (2010b) with the clear indication of good agreement for crack growth path and crack tip speed in all simulations. They also concluded that, on average, the crack angle at initiation remains at about $63.4°$. Comparable results are observed for crack propagation patterns in time, which further indicate the mesh insensitivity of the results.

## 5.5.5  Pre-Cracked Three Point Bending Beam under Impact Loading

A pre-cracked three-point bending beam, previously investigated by Nishioka, Tokudome and Kinoshita (2001) with isotropic properties under a projectile impulse of $5 \text{ m s}^{-1}$, is numerically investigated, as depicted in Figure 5.30. In order to investigate various effects of mixed mode fracture, two eccentricities of loading, defined as $e = 2l_v/l_s$, are considered ($e = 0, 0.1$). The original test was conducted for an isotropic beam, however, two sets of isotropic and orthotropic analyses are performed to assess the performance of the XFEM approach. Isotropic and orthotropic material properties are defined in Table 5.7.

A $120 \times 40$ finite element mesh is chosen to discretize the model and the time-step is considered as $\Delta t = 2 \text{ μs}$.

Figures 5.31 and 5.32 illustrate the results of two eccentricity cases ($e = 0$ and $e = 0.1$), obtained by XFEM (Motamedi and Mohammadi, 2010b, 2012), in comparison with the

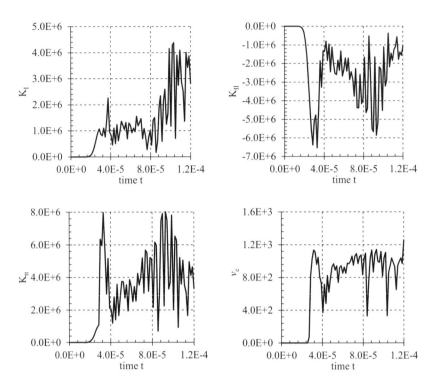

**Figure 5.27**  Variations of stress intensity factors $K_I$, $K_{II}$, $K_{tt}$ and the crack propagation velocity $v_c$ for the Kalthoff experiment with $90°$ orthotropic angle.

reference results by Nishioka, Tokudome and Kinoshita (2001). In the case of zero eccentricity $e = 0$, XFEM results closely match the reference results, whereas slightly different values with more or less similar patterns are observed in the case of eccentric loading ($e = 0.1$). As discussed by Motamedi and Mohammadi (2010b, 2012), the reason may be attributed to the different adopted propagation criteria. This is further ratified by the fact that this difference is minimized (vanished) for the symmetric condition where both propagation criteria predict a similar result.

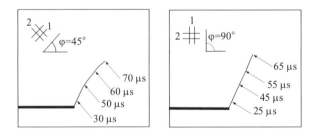

**Figure 5.28**  Crack growth paths for $45°$ and $90°$ orthotropic composite plates with the maximum hoop criterion at four different times (Motamedi and Mohammadi, 2010b). (Reproduced by permission of Springer.)

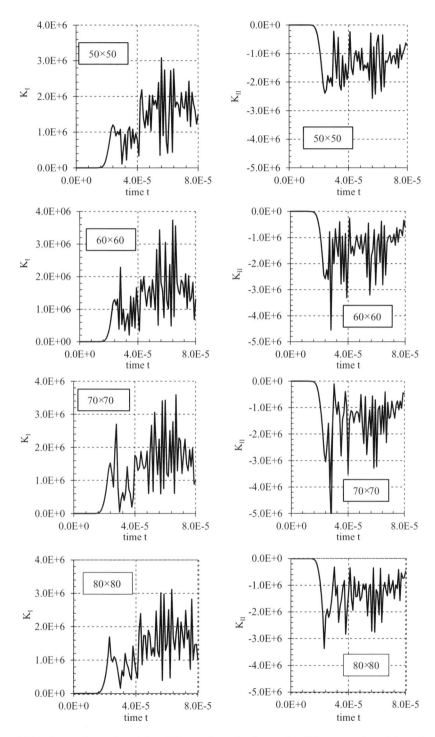

**Figure 5.29**  Comparison of mode I and II stress intensity factors for different meshes of the orthotropic plate with orientation angle of 45°.

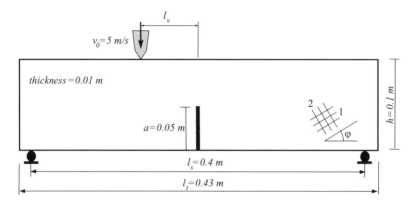

**Figure 5.30**   A three point bending specimen under impact loading (Nishioka, Tokudome and Kinoshita, 2001).

Furthermore, the simulation is extended to orthotropic problems based on the material properties of Table 5.7. Five material orthotropy angles of $\varphi = 0°$, $30°$, $45°$, $60°$ and $90°$ are considered. Figure 5.33 illustrates the variations of crack tip velocity $v_c$ and the dynamic stress intensity factor for the eccentric case of $e = 0.1$ for different orthotropic angles. The radius of the $J$ integral domain is set to $r_J = a/3$ for all simulations.

Figure 5.34 compares the different results, for eccentricities of $e = 0$ and $e = 0.1$, obtained by static and dynamic enrichment functions, (5.101) and (5.113), respectively, as reported by Motamedi and Mohammadi (2012). In the case of $e = 0$, the dynamic stress intensity factors for both cases show similar trends and the existing differences may have been caused by the effect of different incident waves on the elements inside the $J$ integral domain when they reach at different times (Motamedi and Mohammadi, 2012).

More noticeable differences are observed between the results of two enrichments for the eccentric loading. In addition, the crack-tip propagation velocity varies by the size of the $J$ integration domain and the level of finite element refinement in that domain. Also, a slightly faster propagation is observed with the dynamic enrichment simulations. Nevertheless, the general trends and crack propagation paths remain similar. After each propagation step, the energy release rate reduces to a value lower than the critical level, allowing the crack tip to

**Table 5.7**   Isotropic and orthotropic material properties.

|  | Isotropic beam | Orthotropic beam |
| --- | --- | --- |
| $E_1$ (GPa) | 2.94 | 8.6 |
| $E_2$ (GPa) |  | 39 |
| $G_{12}$ (GPA) |  | 3.8 |
| $\nu_{12}$ | 0.3 | 0.061 |
| $\rho$ (kgm$^{-3}$) | 1190 | 2100 |
| $K_{IC}$ (MPa$\sqrt{m}$) | 4.6 | 1 |
| $c_d$ (dilational) (m s$^{-1}$) | 1710 | 2071.3 |
| $c_s$ (shear) (m s$^{-1}$) | 941 | 1545.2 |
| $c_R$ (Rayleigh) (m s$^{-1}$) |  | 1345.6 |

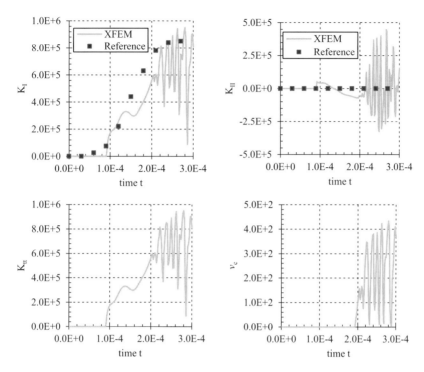

**Figure 5.31** Variations of dynamic stress intensity factors and crack tip velocity for $e = 0$ (Motamedi and Mohammadi, 2010b) in comparison with the reference results of Nishioka, Tokudome and Kinoshita (2001).

rest and remain stationary until the energy accumulates and reaches the critical value again during the upcoming time steps (Motamedi and Mohammadi, 2012).

Figure 5.35 illustrates the crack propagation path at successive times for the eccentric loading. It starts to propagate from the tip of the initial notch and quickly heads towards the point of external loading.

## 5.5.6  Propagating Central Inclined Crack in a Circular Orthotropic Plate

A circular orthotropic plate with a central inclined crack, similar to Figure 5.36 is simulated to study the effect of different crack angles on its dynamic fracture response. The orthotropic material properties are defined in Table 5.8.

The static solution has been studied by Asadpoure, Mohammadi and Vafai (2006), while Motamedi and Mohammdi (2010a, 2012) have reported dynamic XFEM solutions for stationary and propagating cracks.

An unstructured finite element mesh with 877 nodes and 852 four-noded elements is adopted to discretize the model, as depicted in Figure 5.36. $2 \times 2$ and $5 \times 5$ Gauss quadrature rules are employed for ordinary and enriched elements, respectively. The integration domain radius is $r_J = 0.3a$.

The dynamic stress intensity factors $K_I$, $K_{II}$, $K_{tt}$ and crack-tip velocity $v_c$ for different crack angles $30°$, $45°$ and $60°$ are presented in Figures 5.37, 5.38 and 5.39, respectively. According

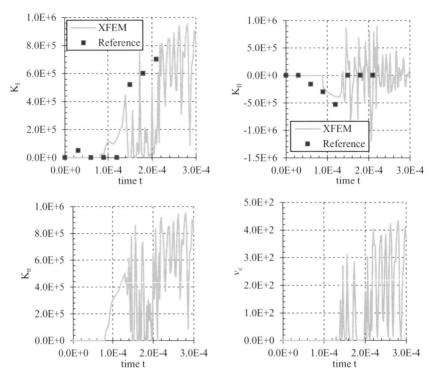

**Figure 5.32** Variations of dynamic stress intensity factors and crack-tip velocity for $e = 0.1$ (Motamedi and Mohammadi, 2010b) in comparison with the reference results of Nishioka, Tokudome and Kinoshita (2001).

**Table 5.8** Orthotropic material properties.

| | |
|---|---|
| $R$(m) | 0.05 |
| $a$(m) | 0.01 |
| $E_1$(GPa) | 8.6 |
| $E_2$(GPa) | 39 |
| $G_{12}$(GPa) | 3.8 |
| $v_{12}$ | 0.061 |
| $\rho$(kgm$^{-3}$) | 2100 |
| $K_{Ic}$(MPa$\sqrt{m}$) | 1 |
| $c_d$ (dilational) (m s$^{-1}$) | 2071.3 |
| $c_s$(shear) (m s$^{-1}$) | 1545.2 |
| $c_R$(Rayleigh) (m s$^{-1}$) | 1345.6 |

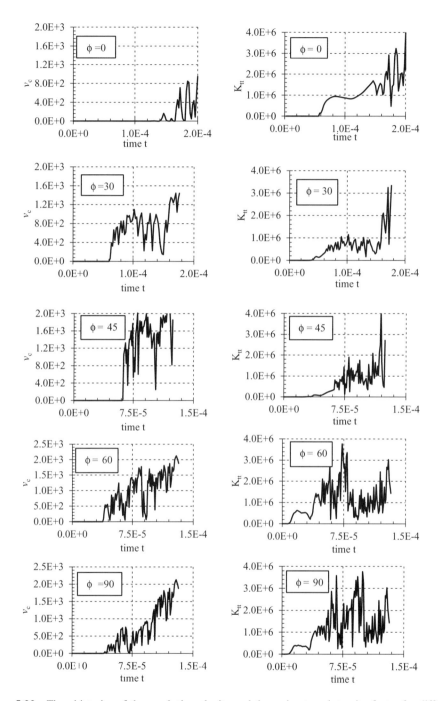

**Figure 5.33**   Time histories of the crack-tip velocity and dynamic stress intensity factor for different orthotropy angles for the eccentric loading ($e = 0.1$).

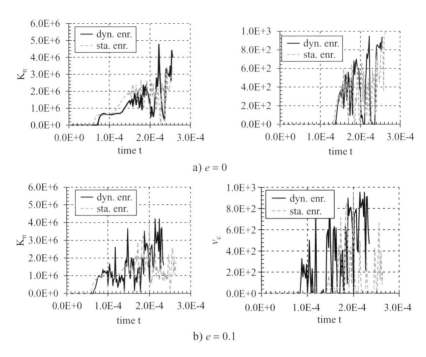

a) $e = 0$

b) $e = 0.1$

**Figure 5.34** Comparison of results obtained from static and dynamic enrichments for cases of $e = 0$ and $e = 0.1$ with the $J$ integration domain of $r_J = a/3$. (Motamedi and Mohammadi, 2012). (Reproduced by permission of Elsevier.)

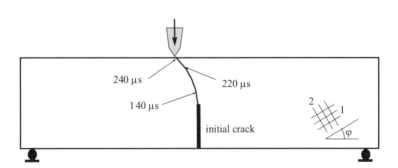

**Figure 5.35** Crack propagation path for the eccentric case of $e = 0.1$ at times $t = 140, 220, 240\,\mu s$.

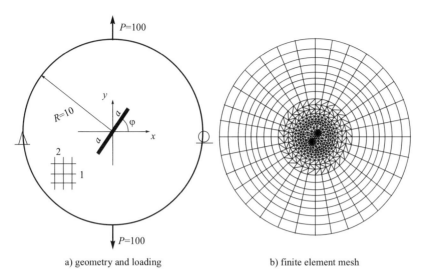

a) geometry and loading        b) finite element mesh

**Figure 5.36** Geometry and boundary conditions for an inclined central crack in a disk subjected to point loads, and the FEM discretization (Asadpoure and Mohammadi, 2007). (Reproduced by permission of John Wiley & Sons, Ltd.)

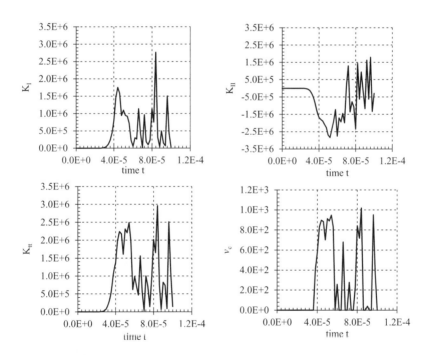

**Figure 5.37** Variations of dynamic stress intensity factors and crack-tip velocity for 30° crack angle in the orthotropic circular plate (Motamedi and Mohammadi, 2012). (Reproduced by permission of Elsevier.)

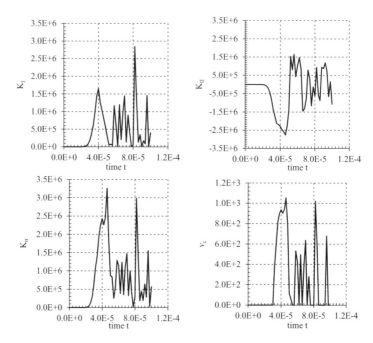

**Figure 5.38**  Variations of dynamic stress intensity factors and crack-tip velocity for a 45° crack angle in the orthotropic circular plate (Motamedi and Mohammadi, 2012). (Reproduced by permission of Elsevier.)

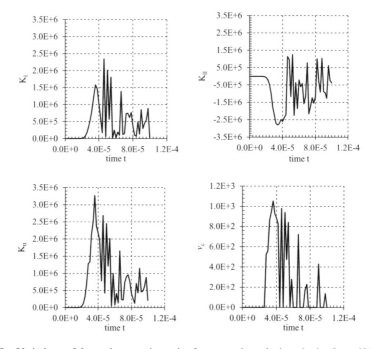

**Figure 5.39**  Variations of dynamic stress intensity factors and crack-tip velocity for a 60° crack angle in the orthotropic circular plate (Motamedi and Mohammadi, 2012). (Reproduced by permission of Elsevier.)

to these figures, the maximum crack propagation velocity and the maximum hoop stress intensity factor remain almost the same for all crack angles, but each at a different time. It is also observed that the propagation initiation time decreases with increase in the crack angle, which can be a direct consequence of reduction in the distance between the loading point and the crack tip. The same normal distance concept may also explain the decreasing trend of total crack lengths 0.0265, 0.0251 and 0.0236 m for increasing crack angles 30°, 45° and 60°, respectively (Motamedi and Mohammadi, 2012).

# 6

# Fracture Analysis of Functionally Graded Materials (FGMs)

## 6.1 Introduction

Advanced material systems in the aerospace and defence industries, microelectronics and piezoelectric applications, biotechnology, medical equipment and so on are required to withstand highly adverse operating conditions. For instance, in high temperature applications, such as combustion chambers, turbine systems and jet engines, the demand for improved thermal efficiency requires the development of increasingly more heat-resistant materials and coatings (Ozturk and Erdogan, 1997). As a result, the use of advanced materials for the purpose of shielding such components against excessive conditions is becoming almost a necessity (Houck, 1987; Batakis and Vogan, 1985; DeMasi, Sheffler and Ortiz, 1989; Meier, Nissley and Sheffler, 1991; Brindley, 1995).

Homogeneous ceramic coatings, however, suffer from a number of shortcomings, such as poor interfacial bonding strength, high thermal and residual stresses and low toughness, which may facilitate the process of unstable cracking. An alternative concept is to use coatings or interfacial zones with graded compositions, also known as functionally graded materials (FGMs), which ensures continuous variation of thermo-mechanical properties (Yamanouchi *et al.*, 1990; Holt *et al.*, 1993; Ilschner and Cherradi, 1995; Nadeau and Ferrari, 1999). Additionally, in contrast to the conventional bimaterial interfaces, the gradual change in FGM material properties seems to improve the resistance to interfacial delamination and fatigue crack growth (Takahashi *et al.*, 1993). It has been shown that grading material properties reduces the magnitude of residual stresses (Pipes and Pagano, 1970, 1974; Pagano, 1974; Lee and Erdogan, 1995), significantly increases the bonding strength (Kurihara, Sasaki and Kawarada, 1990) and improves the toughness and resilience to reduce the concentration of stress in the interface zone (Niino and Maeda, 1990).

Prescribed tailoring of the microstructure enables functionally graded materials to offer an optimized solution for a specific engineering application (Dolbow and Gosz, 2002). Nevertheless, the extent to which material properties can be designed to achieve a specific goal, such as resistance to fracture and failure patterns, is not *a priori* known. Therefore, much of the

*XFEM Fracture Analysis of Composites*, First Edition. Soheil Mohammadi.
© 2012 John Wiley & Sons, Ltd. Published 2012 by John Wiley & Sons, Ltd.

recent research has been devoted to the numerical computation of fracture parameters and the simulation of crack growth in FGMs.

The processing techniques for manufacturing the graded materials usually result in anisotropic components (Sampath *et al.*, 1995; Kaysser and Ilschner, 1995). Therefore, a non-homogeneous orthotropic elastic continuum formulation is essential for any accurate study of FGMs.

From the analytical point of view, variation of the material property affects the fracture mechanics parameters mainly in the asymptotic solution for near tip displacement and stress fields and the stress intensity factors (Erdogan, 1995). Most of the available analytical expressions for stress intensity factors for cracks in FGMs have been limited to semi-infinite or infinite domains and simple load cases. Nevertheless, it has been shown that the asymptotic crack-tip stress and displacement fields for certain classes of FGMs follow the general form of homogeneous materials (Eischen, 1983; Delale and Erdogan, 1983).

Sih and Chen (1980) studied the influence of material inhomogeneity on the behavior of a moving crack in a material whose elastic properties may differ from those of the surrounding material. Theoretical calculations showed that the energy stored in elements ahead of the crack can be raised or lowered, depending on the crack velocity, the crack length and the degree of material innhomogeneity. Also, Delale and Erdogan (1983) investigated the crack problem in isotropic FGMs by exponential variation in material properties.

Analysis of surface cracks in the graded coating bonded to a homogeneous substrate was studied by Kasmalkar (1996) for mode I and Chen and Erdogan (1996) for mixed mode problems. Analytical expressions for mixed-mode stress intensity factors have been obtained by Gu and Asaro (1997) for cases where the crack tip was oriented perpendicular to the material gradient. Also, Ozturk and Erdogan (1997) solved analytically a mode I symmetric crack in an inhomogeneous orthotropic infinite FGM problem where properties of the medium were assumed to vary monotically only in the crack direction. They illustrated that the stress intensity factors were independent of stiffness ratio and concluded that the effect of Poisson's ratio on the stress intensity factors is not generally very significant. Closed form solutions were obtained for the isotropic case. Later, they provided an analytical solution for mixed mode problems (Ozturk and Erdogan, 1997). A more general case of material gradients with respect to crack orientation was obtained for ideal infinite domains by Konda and Erdogan (1994).

The most common way of determining the mixed-mode stress intensity factors in general problems is by proper evaluation of the $J$ integral, usually in the form of equivalent domain integrals over a finite region surrounding the crack tip that avoids precise capture of the singular stress field near the crack tip. Earlier studies showed that, for some inhomogeneous materials and crack-tip orientations, evaluation of the integral on finite contours may exhibit path dependence. Gu and Asaro (1997) and Gu, Dao and Asaro (1999) proposed the use of a very fine mesh near the crack tip to reduce this dependence, while Anlas, Santare and Lambros (2000) developed a modified path-domain form of the $J$ integral. While these techniques were successful to some degree, mainly for mode I problems, the interaction integral method can be used for extraction of mixed mode stress intensity factors in linearly elastic isotropic bodies, or in bodies with smoothly varying elastic moduli.

Kim and Paulino (2002c, 2003a, 2003b) examined various methods using the $J_k$ integral, the modified crack closure method (MCC) and the displacement correlation technique to evaluate mixed mode stress intensity factors in orthotropic FGMs by the finite element approach. Later, Kim and Paulino (2005) further discussed the interaction integral method as a unified framework for evaluating fracture parameters in functionally graded materials. The method was based on a conservation integral involving auxiliary fields.

In the fracture of nonhomogeneous materials, the use of auxiliary fields developed for homogeneous materials results in violation of one of the basic relations of mechanics, that is, equilibrium, compatibility or constitutive, which naturally leads to three independent formulations: nonequilibrium, incompatibile and constant-constitutive-tensor. Each formulation leads to a specific form of the interaction integral in the sense that extra terms are added to compensate for the difference in response between homogeneous and nonhomogeneous materials. The extra terms play a key role in ensuring path independence of the interaction integral (Kim and Paulino, 2005).

The first application of XFEM for FGM problems was reported by Dolbow and Gosz (2002), who further developed the interaction energy integral method in the form of equivalent domain for the computation of mixed mode stress intensity factors at the tips of arbitrarily oriented cracks in FGMs. The necessary auxiliary stress and displacement fields were chosen from the asymptotic near tip fields for a crack in a homogeneous material having the same elastic constants as those of the crack tip in the functionally graded material. The auxiliary strain fields were obtained from the auxiliary stress fields using the constitutive relation for the functionally graded material. As a consequence of the lack of compatibility, extra terms had to be used in the domain integrals.

A number of latest XFEM and PUFEM simulations have focused on thermo-mechanical analysis of orthotropic FGMs (Dag, Yildirim and Sarikaya, 2007), modelling of 3D isotropic FGMs (Ayhan, 2009; Zhang *et al.*, 2011; Moghaddam, Ghajar and Alfano, 2011) and calculation of natural frequencies of cracked isotropic FGMs (Natarajan *et al.*, 2011). Recently, Bayesteh and Mohammadi (2012) have used XFEM with orthotropic crack-tip enrichment functions to analyse several FGM crack stability and propagation problems, based on the use of an incompatible interaction integral to determine the stress intensity factors.

## 6.2  Analytical Solution for Near a Crack Tip

Details of anisotropic elasticity, as discussed in Section 4.2.1, are also valid for inhomogeneous anisotropic materials. In this section, analytical solutions for near a crack tip in functionally graded materials are discussed.

### 6.2.1  Average Material Properties

In order to better describe and compare the behaviour of functionally graded materials, four independent average material properties, which include stiffness parameter $E$, Poison's ratio $v$, stiffness ratio $\delta^4$, and shear modulus $\kappa_0$, can be defined. For the plane stress state they are defined as:

$$E = \sqrt{E_{11}E_{22}} \tag{6.1}$$

$$v = \sqrt{v_{12}v_{21}} \tag{6.2}$$

$$\delta^4 = \frac{E_{11}}{E_{22}} = \frac{v_{12}}{v_{21}} \tag{6.3}$$

$$\kappa_0 = \frac{E}{2G_{12}} - v \tag{6.4}$$

and for a plane strain problem,

$$E = \left[ \frac{E_{11}E_{22}}{(1 - \nu_{13}\nu_{31})(1 - \nu_{23}\nu_{32})} \right]^{\frac{1}{2}} \tag{6.5}$$

$$\nu = \sqrt{ \frac{(\nu_{12} + \nu_{13}\nu_{32})(\nu_{21} + \nu_{23}\nu_{31})}{(1 - \nu_{13}\nu_{31})(1 - \nu_{23}\nu_{32})} } \tag{6.6}$$

$$\delta^4 = \frac{E_{11}(1 - \nu_{23}\nu_{32})}{E_{22}(1 - \nu_{13}\nu_{31})} \tag{6.7}$$

$$\kappa_0 = \frac{E}{2G_{12}} - \nu \tag{6.8}$$

### 6.2.2    Mode I Near Tip Fields in FGM Composites

There is very limited available literature on the analytical solution for fracture analysis in FGM composites. They are usually derived in the form of an infinite series, which cannot be transformed into an appropriate form to derive explicit enrichment functions. Here, the solution for mode I cracks, originally derived by Ozturk and Erdogan (1997), is briefly reviewed.

Consider a typical crack within a linear elastic FGM composite, with orthotropic material axes $x_1$ and $x_2$, as depicted in Figure 6.1. The crack is positioned on the $x_2 = 0$ line within the $-a < x < a$ interval. So, the crack is located in a plane perpendicular to the direction of the property gradient and the principal axes of orthotropy are parallel and perpendicular to the crack.

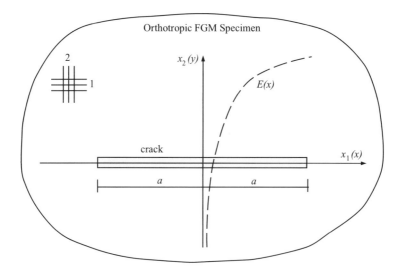

**Figure 6.1**    Definitions for an ideal mode I crack in an FGM material.

The average material properties of Section 6.2.1 are, in general, functions of $x_1$ and $x_2$. Nevertheless, the following simplifications which are in line with usual industrial applications are assumed:

- Variations are assumed to be only in the $x_1$ direction.
- Variations of the Poisson's coefficient are neglected. This is confirmed by a number of previous studies, such as Delale and Erdogan (1983) and Erdogan and Wu (1993), which will be examined by numerical simulations.
- Variations of $E_{11}$, $E_{22}$ and $G_{12}$ are assumed to be proportionally increasing.

As a result, only the elasticity modulus remains a function of $x_1$ and the remaining coefficients are assumed to be constant.

It should be mentioned that the special case of $\kappa_0 = 1$, $\delta = 1$ represents an isotropic material. For a homogenous orthotropic material, $-1 < \kappa_0 < \infty$, and it can be proved that no solution is obtained if $\kappa_0 < -1$. The same conclusions can be made for non-homogenous materials.

New coordinates are defined from the scaling factor $\delta$,

$$x = \frac{x_1}{\sqrt{\delta}} \tag{6.9}$$

$$y = \sqrt{\delta} x_2 \tag{6.10}$$

Similarly, displacement and stress components are scaled:

$$u_x(x, y) = u_1(x_1, x_2)\sqrt{\delta} \tag{6.11}$$

$$u_y(x, y) = \frac{u_2(x_1, x_2)}{\sqrt{\delta}} \tag{6.12}$$

$$\sigma_{xx}(x, y) = \frac{1}{\delta}\sigma_{11}(x_1, x_2) \tag{6.13}$$

$$\sigma_{yy}(x, y) = \delta\sigma_{22}(x_1, x_2) \tag{6.14}$$

$$\sigma_{xy}(x, y) = \sigma_{12}(x_1, x_2) \tag{6.15}$$

Stress–displacement relations can now be written as,

$$\sigma_{xx}(x, y) = \frac{\bar{E}(x, y)}{1 - \nu^2}\left(\frac{\partial u_x(x, y)}{\partial x} + \nu\frac{\partial u_y(x, y)}{\partial y}\right) \tag{6.16}$$

$$\sigma_{yy}(x, y) = \frac{\bar{E}(x, y)}{1 - \nu^2}\left(\frac{\partial u_y(x, y)}{\partial y} + \nu\frac{\partial u_x(x, y)}{\partial x}\right) \tag{6.17}$$

$$\sigma_{xx}(x, y) = \frac{\bar{E}(x, y)}{2(\kappa_0 + \nu)}\left(\frac{\partial u_x(x, y)}{\partial y} + \frac{\partial u_y(x, y)}{\partial x}\right) \tag{6.18}$$

where

$$\bar{E}(x, y) = E(x_1, x_2) \tag{6.19}$$

The equilibrium equations are then transformed to:

$$\frac{\partial^2 u_x}{\partial y^2} + \beta_1 \frac{\partial^2 u_x}{\partial x^2} + \beta_2 \frac{\partial^2 u_y}{\partial x \partial y} + \frac{\beta_1}{\bar{E}} \frac{\partial \bar{E}}{\partial x} \left( \frac{\partial u_x}{\partial x} + v \frac{\partial u_y}{\partial y} \right) + \frac{1}{\bar{E}} \frac{\partial \bar{E}}{\partial y} \left( \frac{\partial u_x}{\partial y} + \frac{\partial u_y}{\partial x} \right) = 0 \quad (6.20)$$

$$\frac{\partial^2 u_y}{\partial x^2} + \beta_1 \frac{\partial^2 u_y}{\partial y^2} + \beta_2 \frac{\partial^2 u_x}{\partial x \partial y} + \frac{\beta_1}{\bar{E}} \frac{\partial \bar{E}}{\partial y} \left( \frac{\partial u_y}{\partial y} + v \frac{\partial u_x}{\partial x} \right) + \frac{1}{\bar{E}} \frac{\partial \bar{E}}{\partial x} \left( \frac{\partial u_x}{\partial y} + \frac{\partial u_y}{\partial x} \right) = 0 \quad (6.21)$$

with

$$\beta_1 = \frac{2(\kappa_0 + v)}{1 - v^2}, \quad \beta_2 = 1 + v\beta_1, \quad (6.22)$$

$E(x, y)$ is assumed to be a monotonically increasing function of $x_1$ as:

$$E(x_1, x_2) = E(x_1) = E_0 e^{\alpha_1 x_1} = \bar{E}(x) = E_0 e^{\gamma x} \quad (6.23)$$

with

$$\gamma = \alpha_1 \sqrt{\delta} \quad (6.24)$$

and the equilibrium equations become

$$\frac{\partial^2 u_x}{\partial y^2} + \beta_1 \frac{\partial^2 u_x}{\partial x^2} + \beta_2 \frac{\partial^2 u_y}{\partial x \partial y} + \beta_1 \gamma \left( \frac{\partial u_x}{\partial x} + v \frac{\partial u_y}{\partial y} \right) = 0 \quad (6.25)$$

$$\frac{\partial^2 u_y}{\partial x^2} + \beta_1 \frac{\partial^2 u_y}{\partial y^2} + \beta_2 \frac{\partial^2 u_x}{\partial x \partial y} + \gamma \left( \frac{\partial u_x}{\partial y} + \frac{\partial u_y}{\partial x} \right) = 0 \quad (6.26)$$

A homogeneous differential equation is then obtained if $\gamma = 0$.

Following the formulation of Ozturk and Erdogan (1997), the solutions are assumed to be in the following general forms (due to symmetry, only $x_2 > 0$ region is considered):

$$u_x(x, y) = \frac{1}{2\pi} \int_{-\infty}^{\infty} \sum_{j=3}^{4} b_j(k) B_j(k) e^{\lambda_j y - ikx} dk \quad (6.27)$$

$$u_y(x, y) = \frac{1}{2\pi} \int_{-\infty}^{\infty} \sum_{j=3}^{4} B_j(k) e^{\lambda_j y - ikx} dk \quad 0 < y < \infty \quad (6.28)$$

$$\sigma_{xx}(x, y) = \frac{\bar{E}(x)}{1 - v^2} \frac{1}{2\pi} \int_{-\infty}^{\infty} \sum_{j=3}^{4} (v\lambda_j - ib_j k) B_j e^{\lambda_j y - ikx} dk \quad (6.29)$$

$$\sigma_{yy}(x, y) = \frac{\bar{E}(x)}{1 - v^2} \frac{1}{2\pi} \int_{-\infty}^{\infty} \sum_{j=3}^{4} (\lambda_j - ivb_j k) B_j e^{\lambda_j y - ikx} dk \quad (6.30)$$

$$\sigma_{xy}(x, y) = \frac{\bar{E}(x)}{2(\kappa_0 + v)} \frac{1}{2\pi} \int_{-\infty}^{\infty} \sum_{j=3}^{4} (\lambda_j b_j - ik) B_j e^{\lambda_j y - ikx} dk, \quad 0 < y < \infty \quad (6.31)$$

where $\lambda_i$ are the roots of the characteristic equation

$$\lambda^4 - (\nu\gamma^2 + 2(k^2 + i\gamma k)\kappa_0)\lambda^2 + (k^2 + i\gamma k)^2 = 0 \tag{6.32}$$

or more explicitly,

$$\lambda_1 = -\lambda_3 = \left\{ \frac{1}{2}\left(\omega^2 + 2\kappa_0\eta\right) + \frac{1}{2}\sqrt{\left(\omega^2 + 2\kappa_0\eta\right)^2 - 4\eta^2} \right\}^{\frac{1}{2}} \tag{6.33}$$

$$\lambda_2 = -\lambda_4 = \left\{ \frac{1}{2}\left(\omega^2 + 2\kappa_0\eta\right) - \frac{1}{2}\sqrt{\left(\omega^2 + 2\kappa_0\eta\right)^2 - 4\eta^2} \right\}^{\frac{1}{2}} \tag{6.34}$$

with

$$\omega^2 = \nu\gamma^2, \quad \eta = k^2 + i\gamma k \tag{6.35}$$

and

$$b_j(k) = \frac{(i\beta_2 k - \gamma\beta_1\nu)}{\lambda_j^2 - \beta_1\eta}\lambda_j \tag{6.36}$$

$B_3(k)$ and $B_4(k)$ should be determined from the boundary conditions. After lengthy calculations, the following set of simultaneous equations is obtained,

$$\sum_{j=3}^{4} (\lambda_j b_j - ik)B_j = 0 \tag{6.37}$$

$$(-ik)(B_3 + B_4) = \int_{-\bar{a}}^{\bar{a}} \frac{\partial u_y(s, 0^+)}{\partial s} e^{iks} ds \tag{6.38}$$

where

$$\bar{a} = \frac{a}{\sqrt{\delta}} \tag{6.39}$$

The final solutions of the displacement field of the crack face $u_2$ and the stress intensity factors $K_I$ at the two crack tips are (Ozturk and Erdogan, 1997),

$$\frac{u_2(x_1, 0^+)}{\omega_0} = -\sqrt{1 - r^2}\left[\sum_{n=1}^{\infty} \frac{A_n}{n} U_{n-1}(r)\right] \tag{6.40}$$

$$\frac{K_I(a)}{K_0} = -e^{\alpha_1 a}\sum_{n=1}^{\infty} A_n \tag{6.41}$$

$$\frac{K_I(-a)}{K_0} = e^{-\alpha_1 a}\sum_{n=1}^{\infty} (-1)^n A_n \tag{6.42}$$

where $p_0$ is a constant corresponding to the amplitude of the external loading and

$$K_0 = p_0\sqrt{a} \tag{6.43}$$

$$r = x_1/a \tag{6.44}$$

$$\omega_0 = \frac{(s_1 + s_2)\delta p_0 a}{E_0} \tag{6.45}$$

$$\kappa_1 = \sqrt{\kappa_0^2 - 1}, \quad s_1 = \sqrt{\kappa_0 + \kappa_1}, \quad s_2 = \sqrt{\kappa_0 - \kappa_1} \tag{6.46}$$

and $A_n$ and $U_n$ are determined from

$$\sum_{n=1}^{\infty} A_n \left[ U_{n-1}(r) + R_n(r) \right] = e^{-\alpha_1 ar} f(r) \tag{6.47}$$

$$U_n(t) = \frac{\sin(n+1)\theta}{\sin\theta}, \quad t = \cos\theta = t_1/a, \quad s = t_1/\sqrt{\delta} \tag{6.48}$$

where

$$R_n(r) = \frac{1}{\pi} \int_{-1}^{1} \frac{a\cos(n\theta)M_1(ra, ta)}{\sqrt{1 - t^2}} dt \tag{6.49}$$

$$f(r) = \frac{p(x_1)}{p_0} \tag{6.50}$$

$p(x_1)$ is the crack boundary conditions $(\sigma_{22}(x_1, 0^+) = p(x_1), -a < x_1 < a)$ and

$$M_1(x_1, t_1) = \frac{1}{2\sqrt{\delta}} L_1\left(\frac{x_1}{\sqrt{\delta}}, \frac{t_1}{\sqrt{\delta}}\right) = \frac{1}{2\sqrt{\delta}} L_1(x, s) \tag{6.51}$$

where $L_1$ is the Fredholm kernel,

$$L_1(x, s) = \int_0^{\infty} 2\{\operatorname{Re}(h(k))\cos(k(s - x)) - \operatorname{Im}(h(k))\sin(k(s - x))\} dk \tag{6.52}$$

with

$$h(k) = \begin{cases} -i\left[\dfrac{(s_1 + s_2)(k + i\gamma)}{\lambda_1 + \lambda_2} - 1\right] & \kappa_0 \neq 1 \\[4mm] -i\left[\dfrac{2(k + i\gamma)}{\sqrt{\nu\gamma^2 + 4(k^2 + i\gamma k)}} - 1\right] & \kappa_0 = 1 \end{cases} \tag{6.53}$$

In real materials, $s_1 + s_2$ is a real number. For inhomogenous isotropic materials with a constant Poisson's ratio:

$$\delta^4 = \frac{E_{11}}{E_{22}} = 1, \quad \kappa_0 = 1, \quad \bar{a} = a, \quad x = x_1, \quad s_1 = 1, \quad s_2 = 1 \tag{6.54}$$

the following simplified Fredholm kernel is obtained,

$$L_1(x, s) = \gamma e^{\xi} \left\{ \nu_0 \frac{|\xi|}{\xi} K_1(\nu_0 |\xi|) + K_0(\nu_0 |\xi|) \right\} - \frac{2}{s - x} \tag{6.55}$$

where $K_0(z)$ and $K_1(z)$ are the modified Bessel functions (Clarke and Hess, 1978; Watson, 1995),

$$K_\alpha(z) = \int_0^\infty e^{-z\cosh(t)} \cosh(\alpha t)\,dt, \quad \alpha = 1, 2 \tag{6.56}$$

and

$$\xi = \frac{\gamma}{2}(s - x) = \frac{\alpha_1}{2}(t_1 - x_1) \tag{6.57}$$

$$\nu_0 = \begin{cases} \sqrt{1 + \nu} & \text{plane stress} \\ \sqrt{\dfrac{1}{1 - \nu}} & \text{plane strain} \end{cases} \tag{6.58}$$

$M_1$ can then be further simplified for isotropic materials,

$$M_1(x_1, t_1) = \frac{\alpha_1 e^\xi}{2} \left\{ \nu_0 \frac{|\xi|}{\xi} K_1\left(\nu_0 |\xi|\right) + K_0\left(\nu_0 |\xi|\right) \right\} - \frac{1}{t_1 - x_1} \tag{6.59}$$

Kernel $M_1(x_1, t_1)$ depends on the non-homogenous parameter $\alpha_1$ and the elastic coefficients $\nu$ and $\kappa_0$, but is independent of $\delta$ and $E$. $s_1$ and $s_2$ depend only on the shear modulus $\kappa_0$. It should also be noted that coefficients $A_n$ depend on the inhomogeneity parameter $\alpha_1$ and the elasticity parameters $\nu$ and $\kappa_0$, but are independent of $\delta$ and $E_0$. Ozturk and Erdogan (1997) have discussed that, under the present assumptions for variations of material properties, the stress intensity factors and $\sigma_{22}(x_1, 0)$ are independent of $\delta$ and $E_0$. As a result a $90°$ rotation of the material axis, that is, replacing $E_{22}$ by $E_{11}$ and $\nu_{12}$ by $\nu_{21}$, does not change $\sigma_{22}(x_1, 0)$ and the stress intensity factors. This is also true in homogeneous orthotropic problems.

## 6.2.3    Stress and Displacement Field (Similar to Homogeneous Orthotropic Composites)

The asymptotic stress and displacement crack-tip fields were derived by Sih, Paris and Irwin (1965) (similar to the solution of homogeneous orthotropic materials). Figure 6.2 defines the global coordinate system $(X, Y)$, the local crack-tip system $(x, y)$ and the local crack tip polar coordinate system $(r, \theta)$ $(x + iy = re^{i\theta})$.

It should be noted that in FGM materials $c_{ij}$ are different from one point to another, which leads to different values of $p_k$, $q_k$ and $s_k$ at different points. For this reason, material properties for the auxiliary field (in the contour integral) and crack-tip enrichment functions are calculated at the crack tip. Consequently, the following replacement is adopted in Section 4.3.1 for FGM materials

$$\chi_k \rightarrow \chi_k^{tip}, \quad \chi_k = c_{ij}, p_k, q_k, s_k \tag{6.60}$$

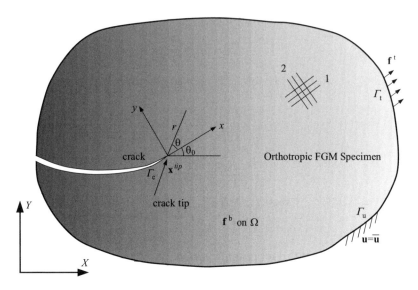

**Figure 6.2**  Global and local crack-tip coordinate systems in Cartesian and polar forms for an arbitrary orthotropic body.

The displacement fields $u_1$ and $u_2$, in the $x$ and $y$ directions, respectively, are defined as,

$$u_1 = K_\mathrm{I} \sqrt{\frac{2r}{\pi}} \mathrm{Re} \left\{ \frac{1}{s_1^\mathrm{tip} - s_2^\mathrm{tip}} \left[ s_1^\mathrm{tip} p_2^\mathrm{tip} g_2(\theta) - s_2^\mathrm{tip} p_1^\mathrm{tip} g_1(\theta) \right] \right\}$$

$$+ K_\mathrm{II} \sqrt{\frac{2r}{\pi}} \mathrm{Re} \left\{ \frac{1}{s_1^\mathrm{tip} - s_2^\mathrm{tip}} \left[ p_2^\mathrm{tip} g_2(\theta) - p_1^\mathrm{tip} g_1(\theta) \right] \right\} \tag{6.61}$$

$$u_2 = K_\mathrm{I} \sqrt{\frac{2r}{\pi}} \mathrm{Re} \left\{ \frac{1}{s_1^\mathrm{tip} - s_2^\mathrm{tip}} \left[ s_1^\mathrm{tip} q_2^\mathrm{tip} g_2(\theta) - s_2^\mathrm{tip} q_1^\mathrm{tip} g_1(\theta) \right] \right\}$$

$$+ K_\mathrm{II} \sqrt{\frac{2r}{\pi}} \mathrm{Re} \left\{ \frac{1}{s_1^\mathrm{tip} - s_2^\mathrm{tip}} \left[ q_2^\mathrm{tip} g_2(\theta) - q_1^\mathrm{tip} g_1(\theta) \right] \right\} \tag{6.62}$$

where

$$g_i(\theta) = \sqrt{\cos(\theta) + s_i^\mathrm{tip} \sin(\theta)} \quad (i = 1, 2) \tag{6.63}$$

$$p_k^\mathrm{tip} = c_{11}^\mathrm{tip} \left( s_k^\mathrm{tip} \right)^2 + c_{12}^\mathrm{tip} - c_{16}^\mathrm{tip} s_k^\mathrm{tip} \quad (k = 1, 2) \tag{6.64}$$

$$q_k^\mathrm{tip} = c_{12}^\mathrm{tip} s_k^\mathrm{tip} + \frac{c_{22}^\mathrm{tip}}{s_k^\mathrm{tip}} - c_{26}^\mathrm{tip} \quad (k = 1, 2) \tag{6.65}$$

$c_{ij}^{tip}$ are the compliance components defined in (4.8) and $s_k^{tip}$ are the roots of the characteristic equation (4.23), all evaluated at the crack tip position. Also, the asymptotic stress components are written as

$$
\sigma_{11} = \frac{K_I}{\sqrt{2\pi r}} \text{Re} \left\{ \frac{s_1^{tip} s_2^{tip}}{s_1^{tip} - s_2^{tip}} \left[ \frac{s_2^{tip}}{g_2(\theta)} - \frac{s_1^{tip}}{g_1(\theta)} \right] \right\}
$$

$$
+ \frac{K_{II}}{\sqrt{2\pi r}} \text{Re} \left\{ \frac{1}{s_1^{tip} - s_2^{tip}} \left[ \frac{(s_2^{tip})^2}{g_2(\theta)} - \frac{(s_1^{tip})^2}{g_1(\theta)} \right] \right\} \tag{6.66}
$$

$$
\sigma_{22} = \frac{K_I}{\sqrt{2\pi r}} \text{Re} \left\{ \frac{1}{s_1^{tip} - s_2^{tip}} \left[ \frac{s_1^{tip}}{g_2(\theta)} - \frac{s_2^{tip}}{g_1(\theta)} \right] \right\}
$$

$$
+ \frac{K_{II}}{\sqrt{2\pi r}} \text{Re} \left\{ \frac{1}{s_1^{tip} - s_2^{tip}} \left[ \frac{1}{g_2(\theta)} - \frac{1}{g_1(\theta)} \right] \right\} \tag{6.67}
$$

$$
\sigma_{12} = \frac{K_I}{\sqrt{2\pi r}} \text{Re} \left\{ \frac{s_1^{tip} s_2^{tip}}{s_1^{tip} - s_2^{tip}} \left[ \frac{1}{g_1(\theta)} - \frac{1}{g_2(\theta)} \right] \right\}
$$

$$
+ \frac{K_{II}}{\sqrt{2\pi r}} \text{Re} \left\{ \frac{1}{s_1^{tip} - s_2^{tip}} \left[ \frac{s_1^{tip}}{g_1(\theta)} - \frac{s_2^{tip}}{g_2(\theta)} \right] \right\} \tag{6.68}
$$

## 6.3　Stress Intensity Factor

### 6.3.1　J Integral

The $J$ integral has been frequently used to compute the stress intensity factors in various mixed mode fracture problems. The general form of path independent $J$ integral can be written as (2.153),

$$
J = \int_\Gamma \left( w_s \delta_{1j} - \sigma_{ij} \frac{\partial u_i}{\partial x_1} \right) n_j \mathrm{d}\Gamma \tag{6.69}
$$

where $\Gamma$ is an arbitrary contour surrounding the crack tip, $n_j$ is the jth component of the outward unit normal to $\Gamma$, $\delta_{ij}$ is the Kronecker delta and the Cartesian coordinate system whose x axis is parallel to the crack surface is considered. $w_s$ is the strain energy density, defined in (2.154).

Using the equivalent domain integral formulation, Eq. (6.69) for inhomogeneous materials is transformed into (Kim and Paulino, 2003a),

$$
J = \int_A (\sigma_{ij} u_{i,1} - w_s \delta_{1j}) q_{,j} \mathrm{d}A + \int_A (\sigma_{ij} u_{i,1} - w_s \delta_{1j})_{,j} q \mathrm{d}A \tag{6.70}
$$

where $q$ is a smooth function varying from $q = 1$ on the interior boundary of $A$ to $q = 0$ on the outer one.

## 6.3.2   Interaction Integral

The interaction or $M$ integral is a powerful and accurate approach, based on superimposition of auxiliary and actual fields, for evaluation of stress intensity factors. Yau, Wang and Corten (1980) used the $M$ integral to evaluate the stress intensity factors in isotropic homogeneous materials. Wang, Yau and Corten (1980) extended the method to orthotropic homogeneous materials. Then, Yau (1979) used the same approach to compute SIFs at the interface of two materials.

The classical form of the interaction integral for computation of mode I and II stress intensity factors is based on the superposition of auxiliary and actual fields,

$$J^s = J + J^{\mathrm{aux}} + M^l \tag{6.71}$$

where $J$ and $J^{\mathrm{aux}}$ are the $J$ integrals corresponding to actual and auxiliary fields, respectively. $J^{\mathrm{aux}}$ is defined by

$$J^{\mathrm{aux}} = \int_A \left( \sigma_{ij}^{\mathrm{aux}} u_{i,1}^{\mathrm{aux}} - w_s^{\mathrm{aux}} \delta_{1j} \right) q_{,j} \mathrm{d}A + \int_A \left( \sigma_{ij}^{\mathrm{aux}} u_{i,1}^{\mathrm{aux}} - w_s^{\mathrm{aux}} \delta_{1j} \right)_{,j} q \mathrm{d}A \tag{6.72}$$

and $M^l$ is the local interaction integral calculated by

$$M^l = \int_A \left\{ \sigma_{ij} u_{i,1}^{\mathrm{aux}} + \sigma_{ij}^{\mathrm{aux}} u_{i,1} - \frac{1}{2} \left( \sigma_{ik} \varepsilon_{ik}^{\mathrm{aux}} + \sigma_{ik}^{\mathrm{aux}} \varepsilon_{ik} \right) \delta_{1j} \right\} q_{,j} \mathrm{d}A$$

$$+ \int_A \left\{ \sigma_{ij} u_{i,1}^{\mathrm{aux}} + \sigma_{ij}^{\mathrm{aux}} u_{i,1} - \frac{1}{2} \left( \sigma_{ik} \varepsilon_{ik}^{\mathrm{aux}} + \sigma_{ik}^{\mathrm{aux}} \varepsilon_{ik} \right) \delta_{1j} \right\}_{,j} q \mathrm{d}A \tag{6.73}$$

The energy release rate in elastic media is calculated as

$$G = J = t_{11} K_{\mathrm{I}}^2 + t_{12} K_{\mathrm{I}} K_{\mathrm{II}} + t_{22} K_{\mathrm{II}}^2 \tag{6.74}$$

where $t_{ij}$ are defined in (4.83)–(4.85). Using the same methodology discussed in Section 4.4.3 leads to the following linear algebraic equations to calculate actual mode I and II stress intensity factors from the local interaction integral $M^l$,

$$M^l \left( K_{\mathrm{I}}^{\mathrm{aux}} = 1, K_{\mathrm{II}}^{\mathrm{aux}} = 0 \right) = 2 t_{11} K_{\mathrm{I}} + t_{12} K_{\mathrm{II}} \tag{6.75}$$

$$M^l \left( K_{\mathrm{I}}^{\mathrm{aux}} = 0, K_{\mathrm{II}}^{\mathrm{aux}} = 1 \right) = t_{12} K_{\mathrm{I}} + 2 t_{22} K_{\mathrm{II}} \tag{6.76}$$

## 6.3.3   FGM Auxillary Fields

Most of the available research for FGMs is based on the assumption of isotropic FGMs. It includes XFEM by Dolbow and Gosz (2002), EFG by Rao and Rahman (2003) and FEM by Kim and Paulino (2003a, 2003b), among others. It is either based on incompatibility or constant constitutive tensor formulations. According to the discussions of Dolbow and Gosz (2002) for isotropic FGMs and Kim and Paulino (2003c) for orthotropic FGMs, the asymptotic crack-tip fields in homogeneous orthotropic media can be used as the auxiliary fields for computation of SIFs in orthotropic FGM problems.

The governing equations for an elasticity problem are equilibrium equations, consistency conditions, and the constitutive relations. Unfortunately, if the analytical asymptotic fields for orthotropic materials are used for definition of auxiliary fields in fracture analysis of FGM composites, one of the main three governing equations may not be satisfied.

Based on which set of equations are ignored in defining the auxiliary fields, three different methods have been proposed: non-equilibrium formulation, incompatibility formulation and constant-constitutive-tensor formulation.

The first method only satisfies the constitutive and compatibility equations

$$\sigma_{ij}^{aux} = D_{ijkl}(\mathbf{x})\varepsilon_{kl}^{aux}, \quad \varepsilon_{ij}^{aux} = \frac{1}{2}\left(u_{i,j}^{aux} + u_{j,i}^{aux}\right), \quad \sigma_{ij,j}^{aux} \neq 0 \tag{6.77}$$

where $D_{ijkl}$ are the components of the material elasticity tensor, defined in (4.3). In the second approach, only the equilibrium and the constitutive equations are satisfied,

$$\sigma_{ij}^{aux} = D_{ijkl}(\mathbf{x})\varepsilon_{kl}^{aux}, \quad \varepsilon_{ij}^{aux} \neq \frac{1}{2}\left(u_{i,j}^{aux} + u_{j,i}^{aux}\right), \quad \sigma_{ij,j}^{aux} = 0 \tag{6.78}$$

and the constant-constitutive-tensor only satisfies equilibrium and compatibility

$$\sigma_{ij}^{aux} \neq D_{ijkl}(\mathbf{x})\varepsilon_{kl}^{aux}, \quad \varepsilon_{ij}^{aux} = \frac{1}{2}\left(u_{i,j}^{aux} + u_{j,i}^{aux}\right), \quad \sigma_{ij,j}^{aux} = 0 \tag{6.79}$$

Table 6.1 briefly compares the three approaches.

### 6.3.3.1   Non-Equilibrium Formulation

In this method, consistency and constitutive laws are satisfied, while equilibrium is violated. From the constitutive law,

$$\sigma_{ij}^{aux} = D_{ijkl}(\mathbf{x})\varepsilon_{kl}^{aux} \tag{6.80}$$

and its differentiation,

$$\begin{aligned}
\sigma_{ij,j}^{aux} &= D_{ijkl,j}(\mathbf{x})\varepsilon_{kl}^{aux} + D_{ijkl}(\mathbf{x})\varepsilon_{kl,j}^{aux} \\
&= \left[D_{ijkl}^{tip}\varepsilon_{kl,j}^{aux}\right] + \left[D_{ijkl,j}(\mathbf{x})\varepsilon_{kl}^{aux}\right] + \left[\left(D_{ijkl}(\mathbf{x}) - D_{ijkl}^{tip}\right)\varepsilon_{kl,j}^{aux}\right]
\end{aligned} \tag{6.81}$$

**Table 6.1**   Auxiliary fields for evaluation of the interaction integral, reproduced from Kim and Paulino (2003c).

| Auxiliary fields | Formulation | | |
| --- | --- | --- | --- |
| | Non-equilibrium | Incompatible | Constant $D_{ijkl}^{tip}$ |
| displacement | $u^{aux}$ | $u^{aux}$ | $u^{aux}$ |
| second variable | $\varepsilon_{ij}^{aux} = \frac{1}{2}\left(u_{i,j}^{aux} + u_{j,i}^{aux}\right)$ | $\sigma_{ij}^{aux}$ | $\varepsilon_{ij}^{aux} = \frac{1}{2}\left(u_{i,j}^{aux} + u_{j,i}^{aux}\right)$ |
| third variable | $\sigma_{ij}^{aux} = D_{ijkl}(\mathbf{x})\varepsilon_{kl}^{aux}$ | $\varepsilon_{ij}^{aux} = D_{ijkl}^{-1}(\mathbf{x})\sigma_{kl}^{aux}$ | $\sigma_{ij}^{aux} = D_{ijkl}^{tip}\varepsilon_{kl}^{aux}$ |
| violated equation | $\sigma_{ij,j}^{aux} \neq 0$ | $\varepsilon_{ij}^{aux} \neq \frac{1}{2}\left(u_{i,j}^{aux} + u_{j,i}^{aux}\right)$ | variable $D_{ijkl}(\mathbf{x})$ |

where the first bracket vanishes. Clearly, it does not satisfy the equilibrium equation $\sigma_{ij,j}^{\text{aux}} \neq 0$. Such a choice for the auxiliary field was adopted by Dolbow and Gosz (2002) to solve isotropic FGM problems.

The resultant $M$ integral is then defined as,

$$M^l = \int_A \left\{ \sigma_{ij} u_{i,1}^{\text{aux}} + \sigma_{ij}^{\text{aux}} u_{i,1} - \sigma_{ik} \varepsilon_{ik}^{\text{aux}} \delta_{1j} \right\} q_{,j} \mathrm{d}A$$

$$+ \int_A \left\{ \sigma_{ij,j}^{\text{aux}} u_{i,1} - \mathsf{D}_{ijkl,1} \varepsilon_{kl} \varepsilon_{ij}^{\text{aux}} \right\} q \mathrm{d}A \tag{6.82}$$

where the first term of the second integral ruins the equilibrium due to inhomogeneity of the FGM composite material.

### 6.3.3.2 Incompatible Formulation

In this method, equilibrium equations and constitutive laws are satisfied, but the compatibility is violated. The resultant $M$ integral is then defined as,

$$M^l = \int_A \left\{ \sigma_{ij} u_{i,1}^{\text{aux}} + \sigma_{ij}^{\text{aux}} u_{i,1} - \sigma_{ik} \varepsilon_{ik}^{\text{aux}} \delta_{1j} \right\} q_{,j} \mathrm{d}A$$

$$+ \int_A \left\{ \sigma_{ij} \left( u_{i,1j}^{\text{aux}} - \varepsilon_{ij,1}^{\text{aux}} \right) - \mathsf{D}_{ijkl,1} \varepsilon_{kl} \varepsilon_{ij}^{\text{aux}} \right\} q \mathrm{d}A \tag{6.83}$$

Again, the first term of the second integral, which is due to inhomogeneity of FGM, leads to inconsistency of the formulation.

### 6.3.3.3 Constant-Constitutive-Tensor

Finally, the third approach satisfies equilibrium and consistency conditions, but the constitutive laws are not satisfied. The $M$ integral of this formulation can be written as,

$$M^l = \int_A \left\{ \sigma_{ij} u_{i,1}^{\text{aux}} + \sigma_{ij}^{\text{aux}} u_{i,1} - \frac{1}{2} \left( \sigma_{ik} \varepsilon_{ik}^{\text{aux}} + \sigma_{ik}^{\text{aux}} \varepsilon_{ik} \right) \delta_{1j} \right\} q_{,j} \mathrm{d}A$$

$$+ \int_A \frac{1}{2} \left\{ \sigma_{ij} \varepsilon_{ij,1}^{\text{aux}} - \sigma_{ij,1} \varepsilon_{ij}^{\text{aux}} + \sigma_{ij}^{\text{aux}} \varepsilon_{ij,1} - \sigma_{ij,1}^{\text{aux}} \varepsilon_{ij} \right\} q \mathrm{d}A \tag{6.84}$$

It should be mentioned that, in this method, derivatives of real stress and strain fields should be computed. As a result, less accurate results are expected even though more computational effort is required. In contrast, the first two approaches are faster and easier ways of computing the $M$ integral, and both require similar derivatives of the material modulus.

While the constant-constitutive-tensor formulation needs derivatives of stress and strain in actual fields, which leads to inaccuracies in the $C^0$ finite element formulation, the two other cases have more or less the same level of accuracy (Kim and Paulino, 2005).

### 6.3.3.4  Further Simplifications

The incompatibility formulation, which requires less complicated derivatives, is further examined. Beginning with the incompatible formulation (6.83), the following relations (Kim and Paulino, 2003c)

$$\sigma_{ij}^{\text{aux}} u_{i,1j} = \sigma_{ik}^{\text{aux}} \varepsilon_{ij,1} \tag{6.85}$$

$$\sigma_{ij} \varepsilon_{ij}^{\text{aux}} = \sigma_{ik}^{\text{aux}} \varepsilon_{ij} \tag{6.86}$$

$$\mathsf{D}_{ijkl,1} \varepsilon_{ij}^{\text{aux}} \varepsilon_{kl} = \mathsf{D}_{ijkl,1} \varepsilon_{ij} \varepsilon_{kl}^{\text{aux}} \tag{6.87}$$

can be used to simplify $M^l$

$$M^l = \int_A \left\{ \sigma_{ij} u_{i,1}^{\text{aux}} + \sigma_{ij}^{\text{aux}} u_{i,1} - \sigma_{ik} \varepsilon_{ik}^{\text{aux}} \delta_{1j} \right\} q_{,j} dA$$

$$+ \int_A \left\{ \sigma_{ij}^{\text{aux}} \left( u_{i,1j}^{\text{aux}} - \varepsilon_{ij,1}^{\text{aux}} \right) - \mathsf{D}_{ijkl,1} \varepsilon_{kl} \varepsilon_{ij}^{\text{aux}} \right\} q \, dA \tag{6.88}$$

Evaluation of $\sigma_{ij} \varepsilon_{ij,1}^{\text{aux}}$ can be simplified by the method proposed by Kim and Paulino (2004),

$$\sigma_{ij} \varepsilon_{ij,1}^{\text{aux}} = \sigma_{ij} \left[ \mathsf{C}_{ijkl}(\mathbf{x}) \sigma_{kl}^{\text{aux}} \right]_{,1}$$

$$= \sigma_{ij} \left\{ \mathsf{C}_{ijkl,1}(\mathbf{x}) \sigma_{kl}^{\text{aux}} + \mathsf{C}_{ijkl}(\mathbf{x}) \sigma_{kl,1}^{\text{aux}} \right\}$$

$$= \sigma_{ij} \mathsf{C}_{ijkl,1}(\mathbf{x}) \sigma_{kl}^{\text{aux}} + \sigma_{ij} \left( \mathsf{C}_{ijkl}(\mathbf{x}) - \mathsf{C}_{ijkl}^{\text{tip}} \right) \sigma_{kl,1}^{\text{aux}} + \sigma_{ij} \underbrace{\mathsf{C}_{ijkl}^{\text{tip}} \sigma_{kl,1}^{\text{aux}}}_{\varepsilon_{ij,1}^{\text{tip}}} \tag{6.89}$$

$$= \sigma_{ij} \mathsf{C}_{ijkl,1}(\mathbf{x}) \sigma_{kl}^{\text{aux}} + \sigma_{ij} \left( \mathsf{C}_{ijkl}(\mathbf{x}) - \mathsf{C}_{ijkl}^{\text{tip}} \right) \sigma_{kl,1}^{\text{aux}} + \sigma_{ij} u_{i,1j}^{\text{aux}}$$

where $\mathsf{C}_{ijkl}$ are the components of the compliance tensor, defined in (4.4). Therefore,

$$\sigma_{ij} \left( u_{i,1j}^{\text{aux}} - \varepsilon_{ij,1}^{\text{aux}} \right) = -\sigma_{ij} \mathsf{C}_{ijkl,1}(\mathbf{x}) \sigma_{kl}^{\text{aux}} - \sigma_{ij} \left( \mathsf{C}_{ijkl}(\mathbf{x}) - \mathsf{C}_{ijkl}^{\text{tip}} \right) \sigma_{kl,1}^{\text{aux}} \tag{6.90}$$

The term $\sigma_{ij} \mathsf{C}_{ijkl,1}(\mathbf{x}) \sigma_{kl}^{\text{aux}}$ vanishes if $r \to 0$, as a result: (Kim and Paulino, 2004)

$$\left\{ \sigma_{ij} \left( u_{i,1j}^{\text{aux}} - \varepsilon_{ij,1}^{\text{aux}} \right) \right\}_{r \to 0} \cong -\sigma_{ij} \left( \mathsf{C}_{ijkl}(\mathbf{x}) - \mathsf{C}_{ijkl}^{\text{tip}} \right) \sigma_{kl,1}^{\text{aux}} \tag{6.91}$$

and

$$M^l = \int_A \left\{ \sigma_{ij} u_{i,1}^{\text{aux}} + \sigma_{ij}^{\text{aux}} u_{i,1} - \sigma_{ik} \varepsilon_{ik}^{\text{aux}} \delta_{1j} \right\} q_{,j} dA$$

$$+ \int_A \left\{ -\sigma_{ij} \left( \mathsf{C}_{ijkl}(\mathbf{x}) - \mathsf{C}_{ijkl}^{\text{tip}} \right) \sigma_{kl,1}^{\text{aux}} - \mathsf{D}_{ijkl,1} \varepsilon_{kl} \varepsilon_{ij}^{\text{aux}} \right\} q \, dA \tag{6.92}$$

### *6.3.4 Isoparametric FGM*

Any FGM material properties $P$, such as modules of elasticity $E_{11}$, $E_{22}$, shear modules $G_{12}$ and Poisson's ratio $\nu_{12}$ vary at different points of the domain in the form of

$$P(x, y) = \varphi(x, y) \tag{6.93}$$

where $\varphi(x, y)$ is a predefined function, usually exponential or linear. Such variable properties lead to variation of constitutive tensors in different points of the inhomogeneous domain. While in the standard FGM modelling, the material properties (6.93) are defined on each Gauss point, in the isoparametric graded finite element method (Kim and Paulino, 2002b), material properties are interpolated from nodal values by FEM shape functions,

$$P(x, y) = \sum_{i=1}^{n_e} N_i(x, y) \widehat{\bar{P}}_i \tag{6.94}$$

where $\bar{P}_i$ is the nodal value of property $P$. Derivatives of $P$ can then be defined as,

$$P_{,j}(x, y) = \sum_{i=1}^{n_e} N_{,j}(x, y) \bar{P}_i, \quad j = 1, 2 \tag{6.95}$$

Therefore, derivatives of $\mathsf{D}_{ijkl}$ or $\mathsf{C}_{ijkl}$ in the interaction integral (6.82), (6.83) and (6.84) can be computed accordingly.

Also, the roots of equation (4.22) for evaluation of auxiliary fields should be calculated at the crack tip, as described in relation (6.60). Similarly, values of $s_{kx}$ and $s_{ky}$ in enrichment equations (6.104) and (6.105) can also be calculated at the crack tip, but if a large area is to be enriched, $s_{kx}$ and $s_{ky}$ should then be calculated at enriched nodes and interpolated by FEM shape functions (6.94) and (6.95).

## 6.4   Crack Propagation in FGM Composites

In homogeneous isotropic materials, various crack propagation criteria are available in the literature, such as the maximum hoop stress (Erdogan and Sih, 1963), the maximum strain energy release rate (Hussain, Pu and Underwood, 1974) and the minimum strain energy density (Sih, 1974). Also, various homogeneous orthotropic mixed mode crack propagation criteria are available, such as the maximum circumferential stress by Saouma, Ayari and Leavell (1987), the minimum strain energy density (Ye and Ayari, 1994) and the maximum circumferential strain criteria (Ayari and Ye, 1995). Unfortunately, little literature is available for crack propagation in inhomogeneous media. Most of the studies in this context have assumed equivalent homogeneous isotropic or orthotropic criteria. In fact, the material behaviour has been idealized as a homogeneous orthotropic continuum to simplify the process of crack modelling. For details see Section 4.4.4.

## 6.5  Inhomogeneous XFEM

The general methodology of the orthotropic extended finite element method (Section 4.5) can be extended to inhomogeneous problems. The crack-tip enrichment functions are extracted from the analytical solutions (6.61) and (6.62).

### 6.5.1  Governing Equation

Consider an FGM body subjected to body and traction forces, $\mathbf{f}^b$ and $\mathbf{f}^t$, respectively, as depicted in Figure 6.2. The strong form of the equilibrium equation in terms of the stress tensor $\boldsymbol{\sigma}$ can be written as:

$$\nabla.\boldsymbol{\sigma} + \mathbf{f}^b = 0 \quad \text{in } \Omega \tag{6.96}$$

with the following boundary conditions:

$$\boldsymbol{\sigma} \cdot \mathbf{n} = \mathbf{f}^t \quad \text{on } \Gamma_t : \quad \text{external traction} \tag{6.97}$$

$$\mathbf{u} = \bar{\mathbf{u}} \quad \text{on } \Gamma_u : \quad \text{prescribed displacement} \tag{6.98}$$

$$\boldsymbol{\sigma} \cdot \mathbf{n} = 0 \quad \text{on } \Gamma_c : \quad \text{traction free crack} \tag{6.99}$$

where $\Gamma_t$, $\Gamma_u$ and $\Gamma_c$ are traction, displacement and crack boundaries, respectively.

The variational formulation of the boundary value problem can be defined as:

$$\int_\Omega \boldsymbol{\sigma} \cdot \delta\boldsymbol{\varepsilon} \, d\Omega = \int_\Omega \mathbf{f}^b \cdot \delta\mathbf{u} \, d\Omega + \int_\Gamma \mathbf{f}^t \cdot \delta\mathbf{u} \, d\Gamma \tag{6.100}$$

### 6.5.2  XFEM Approximation

In the extended finite element method, the effects of crack surfaces and crack tips are utilized by the approximation (3.28) to calculate the displacement for a point $\mathbf{x}$ located within the domain (for details see Section 3.4.3):

$$\mathbf{u}^h(\mathbf{x}) = \mathbf{u}(\mathbf{x}) + \mathbf{u}^H(\mathbf{x}) + \mathbf{u}^{\text{tip}}(\mathbf{x}) + \mathbf{u}^{\text{tra}}(\mathbf{x}) \tag{6.101}$$

or more explicitly

$$\mathbf{u}^h(\mathbf{x}) = \left[ \sum_{j=1}^n N_j(\mathbf{x})\mathbf{u}_j \right] + \left[ \sum_{h=1}^{mh} N_h(\mathbf{x}) \left( H(\mathbf{x}) - H(\mathbf{x}_h) \right) \mathbf{a}_h \right]$$

$$+ \left[ \sum_{k=1}^{mt} N_k(\mathbf{x}) \left( \sum_{l=1}^{mf} [F_l(\mathbf{x}) - F_l(\mathbf{x}_k)] \mathbf{b}_k^l \right) \right]$$

$$+ \left[ \sum_{m=1}^{mst} \bar{N}_m(\mathbf{x})\mathbf{c}_m + \sum_{n=1}^{msh} \bar{N}_n(\mathbf{x})H(\mathbf{x})\mathbf{d}_n \right] \tag{6.102}$$

where $\mathbf{u}_j$ is the displacement vector at node $j$, $\mathbf{b}_k^l$ are the added DOFs for tip enrichments $F_l$, $\mathbf{a}_h$ are the added DOFs for the crack discontinuity approximation by the Heaviside function

$H$, and $\mathbf{c}_i$ and $\mathbf{d}_j$ are added DOFs which correspond to hierarchical nodes in the sets *mst* and *msh*, respectively. $N_i$ are the standard finite element shape functions and $\bar{N}_i$ are the hierarchical shape functions for the transition domain.

### 6.5.2.1 Orthotropic Crack-Tip Enrichment Functions

General orthotropic crack-tip enrichment functions (4.112) in the local crack-tip polar coordinate system $(r, \theta)$ are defined as,

$$\{F_l(r, \theta)\}_{l=1}^4 = \left\{ \sqrt{r}\cos\frac{\theta_1}{2}\sqrt{g_1(\theta)}, \ \sqrt{r}\cos\frac{\theta_2}{2}\sqrt{g_2(\theta)}, \ \sqrt{r}\sin\frac{\theta_1}{2}\sqrt{g_1(\theta)}, \ \sqrt{r}\sin\frac{\theta_2}{2}\sqrt{g_2(\theta)} \right\}$$

$$(6.103)$$

where functions $g_k(\theta)$ and $\theta_k$, in their general form for all orthotropic composites, are defined as:

$$g_k(\theta) = \sqrt{(\cos\theta + s_{kx}\sin\theta)^2 + (s_{ky}\sin\theta)^2} \qquad (6.104)$$

$$\theta_k = \arctan\left(\frac{s_{ky}\sin\theta}{\cos\theta + s_{kx}\sin\theta}\right) \qquad (6.105)$$

where $s_{kx}$ and $s_{ky}$ have been defined in Section 4.3.1.

A number of previous studies, such as Dolbow and Gosz (2002), have used the standard (isotropic homogenous) crack-tip enrichments to analyse FGMs,

$$\{F_l(r, \theta)\}_{l=1}^4 = \left\{ \sqrt{r}\cos\frac{\theta}{2}, \ \sqrt{r}\sin\frac{\theta}{2}, \ \sqrt{r}\sin\theta\cos\frac{\theta}{2}, \ \sqrt{r}\sin\theta\sin\frac{\theta}{2} \right\} \qquad (6.106)$$

### 6.5.2.2 Transition Domain

The term $\mathbf{u}^{\text{tra}}(\mathbf{x})$ has been added to the original XFEM approximation (4.103) to overcome the incompatibility between the enriched and non-enriched domains by introduction of a transition (blending) zone. Two types of hierarchical nodes exist: hierarchical nodes that correspond to edges connecting non-enriched and tip enriched nodes (defined by *mst*), and those edges that connect tip and Heaviside enriched nodes (defined by *mth*). Tarancon *et al.* (2009) have discussed the definition of hierarchical shape functions $\bar{N}_i$, as proposed by Szabo and Babuska (1991). For example, for nodes $t_1$ and $t_2$ in Figure 6.3, the hierarchical shape functions in terms of the isoparametric coordinates $\xi, \eta$ are defined as

$$\bar{N}_{p_1}(\xi, \eta) = -\frac{1}{2}\sqrt{\frac{3}{2}}(1 - \xi^2)\frac{1 - \eta}{2} \qquad (6.107)$$

$$\bar{N}_{p2}(\xi, \eta) = -\frac{1}{2}\sqrt{\frac{3}{2}}(1 - \eta^2)\frac{1 - \xi}{2} \qquad (6.108)$$

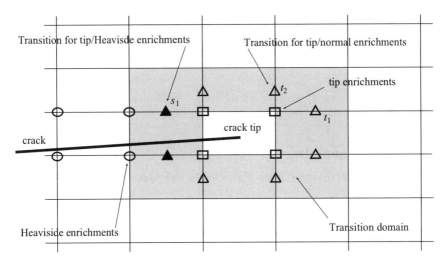

**Figure 6.3**    Added hierarchical nodes in the blending elements (Tarancon *et al.*, 2009).

## 6.5.3   XFEM Discretization

Discretization of Eq. (6.96) using the XFEM approximation (6.101) results in a discrete system of linear equilibrium equations:

$$\mathbf{Ku}^h = \mathbf{f} \tag{6.109}$$

where $\mathbf{u}^h$ is the vector of degrees of nodal freedom (for both classical and enriched ones),

$$\mathbf{u}^h = \{\mathbf{u} \quad \mathbf{a} \quad \mathbf{b}_\alpha \quad \mathbf{c} \quad \mathbf{d}\}^T \tag{6.110}$$

The global stiffness matrix $\mathbf{K}$ is assembled from the stiffness of each element $\mathbf{K}_{ij}^e$ with components $\mathbf{K}_{ij}^{rs}$

$$\mathbf{K}_{ij}^{rs} = \int_{\Omega^e} \left(\mathbf{B}_i^r\right)^T \mathbf{DB}_j^s \, d\Omega \quad (r, s = \mathbf{u}, \mathbf{a}, \mathbf{b}, \mathbf{c}, \mathbf{d}) \tag{6.111}$$

where $\mathbf{B}_i^{\mathbf{u}}$, $\mathbf{B}_i^{\mathbf{a}}$ and $\mathbf{B}_i^{\mathbf{b}}$ are derivatives of shape functions defined in Eqs. (3.49)–(3.55), and $\mathbf{B}_i^{\mathbf{c}}$ and $\mathbf{B}_i^{\mathbf{d}}$ are defined as,

$$\mathbf{B}_i^{\mathbf{c}} = \begin{bmatrix} \bar{N}_{i,x} & 0 \\ 0 & \bar{N}_{i,y} \\ \bar{N}_{i,y} & \bar{N}_{i,x} \end{bmatrix} \tag{6.112}$$

$$\mathbf{B}_i^{\mathbf{d}} = \begin{bmatrix} \left(\bar{N}_i H\right)_{,x} & 0 \\ 0 & \left(\bar{N}_i H\right)_{,y} \\ \left(\bar{N}_i H\right)_{,y} & \left(\bar{N}_i H\right)_{,x} \end{bmatrix} \tag{6.113}$$

$\mathbf{f}$ is the nodal force vector,

$$\mathbf{f}_i^e = \left\{\mathbf{f}_i^{\mathbf{u}} \quad \mathbf{f}_i^{\mathbf{a}} \quad \mathbf{f}_i^{\mathbf{b}_\alpha} \quad \mathbf{f}_i^{\mathbf{c}} \quad \mathbf{f}_i^{\mathbf{d}}\right\}^T \tag{6.114}$$

where $\mathbf{f}_i^{\mathbf{u}}$, $\mathbf{f}_i^{\mathbf{a}}$ and $\mathbf{f}_i^{\mathbf{b}\alpha}$ are defined in (4.109)–(4.111) and the new components are:

$$\mathbf{f}_i^{\mathbf{c}} = \int_{\Gamma_t} \bar{N}_i \mathbf{f}^t \, d\Gamma + \int_{\Omega^e} \bar{N}_i \mathbf{f}^b \, d\Omega \qquad (6.115)$$

$$\mathbf{f}_i^{\mathbf{d}} = \int_{\Gamma_t} \bar{N}_i H \mathbf{f}^t \, d\Gamma + \int_{\Omega^e} \bar{N}_i H \mathbf{f}^b \, d\Omega \qquad (6.116)$$

## 6.6  Numerical Examples

### 6.6.1  Plate with a Centre Crack Parallel to the Material Gradient

In the first example, a plane stress FGM square plate with a centre crack parallel to the material gradient is considered, as depicted in Figure 6.4. A constant tensile strain loading, defined in terms of intact material properties, is applied by means of its equivalent normal stress $\sigma_{22}(x, 10) = \bar{\varepsilon} E(x, 10)$. The boundary conditions and the finite element mesh of 1369 four-node elements and 1444 nodes are depicted in Figure 6.4.

Material properties $E_{11}$, $E_{22}$ and $G_{12}$ are assumed to vary similarly with an exponential law along the $x$ direction,

$$E_{11}(x) = E_{11}^0 e^{\beta x}, \quad E_{22}(x) = E_{22}^0 e^{\beta x}, \quad G_{12}(x) = G_{12}^0 e^{\beta x} \qquad (6.117)$$

Different values for effective material parameters $v$ and $\kappa_0$ (defined in equations (6.1)–(6.8)) and the gradient coefficient $\beta a$ are examined:

$$v = .1, .2, .3, .4, .5, .7, .9 \qquad (6.118)$$

$$\kappa_0 = 0.5, 5 \qquad (6.119)$$

$$\beta a = 0, .25, .5, .75, 1, 1.25, 1.5 \qquad (6.120)$$

The predicted normalized stress intensity factors at both crack tips for various material gradations have been compared with the reference values reported by Kim and Paulino based

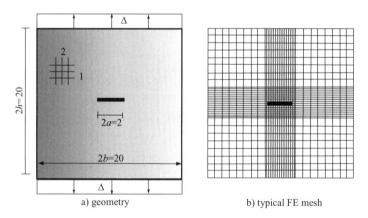

a) geometry                              b) typical FE mesh

**Figure 6.4**  FGM plate with a centre crack parallel to the material gradient under constant strain loading, and the adaptively structured finite element mesh.

**Table 6.2**   The effect of material gradient $\beta a$ on the normalized stress intensity factors of left and right crack tips ($\nu = .3$)(Bayesteh and Mohammadi, 2012).

| $\dfrac{\bar{K}_{\mathrm{I}}^{\mathrm{left}}}{\bar{K}_{\mathrm{I}}^{\mathrm{right}}}$ | $\beta a$ | | | | | | |
|---|---|---|---|---|---|---|---|
| | 0.00 | 0.25 | 0.50 | 0.75 | 1.00 | 1.25 | 1.50 |
| FEM-MCC$^{a}$ ($\kappa_0 = 0.5$) | 0.999 | 0.823 | 0.668 | 0.536 | 0.429 | — | — |
| | 0.997 | 1.204 | 1.437 | 1.706 | 2.064 | — | — |
| FEM-M Integral$^{b}$ ($\kappa_0 = 0.5$) | 0.997 | 0.825 | 0.671 | 0.540 | 0.434 | — | — |
| | 0.999 | 1.210 | 1.448 | 1.726 | 2.064 | — | — |
| Isotropic XFEM $^{c}$ ($\kappa_0 = 0.5$) | 1.020 | — | — | — | — | — | — |
| | 1.020 | — | — | — | — | — | — |
| XFEM ($\kappa_0 = 0.5$) | 0.998 | 0.834 | 0.678 | 0.543 | 0.436 | 0.350 | 0.280 |
| | 0.998 | 1.206 | 1.423 | 1.663 | 1.955 | 2.305 | 2.718 |
| XFEM ($\kappa_0 = 5$) | 1.001 | 0.829 | 0.665 | 0.524 | 0.414 | 0.327 | 0.257 |
| | 1.001 | 1.218 | 1.445 | 1.699 | 2.014 | 2.402 | 2.877 |

$^{a}$Kim and Paulino (2003c).
$^{b}$Kim and Paulino (2002a).
$^{c}$Asadpoure and Mohammadi (2007).

on the MCC method (Kim and Paulino, 2002a) and the interaction integral method (Kim and Paulino, 2003c) in Table 6.2. The normalized stress intensity factor is defined as

$$\bar{K}_{\mathrm{I}} = \frac{K_{\mathrm{I}}}{K_0} \tag{6.121}$$

where $K_{\mathrm{I}}$ is the stress intensity factor, and

$$K_0 = \bar{\varepsilon}\bar{E}^0\sqrt{\pi a}, \quad \bar{E}^0 = \frac{E^0}{\delta^2}, \quad E^0 = \sqrt{E_{11}^0 E_{22}^0} \tag{6.122}$$

The reference finite element results were based on a mesh of 8-node elements and more than 5800 nodes, nearly four times the number of nodes of XFEM simulations by Bayesteh and Mohammadi (2012) for inhomogeneous cases and Asadpoure and Mohammadi (2007) for the case of homogeneous orthotropic media ($\beta a = 0$). Very close agreement is observed for both crack tips, which is an indication of the efficiency of XFEM in achieving similar accuracies with far lower degrees of freedom.

In addition, it is important to note that the differences between the stress intensity factors at the left and right crack tips become larger for higher gradient coefficients $\beta a$. For instance, while the difference for the case of $\beta a = 0.25$ is about 40%, the right crack-tip stress intensity factor becomes more than ten times its value at the left tip.

The problem is also solved for different constant effective Poisson's ratios, as it is well known that its functional variation has little effect on the response of an FGM specimen. Table 6.3 compares the predicted normalized stress intensity factors at both crack tips for the

**Table 6.3**  The effect of Poisson's ratio on normalized stress intensity factor for the left and right crack tips under constant strain loading (Bayesteh and Mohammadi, 2012).

| $\dfrac{\bar{K}_I^{\text{left}}}{\bar{K}_I^{\text{right}}}$ | $\nu$ | | | | | | |
|---|---|---|---|---|---|---|---|
|  | 0.1 | 0.2 | 0.3 | 0.4 | 0.5 | 0.7 | 0.9 |
| FEM- MCC[a] | 0.667 | 0.669 | 0.671 | 0.673 | 0.675 | 0.679 | 0.683 |
|  | 1.430 | 1.433 | 1.437 | 1.441 | 1.444 | 1.451 | 1.456 |
| FEM[b] | 0.665 | 0.668 | 0.671 | 0.673 | 0.676 | 0.680 | 0.685 |
|  | 1.418 | 1.423 | 1.428 | 1.433 | 1.437 | 1.445 | 1.452 |
| XFEM | 0.678 | 0.678 | 0.678 | 0.677 | 0.677 | 0.674 | 0.666 |
|  | 1.406 | 1.415 | 1.423 | 1.431 | 1.438 | 1.451 | 1.452 |

[a]Kim and Paulino (2003c).
[b]Ozturk and Erdogan (1997).

gradient coefficient $\beta a = 0.5$ with the reference finite element results by Kim and Paulino (2003c) and Ozturk and Erdogan (1997). Again, very close agreement is observed at both crack tips and for all Poisson ratios.

Another important parameter that affects the post-processing solution of stress intensity factors is the size or radius of the contour integral, $r_J/a$. Figure 6.5 compares the variations of the normalized mode I stress intensity factor $\bar{K}_I$ at both crack tips with respect to $r_J/a$ for two different material gradient coefficients. It is clearly observed that $\bar{K}_I$ is almost insensitive to the size or radius of the $J$ integral. Similar conclusions have been reported by Bayesteh and Mohammadi (2012) for other material properties.

Since there is no available analytical solution in the literature for $\beta a = 0.75$ (to the best knowledge of the author) and the reference values of $\bar{K}$ presented by Kim and Paulino (2002a, 2003c) differ, a very fine model of 154 904 DOFs is utilized as the exact solution. Accordingly, convergence of the error of the normalized stress intensity factors for $\beta a = 0$ and

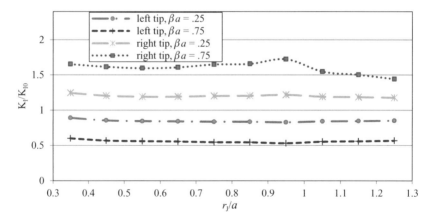

**Figure 6.5**  The effect of the radius of contour integral on the normalized stress intensity factors of left and right crack tips.

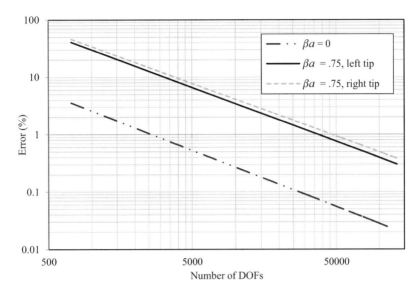

**Figure 6.6**   Convergence of error of normalized stress intensity factors.

$\beta a = 0.75$ (using a constant enriched area of radius $r = 0.15a$) is compared in Figure 6.6 on a logarithmic scale,

$$\text{Error} = \frac{\bar{K} - \bar{K}_{\text{exact}}}{\bar{K}_{\text{exact}}} \times 100 \qquad (6.123)$$

Clearly a good convergence rate, almost similar for all cases, is observed. A similar convergence rate has been reported by Bayesteh and Mohammadi (2012) for the error $\bar{K}$ of the right crack tip with $\beta a = 0$. In this case, however, the absolute error is far lower than the graded material $\beta a = 0.75$.

Finally, Figure 6.7 compares the stress contours of two different cases of $\beta a = 1$ and $\beta a = 0.2$. A strong tendency is observed towards the right edge for $\beta a = 1$.

### 6.6.2   Proportional FGM Plate with an Inclined Central Crack

A proportional FGM plate with an inclined central crack is considered to investigate the efficiency of XFEM in dealing with mixed mode fracture problems in functionally graded materials. Figure 6.8 depicts the geometry and the finite element mesh of the plane stress plate which is subjected to an equivalent traction $\sigma_{22}(x, 10) = \bar{\varepsilon}E(x, 10)$ of the applied constant strain $\bar{\varepsilon}$. The finite element model is constructed from an adaptive structured mesh of 1024 and 1444 nodes for the isotropic and orthotropic cases, respectively.

An exponential proportional variation of orthotropic material properties is considered.

$$E_{11}(x) = E_{11}^0 e^{\beta x}, \quad E_{22}(x) = E_{22}^0 e^{\beta x}, \quad G_{12}(x) = G_{12}^0 e^{\beta x} \qquad (6.124)$$

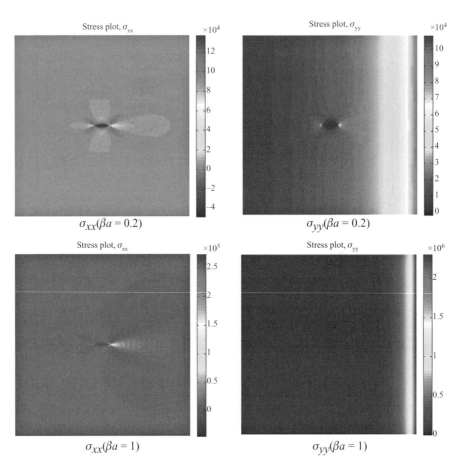

**Figure 6.7**  $\sigma_{xx}$ and $\sigma_{yy}$ stress contours for two different values of $\beta a = 1$ and $\beta a = 0.2$.

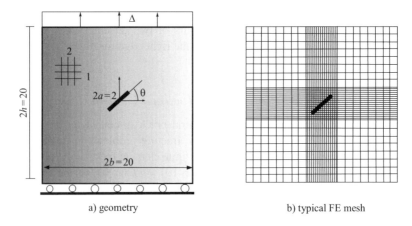

a) geometry                                          b) typical FE mesh

**Figure 6.8**  The geometry and typical finite element mesh of a tensile plate with an inclined central crack.

**Table 6.4** Comparison of stress intensity factors of the right crack tip for the isotropic case (Bayesteh and Mohammadi, 2012).

| Method | | $\theta°$ | | | | |
|---|---|---|---|---|---|---|
| | | 0 | 18 | 36 | 54 | 72 |
| Present XFEM | $K_{\mathrm{I}}$ | 2.532 | 2.278 | 1.636 | 0.859 | 0.241 |
| | $K_{\mathrm{II}}$ | 0 | 0.609 | 0.969 | 0.963 | 0.576 |
| Singular FEM | $K_{\mathrm{I}}$ | 2.523 | 2.275 | 1.635 | 0.865 | 0.257 |
| Kim and Paulino, 2005 | $K_{\mathrm{II}}$ | 0 | 0.612 | 0.975 | 0.946 | 0.558 |
| Isotropic XFEM | $K_{\mathrm{I}}$ | 2.528 | 2.270 | 1.635 | 0.862 | 0.255 |
| Dolbow and Gosz, 2002 | $K_{\mathrm{II}}$ | 0 | 0.626 | 0.993 | 0.957 | 0.560 |
| Semi-analytical | $K_{\mathrm{I}}$ | 2.524 | 2.278 | 1.640 | 0.869 | 0.259 |
| Konda and Erdogan, 1994 | $K_{\mathrm{II}}$ | 0 | 0.610 | 0.971 | 0.943 | 0.557 |

with

$$E_{11}^0 = 10^4, \quad E_{22}^0 = 10^3, \quad G_{12}^0 = 1216, \quad v_{12} = 0.3 \qquad (6.125)$$

The isotropic case with $E^0 = 1$ and $v(x) = 0.3$ was previously studied by a semi-analytical method (Konda and Erdogn, 1994), the extended finite element method (Dolbow and Gosz, 2002) and the singular finite element method using a mesh almost 5 times finer than the present XFEM model (Kim and Paulino, 2005).

The normalized stress intensity factors of the isotropic and orthotropic cases at both crack tips are compared in Tables 6.4 and 6.5, respectively. The results for both modes agree well for the isotropic case. In contrast, despite close agreement between XFEM and the reference for mode I stress intensity factors, predictions of the present XFEM for mode II stress intensity factor in the orthotropic case are almost twice the values presented by Kim and Paulino (2005). Further study is required to examine the potential reasons for the existing differences of mode II SIF of the present and reference results.

**Table 6.5** Comparison of stress intensity factors for the orthotropic case (Bayesteh and Mohammadi, 2012).

| Method | | | $\theta°$ | | | | |
|---|---|---|---|---|---|---|---|
| | | | 0 | 18 | 36 | 54 | 72 |
| XFEM | Right tip | $K_{\mathrm{I}}$ | 2531 | 2341 | 1739 | 1078 | 361 |
| | | $K_{\mathrm{II}}$ | 0 | 944 | 1539 | 1516 | 879 |
| | Left tip | $K_{\mathrm{I}}$ | 1178 | 1092 | 904 | 679 | 476 |
| | | $K_{\mathrm{II}}$ | 0 | 688 | 1123 | 1190 | 734 |
| Singular FEM | Right tip | $K_{\mathrm{I}}$ | 2531 | 2344 | 1803 | 1064 | 383 |
| Kim and Paulino, 2002a | | $K_{\mathrm{II}}$ | 0 | 389 | 729 | 794 | 515 |
| | Left tip | $K_{\mathrm{I}}$ | 1181 | 1063 | 736 | 321 | 9 |
| | | $K_{\mathrm{II}}$ | 0 | 430 | 737 | 775 | 500 |

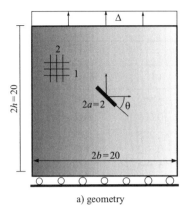
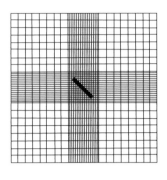

a) geometry                              b) typical FE mesh

**Figure 6.9**  The geometry of a tensile plate with an inclined central crack ($\theta = 36°$) and the finite element mesh.

More fundamental differences are observed between the isotropic and orthotropic problems. For instance, in both cases, variations of the mode II stress intensity factor are very limited, whereas the mode I stress intensity factor in the isotropic case reduces by a factor of more than 10 as the crack angle is changed from 0° to 72°. This trend, however, is not observed for variations of the mode I stress intensity factor at the right and left crack tips in the orthotropic case.

### 6.6.3  Non-Proportional FGM Plate with a Fixed Inclined Central Crack

The orthotropic plate of Figure 6.9, which is geometrically similar to the previous example for $\theta = 36°$, is now analysed in a mixed mode fracture of non-proportional FGMs. The geometry of the model and the adaptive structured finite element mesh are depicted in Figure 6.9.

The material properties are assumed to follow different (non-proportional) exponential laws:

$$E_{11}(x) = E_{11}^0 e^{\alpha x}, \quad E_{22}(x) = E_{22}^0 e^{\beta x}, \quad G_{12}(x) = G_{12}^0 e^{\gamma x} \tag{6.126}$$

with

$$E_{11}^0 = 0.75, \quad E_{22}^0 = 1, \quad G_{12}^0 = 0.5, \quad \nu_{12} = 0.3 \tag{6.127}$$

$$\alpha a = 0.2, \quad \beta a = 0.25, \quad \gamma a = 0.15 \tag{6.128}$$

Table 6.6 compares the results of the normalized mode I and II stress intensity factors, $\bar{K}_I = K_I/K_0$ and $\bar{K}_{II} = K_{II}/K_0$, respectively, the complex stress intensity factors $K$ and the phase angle $\psi$ for non-proportional and various proportional modellings. The reference stress intensity factor is $K_0 = \bar{\varepsilon} E_{22}^0 \sqrt{\pi a}$ and the complex stress intensity factor and the phase angle are defined as,

$$K = |K| e^{i\psi}, \quad |K| = \sqrt{K_I^2 + K_{II}^2} \tag{6.129}$$

$$\psi = \tan^{-1}\left(\frac{K_{II}}{K_I}\right) \tag{6.130}$$

Kim and Paulino (2003d) used a finite element model with almost 2400 nodes to compute the values of normalized mode I and II stress intensity factors for the non-proportional case as

**Table 6.6** Comparison of major fracture mechanics parameters for various material properties (Bayesteh and Mohammadi, 2012) (Reproduced by permissions of Elsevier.).

| $\alpha$ | $\beta$ | $\gamma$ | $\bar{K}_I = K_I/K_0$ | $\bar{K}_{II} = K_{II}/K_0$ | $K$ | $\psi$ |
|---|---|---|---|---|---|---|
| 0.2 | 0.25 | 0.15 | 0.757 | −0.537 | 1.644 | −35.3 |
| 0 | 0 | 0 | 0.661 | −0.492 | 1.460 | −36.7 |
| 0.2 | 0.2 | 0.2 | 0.751 | −0.530 | 1.629 | −35.3 |
| 0.4 | 0.4 | 0.4 | 0.845 | −0.561 | 1.797 | −33.6 |
| 0.6 | 0.6 | 0.6 | 0.942 | −0.587 | 1.967 | −31.9 |
| 0.8 | 0.8 | 0.8 | 1.050 | −0.616 | 2.157 | −30.4 |
| 1 | 1 | 1 | 1.172 | −0.649 | 2.375 | −29.0 |

0.756 and 0.572, respectively, which closely match the XFEM predictions (on a mesh with half number of nodes) with differences of less than 0.1% and 5% for modes I and II, respectively. Since there is no analytical solution for this problem, no specific conclusion can be made to determine the more accurate approach.

Figure 6.10 illustrates variations of complex stress intensity factors for different radius of $J$ integral contour, which shows the insensitivity of the results with respect to $r_J$. Similar conclusions have been reported by Bayesteh and Mohammadi (2012) for the insensitivity of the energy release rate.

### 6.6.4   Rectangular Plate with an Inclined Crack (Non-Proportional Distribution)

This problem, which has already been investigated by several authors, including Sih, Paris and Irwin (1965), Wang, Yau and Corten (1980) and Atluri, Kobayashi and Nakagaki (1975a, 1975b) for homogeneous orthotropic cases (no variation in material properties), Kim and

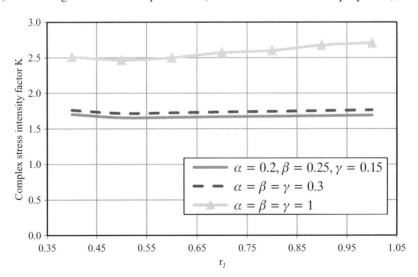

**Figure 6.10**   The complex stress intensity factor versus the radius of the $J$ integral contour (Bayesteh and Mohammadi, 2012).

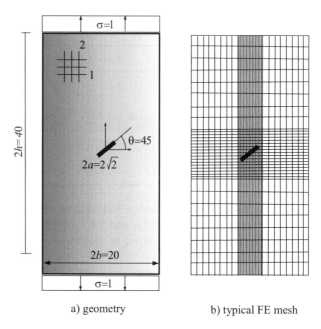

a) geometry                                    b) typical FE mesh

**Figure 6.11**   Geometry and finite element mesh of rectangular FGM plate.

Paulino (2002a) and Bayesteh and Mohammadi (2012) for various orthotropic FGM cases, is revisited. The problem, defined in Figure 6.11, is to analyse an inclined crack in a rectangular plate with the exponential form of the orthotropic FGM properties,

$$E_{11}(x) = E_{11}^0 e^{\alpha x}, \quad E_{22}(y) = E_{22}^0 e^{\beta x}, \quad G_{12}(x) = G_{12}^0 e^{\gamma x} \tag{6.131}$$

with the following reference material properties

$$E_{11}^0 = 3.5 \times 10^6, \quad E_{22}^0 = 12 \times 10^6, \quad G_{12}^0 = 3 \times 10^6, \quad \nu_{12} = 0.204 \tag{6.132}$$

Three sets of proportional and non-proportional material properties are considered:

Set 1: Homogeneous orthotropic distribution,

$$(\alpha, \beta, \gamma) = (0, 0, 0)$$

Set 2: Orthotropic FGM with proportional material variation,

$$(\alpha, \beta, \gamma) = (0.2, 0.2, 0.2)$$
$$(\alpha, \beta, \gamma) = (0.5, 0.5, 0.5)$$

Set 3: Orthotropic FGM with non-proportional material variation,

$$(\alpha, \beta, \gamma) = (0.5, 0.4, 0.3)$$

$$(\alpha, \beta, \gamma) = (0.3, 0.2, 0.1)$$

$$(\alpha, \beta, \gamma) = (0.8, 0.5, 0.2)$$

$$(\alpha, \beta, \gamma) = (0.8, 0.7, 0.6)$$

The finite element mesh is constructed from an adaptive structured grid of 2552 nodes and 2451 elements, as depicted in Figure 6.11b. Table 6.7 compares the results of left and right crack-tip stress intensity factors for several analyses by XFEM and the existing references for the three material sets. The results of the main reference (Kim and Paulino, 2002a) were based on a finite element of 3694 nodes. Interestingly, very close agreement is observed for the stress intensity factors of both crack tips in a large variety of material properties.

Bayesteh and Mohammadi (2012) have investigated the mesh independency of XFEM simulations for the specific non-proportional case of $(\alpha, \beta, \gamma) = (0.5, 0.4, 0.3)$. Table 6.8 compares the mode I and II stress intensity factors for different types of finite element meshes and clearly indicates the mesh insensitivity of both stress intensity factors.

Finally, Figure 6.12 illustrates the $\sigma_{22}$ stress contours on a deformed shape of FGM plate (with a magnification factor of $10^5$) for four different inhomogeneity coefficients $(\alpha, \beta, \gamma) = (0, 0, 0), (0.2, 0.2, 0.2), (0.5, 0.5, 0.5)$ and $(0.5, 0.4, 0.3)$. It is observed that by increasing the inhomogeneity factor, the deformation of the plate becomes more disturbed, as the stiffer part of the plate undergoes smaller deformation and endures higher stresses, the softer part experiences substantially larger deformations under much lower stresses. This also confirms the expected differences between the values of the stress intensity factors for the two crack tips.

## 6.6.5   Crack Propagation in a Four-Point FGM Beam

### 6.6.5.1   Isotropic Case

In a series of experimental tests and numerical simulations, Rousseau and Tippur (2000) investigated the mixed-mode fracture performance of four-point bending beams with cracks perpendicular to a glass-epoxy material gradient, as depicted in Figure 6.13. A dimensionless normalized parameter $\xi$ is used to specify the position of the crack with respect to the left edge of the graded portion (see Figure 6.13). Two main crack positions are examined: $\xi = 0.17$ and $\xi = 0.58$.

Rousseau and Tippur (2000) and Kim and Paulino (2007) have defined the main material properties in the graded part of the beam, including the modulus of elasticity, Poisson's ratio and the fracture toughness, as presented in Table 6.9.

Bayesteh and Mohammadi (2012) have adopted a fixed unstructured finite element mesh of Figure 6.13, constructed from 2547 nodes and 4848 triangular elements, to simulate this problem for all crack positions, different material properties and through all crack propagation patterns. Also, a fixed circular area with radius $r = 0.2a$ is enriched around the crack tip in all quasi-static propagation steps. The stress intensity factors are computed by the domain interaction integral approach. A constant crack increment $\Delta a = 2$ mm is used in all propagation steps, unless stated otherwise.

**Table 6.7** Right and left tips stress intensity factors for mixed mode crack in a rectangular plate (Bayesteh and Mohammadi, 2012).

| Material Properties | | | Stress Intensity Factors | | | | |
|---|---|---|---|---|---|---|---|
| | | | Left tip | | Right tip | | |
| $\alpha$ | $\beta$ | $\gamma$ | $K_\mathrm{I}$ | $K_\mathrm{II}$ | $K_\mathrm{I}$ | $K_\mathrm{II}$ | Reference |
| 0 | 0 | 0 | 1.054 | 1.054 | 1.054 | 1.054 | Sih, Paris and Irwin (1965) |
| | | | 1.020 | 1.080 | 1.020 | 1.080 | Atluri, Kobayashi and Nakagaki (1975a, 1975b) |
| | | | 1.023 | 1.049 | 1.023 | 1.049 | Wang, Yau and Corten (1980) |
| | | | 1.067 | 1.044 | 1.067 | 1.044 | (MCC) Kim and Paulino (2002a) |
| | | | 1.077 | 1.035 | 1.077 | 1.035 | (DCT) Kim and Paulino (2002a) |
| | | | 1.047 | 0.898 | 1.047 | 0.898 | XFEM-non-equilibrium Mahmoudi (2009) |
| | | | 1.010 | 0.910 | 1.010 | 0.933 | XFEM-incompatible Bayesteh, Mohammadi (2012) |
| 0.2 | 0.2 | 0.2 | 1.403 | 1.288 | 1.762 | 1.439 | (MCC) Kim and Paulino (2002a) |
| | | | 1.419 | 1.284 | 1.769 | 1.419 | (DCT) Kim and Paulino (2002a) |
| | | | 1.315 | 1.021 | 1.828 | 1.329 | XFEM-non-equilibrium Mahmoudi (2009) |
| | | | 1.356 | 1.198 | 1.794 | 1.284 | XFEM-incompatible Bayesteh, Mohammadi (2012) |
| 0.5 | 0.5 | 0.5 | 1.025 | 0.611 | 2.367 | 1.08 | XFEM-non-equilibrium Mahmoudi (2009) |
| | | | — | — | 2.384 | 1.581 | (MCC) Kim and Paulino (2002a) |
| 0.5 | 0.4 | 0.3 | — | — | 2.387 | 1.553 | (DCT) Kim and Paulino (2002a) |
| | | | 1.452 | 1.069 | 2.355 | 1.448 | XFEM-incompatible Bayesteh, Mohammadi (2012) |
| 0.3 | 0.2 | 0.1 | 1.393 | 1.219 | 1.807 | 1.259 | XFEM-incompatible Bayesteh, Mohammadi (2012) |
| 0.8 | 0.5 | 0.2 | 1.321 | 0.887 | 2.736 | 0.980 | XFEM-incompatible Bayesteh, Mohammadi (2012) |
| 0.8 | 0.7 | 0.6 | 0.645 | 0.313 | 1.763 | 0.433 | XFEM-incompatible Bayesteh, Mohammadi (2012) |

**Table 6.8**  The effect of different finite element meshes for the non-proportional $(\alpha, \beta, \gamma) = (0.5, 0.4, 0.3)$ FGM on the right crack tip stress intensity factors (Bayesteh and Mohammadi, 2012).

| Element Type | Number of elements | Number of nodes | Number of DOFs | $K_{\mathrm{I}}$ | $K_{\mathrm{II}}$ |
|---|---|---|---|---|---|
| Three-node (T3) | 928 | 499 | 1056 | 2.501 | 1.284 |
| | 3618 | 1878 | 3842 | 2.467 | 1.382 |
| | 14302 | 7288 | 14674 | 2.490 | 1.362 |
| | 5628 | 5780 | 11716 | 2.470 | 1.346 |
| Four-node (Q4) | 2451 | 2552 | 5232 | 2.562 | 1.332 |
| | 11817 | 12036 | 24276 | 2.545 | 1.326 |

Table 6.10 compares the XFEM predictions for the critical load $P_{cr}$, the stress intensity factors $K_{\mathrm{I}}$ and $K_{\mathrm{II}}$, the complex stress intensity factor $\bar{K}$, the phase angle $\psi$ and the initial angle of crack propagation $\alpha$ with available test and numerical results by Rousseau and Tippur (2000), FEM results of Kim and Paulino (2007) and Comi and Mariani (2007) without crack-tip enrichments. The reference XFEM results were available only for the initial step. A close agreement is observed in all cases.

The crack trajectory for different crack increment sizes $\Delta a$ has been compared in Figure 6.14 with the experimental results of Rousseau and Tippur (2000) and the finite element predictions, reported by Kim and Paulino (2007). An excellent agreement is observed between the XFEM predictions and the reference results, with almost no sensitivity with respect to the crack increment size. The reason may be attributed to the fact that after an initial propagation step, the general state of mixed mode crack propagation is mainly changed to a pure self-similar crack propagation mode, which is less affected by the crack increment length (Bayesteh and Mohammadi, 2012).

Bayesteh and Mohammadi (2012) have reported that almost identical results are obtained from the maximum hoop stress and maximum energy release rate propagation criteria on the crack trajectory and the critical values of $P_{cr}$ for each crack length.

The effect of $\Delta a$ on the critical load for crack position $\xi = 0.58$ is depicted in Figure 6.15, which shows identical critical loads for all crack increments $\Delta a = 1\,\mathrm{mm}$, $\Delta a = 2\,\mathrm{mm}$ and $\Delta a = 3\,\mathrm{mm}$. Similar conclusions have been made by Bayesteh and Mohammadi (2012) for the case of crack position $\xi = 0.17$.

### 6.6.5.2  Orthotropic FGM

For the final part of the simulations, the FGM beam is assumed to be made of an orthotropic FGM material with $\kappa_0 = .5$ and $E$, $\nu$ and $K_{\mathrm{Ic}}^1$ according to Table 6.9, and various orthotropic stiffness ratios are considered: $\lambda = E_2/E_1 = 5$, 1 and 0.2. It is also assumed that the ratio of the critical stress intensity factors in the two orthotropic directions is $K_{\mathrm{Ic}}^2 = K_{\mathrm{Ic}}^1 E_1/E_2 = K_{\mathrm{Ic}}^1/\lambda$ (see Section 4.4.4.1). The initial crack position is set to $\xi = 0.17$.

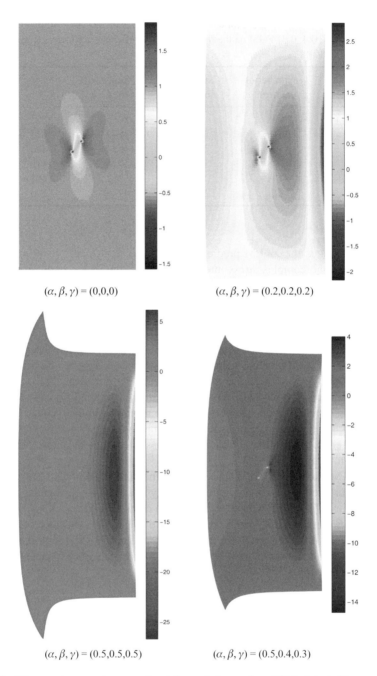

$(\alpha, \beta, \gamma) = (0,0,0)$       $(\alpha, \beta, \gamma) = (0.2, 0.2, 0.2)$

$(\alpha, \beta, \gamma) = (0.5, 0.5, 0.5)$     $(\alpha, \beta, \gamma) = (0.5, 0.4, 0.3)$

**Figure 6.12** The $\sigma_{22}$ stress contours on the deformed shape of an FGM plate (with a magnification factor of $10^5$) for four different inhomogeneity coefficients $(\alpha, \beta, \gamma)$(Mahmoudi, 2009; Mohammadi and Bayesteh, 2012).

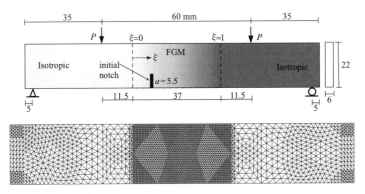

**Figure 6.13** The geometry, boundary conditions and finite element mesh of the FGM beam (Bayesteh and Mohammadi, 2012). (Reproduced by permission of Elsevier.)

**Table 6.9** Modulus of elasticity, Poisson's ratio and critical SIF in graded domain of glass-filled epoxy (Rousseau and Tippur, 2000; Kim and Paulino, 2007).

| $\xi$ | $E$ (MPa) | $\nu$ | $K_{Ic}$ (MPa$\sqrt{m}$) |
|---|---|---|---|
| $0 \leq$ | 3000 | 0.35 | 1.2 |
| 0.17 | 3300 | 0.34 | 2.1 |
| 0.33 | 5300 | 0.33 | 2.7 |
| 0.58 | 7300 | 0.31 | 2.7 |
| 0.83 | 8300 | 0.30 | 2.6 |
| $1 \geq$ | 8600 | 0.29 | 2.6 |

**Table 6.10** Comparison of $P_{cr}$, stress intensity factors and phase angle $\psi$ for the initial crack (Bayesteh and Mohammadi, 2012).

| $\xi$ | | Test[a] | CGS[a] | FEM[a] | FEM[b] | XFEM[c] | XFEM |
|---|---|---|---|---|---|---|---|
| $n_{elements}$ | | — | — | 10000 | 1234 | 13084 | 4848 |
| $n_{nodes}$ | | — | — | 30000 | 3725 | 6610 | 2547 |
| $P_{cr}$ | 0.17 | — | — | — | 249.3 | — | 250 |
| | 0.58 | — | — | — | 298 | — | 300 |
| $K_I$ | 0.17 | — | — | — | 2.088 | — | 2.087 |
| | 0.58 | — | — | — | 2.695 | — | 2.694 |
| $K_{II}$ | 0.17 | — | — | — | −0.127 | — | −0.117 |
| | 0.58 | — | — | — | −0.094 | — | −0.085 |
| $\bar{K}$ | 0.17 | — | 1.07 | 1.06 | — | — | 1.027 |
| | 0.58 | — | 1.08 | 1.11 | — | — | 1.11 |
| $\psi$ | 0.17 | — | −2.2 | −4.5 | −3.480 | — | −3.197 |
| | 0.58 | — | 1.0 | −1.9 | −1.997 | — | −1.81 |
| $\alpha$ | 0.17 | 7° | — | — | 6.98 | 7.22° | 7.01° |
| | 0.58 | 4° | — | — | 4.01 | 4.07° | 3.99° |

[a]Rousseau and Tippur (2000).
[b]Kim and Paulino (2007).
[c]Comi and Mariani (2007).

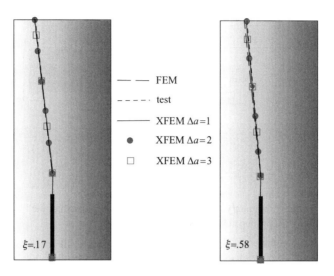

**Figure 6.14**  Comparison of the crack trajectory for two different positions of the initial notch at $\xi = 0.17$ and $\xi = 0.58$ with the experimental trajectory (Rousseau and Tippur, 2000) and FEM predictions by Kim and Paulino (2007) (Bayesteh and Mohammadi, 2012).

The crack propagation paths for a variety of orthotropic stiffness ratios $\lambda = 5, 1, 0.2$ are depicted in Figure 6.16. The predicted paths indicate the large effect of orthotropic properties on crack propagation trajectories. For the case of $\lambda = 5$, where the stiffer material axis is in the vertical direction, crack propagation as expected tends towards a vertical path. On the other hand, it is clearly observed that by decreasing $\lambda$ (i.e. the stronger material axis towards the $x$ direction), the crack has a trend to propagate towards the horizontal $x$ axis. It should be mentioned that all simulations have been performed on a fixed finite element mesh.

Similar conclusions can be reached by varying the critical load $P_{cr}$ with respect to the total crack length for different values of $\lambda$, as depicted in Figure 6.17. It is clearly observed that the

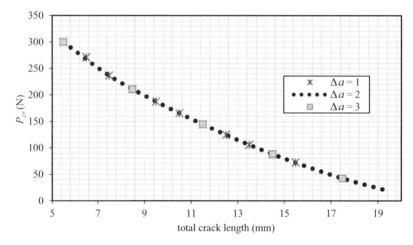

**Figure 6.15**  Comparison of the effect of the crack increment length on the critical load for $\xi = 0.58$.

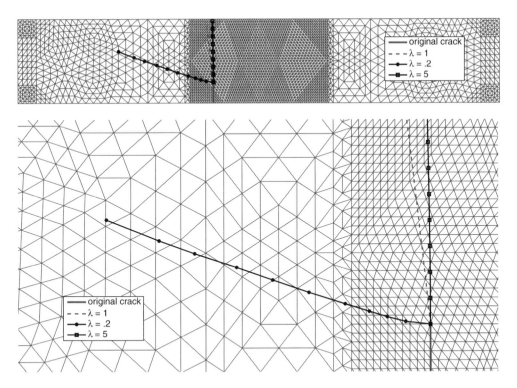

**Figure 6.16** Overall and detailed illustrations of crack propagation paths for $\lambda = 5, 1, 0.2$ (Bayesteh and Mohammadi, 2012). (Reproduced by permission of Elsevier.)

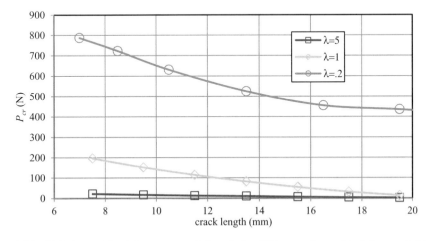

**Figure 6.17** Variations of load $P$ versus crack length for different values of $\lambda$ (after crack kinking) (Bayesteh and Mohammadi, 2012). (Reproduced by permission of Elsevier.)

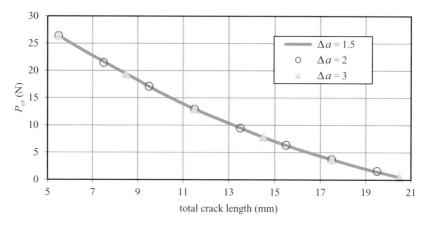

**Figure 6.18** Comparison of the effect of crack increment length on the critical load for the orthotropic FGM beam.

least load bearing capacity is obtained for $\lambda = 5$, where the principle orthotropic axis in the vertical direction, and the bending of the beam should be endured by its weakest longitudinal (horizontal) axis. In contrary, $\lambda = 0.2$ represents a nearly horizontal principle axis, which is best suited to the present bending problem.

Finally, Figure 6.18 illustrates the effect of the length of the crack increment $\Delta a$ on $P_{cr}$. The identical results prove the insensitivity of the present quasi-static XFEM simulations with respect to the crack increment sizes for orthotropic FGM crack propagations.

# 7

# Delamination/Interlaminar Crack Analysis

## 7.1 Introduction

Delamination has been a major source of concern in the performance and safety of composite laminates. This type of failure, also known as interlaminar debonding or interface cracking, is among the most commonly encountered failure modes in composite laminates which, individually or in combination with other types of failures, can cause extensive damage, stiffness and load bearing capacity reduction, and even structural dis-integrity due to reduced ductility or brittle and sudden fracture mechanisms. Such a paramount importance, therefore, has made the subject of prediction of the behaviour of laminates subjected to potential delamination a major topic in computational mechanics for a wide range of applications in sensitive industries, such as aerospace, defence and nanotechnology, as well as the more usual engineering applications, such as the use of FRP for structural strengthening, seismic retrofitting, and repair applications (Meier, 1995; Rizkalla, Hassan and Hassan, 2003; Lau and Zhou, 2001; Teng *et al.*, 2002).

Delamination can occur due to high concentrations of normal and shear stresses caused by geometrical effects, such as curved sections, sudden changes of cross-sections, and free edges, or from the mechanical, manufacturing, transportation and service effects, such as temperature, moisture, matrix shrinkage and general loading conditions. Sudden concentrated loadings, such as impact and explosion, can also be a major source of delamination in composite structures which may cause substantial internal damage at the interface of two adjacent plies without any apparent destruction (Mohammadi and Moosavi, 2007; Mohammadi and Forouzan-Sepehr, 2003).

Analysis of delamination can be classified into two main types; analytical methods and computational approaches. The first includes the analytical solutions for ideal cases of semi-infinite bimaterial problems, and comprises static and dynamic solutions for isotropic, orthotropic and functionally graded materials. Williams (1959) discovered the oscillatory near-tip behaviour for a traction-free interface crack between two dissimilar isotropic elastic materials. Later, England (1965) reported a violation of the basic open-crack assumption by predicting potential interpenetration or overlapping of crack surfaces in bimaterials, which was an indication of

*XFEM Fracture Analysis of Composites*, First Edition. Soheil Mohammadi.
© 2012 John Wiley & Sons, Ltd. Published 2012 by John Wiley & Sons, Ltd.

the existence of a contact zone near the crack tip. This contradiction was further studied by Comninou (1977) and Comninou and Schmuser (1979) by examining the stress singularities near the tip of an interfacial crack by assuming that the crack surfaces were in contact near the tip.

Derivation of bimaterial stress intensity factors $K_1$ and $K_2$ was performed by Erdogan (1963), Rice and Sih (1965) and Malysev and Salganik (1965), based on the fundamental concepts of fracture mechanics. Unlike the homogeneous case, they concluded that the individual stress intensity factors were not associated with the opening and shearing modes of fracture, respectively. Later, Sun and Jih (1987) discussed the definition of the total strain energy release rate of an interfacial crack between isotropic solids and the non-existence of separable strain energy release rates for mode I ($G_1$) and mode II ($G_2$) due to their oscillatory natures. Similar discussions were reported by Sun and Jih (1987), Hutchinson, Mear and Rice (1987) and Rice (1988).

The study of interface cracks was not limited to isotropic bimaterials and several others, almost simultaneously, discussed the interface crack between two anisotropic materials (Gotoh, 1967; Clements, 1971; Willis, 1971). Then, mixed mode analysis of interface cracks in anisotropic composites was reported by Wang and Choi (1983a, 1983b), followed by Ting (1986), Tewary, Wagoner and Hirth (1989), Wu (1990), Gao, Abbudi and Barnett (1992) and Hwu (1993a, 1993b), among others, discussing the bimaterial oscillatory index and its dependence on material orientations.

A major step forward was taken by Bassani and Qu (1989) by proving a necessary and sufficient non-oscillatory condition for separately defining the three fracture modes. They also showed that the Irwin-type energy release rate was simply the average of the corresponding results for the two homogeneous materials. Similar methodologies were reportedly used by Sun and Manoharan (1989), Sou (1990), Yang, Sou and Shih (1991), Hwu (1993b), Qian and Sun (1998), Lee (2000) and Hemanth et al. (2005) to evaluate mixed mode strain energy release rates and other fracture mechanics parameters for an interfacial crack between two orthotropic layers.

In a completely different analytical approach, stress-based methods have also been developed to assess the performance of debonded composite laminates and to discuss potential delamination propagation (Mohammadi and Forouzan-Sepehr, 2003; Mohammadi and Moosavi, 2007; Rabinovitch, 2008). In this approach, evaluation of the interfacial stress distribution is based on an elastic assumption and the delamination criterion is a simple comparison of the computed stress state with the adhesive strength (Parker, 1981; O'Brien, 1985; Rowlands, 1985). For example, Roberts (1989) presented a general formulation for the analysis of composite beams with partial interaction and proposed some models for the interface shear stress concentration near the ends of the epoxy-bonded external plates. Also, Taljsten (1997) used the linear elastic theory to derive shear and peeling stresses and to calculate the critical stress levels at the end of an outer reinforcement plate (Esna Ashari and Mohammadi, 2012).

In the second category of analysis of bimaterial interface cracks, a vast variety of computational techniques have been adopted for analysis of multilayer orthotropic composites and the study of interlaminar crack stability and propagation. The finite element method (FEM), which has been the basis for hundreds/thousands of studies of fracture mechanics related problems for the past 50 years, is probably the first choice of analysis for general engineering problems, including fracture, unless a better solution is proposed. In addition, the boundary element method (BEM), the discrete element method (DEM) (Mohammadi, 2003a), meshless methods such as the element-free Galerkin method (EFG) (Belytschko, Organ and Krongauz, 1995; Ghorashi, Mohammadi and Sabbagh-Yazdi, 2011) and the meshless local Petrov–Galerkin approach (MLPG), the extended finite element method (XFEM) and the extended isogeo-metric analysis (XIGA) (Ghorashi, Valizadeh and Mohammadi, 2012) are probably the main classes of computational techniques that are currently available for efficient analysis of various crack problems, including interface cracks.

Most of the performed numerical simulations prior to the end of twentieth century were related to the finite element method. Interlaminar crack simulation in FEM has been performed with a number of assumptions on the way a crack is represented. They mainly include contin-uous smeared crack models, discrete inter-element crack models, cracked interface elements and the discrete element based approach, which may use general contact mechanics algo-rithms to simulate progressive delamination and fracture problems (Mohammadi, Owen and Peric, 1997; Mohammadi, 2008). These models may represent a cohesive interface approach, which employs a layer of interface elements (Sprenger, Gruttmann and Wagner, 2000) or a general contact interface between the materials to analyse partially delaminated layered com-posites (Mohammadi and Moosavi, 2007; Mohammadi and Forouzan-Sepehr, 2003; Wong and Vecchio, 2003; Wu, Yuan and Niu, 2002; Wu and Yin, 2003; Lu et al., 2005; Wang, 2006; Teng, Yuan and Chen, 2006).

On the other hand, a large number of studies have used computationally evaluated fracture mechanics parameters, such as the stress intensity factors, the energy release rate or the $J$ integral, in comparison with the relevant critical values to assess the crack stability (Pesic and Pilakoutas, 2003; Colombi, 2006; Lu et al., 2006; Bruno, Carpino and Greco, 2007). For instance, the initiation and propagation of the delamination, based on this concept, occurs if the amount of energy released by the system due to an infinitesimal crack growth is larger than the specific fracture energy of the material. Most of the existing FEM-based LEFM debonding analyses differ mainly in the way the fracture mechanics parameters are evaluated. For example, Rabinovitch and Frosting (2001) and Rabinovitch (2004) used a high-order theory for the stress analysis of the FRP-strengthened beam, in combination with the $J$ integral formulation or numerical differentiation of the total potential energy, for evaluation of the energy release rate. Yang, Peng and Kwan (2006) adopted FEM and the virtual crack extension method while Greco, Lonetti and Blasi (2007) used a layered shear deformation model for the stress analysis in combination with the virtual crack closure method.

The extended finite element method evolved from the basis of the finite element method by the merit of enhancing its deficiencies in the simulation of crack problems; to avoid local mesh generations even for crack propagation problems and to reproduce the singular nature of the stress field around a crack tip. The first attempt to analyse delamination in isotropic materials was performed by Remmers, Wells and de Borst (2003) who presented a partition of unity finite element based on a solid-like shell element for simulation of delamination growth in thin layered structures. Then, Sukumar et al. (2004), developed partition of unity-based enrichment functions for bimaterial interface cracks between two isotropic media. Also, Nagashima and Suemasu (2004, 2006) applied XFEM to analyse interface cracks between dissimilar isotropic materials. They also studied the behaviour of orthotropic composite ma-terials by adopting the near-tip enrichment functions for homogeneous isotropic cracks, and concluded that the isotropic enrichment functions cannot represent the asymptotic solution for an interface crack in orthotropic materials. At the same time, Hettich and Ramm (2006) simulated the delamination crack as a jump in the displacement field without using any crack-tip enrichment.

Development of the set of orthotropic crack-tip enrichments by Asadpoure, Mohammadi and Vafai (2006, 2007) and Asadpoure and Mohammadi (2007) for static problems, Motamedi and Mohammadi (2010a, 2010b, 2012) for dynamic cracks and Esna Ashari and Mohammadi (2011) for orthotropic bimaterial interface cracks have further extended the method well to the anisotropic fracture mechanics problems.

In this chapter, details of XFEM for modelling interfacial cracks between two orthotropic media by the set of bimaterial orthotropic enrichment functions are discussed. The chapter is mainly based on the comprehensive work of Esna Ashari (2009) and Esna Ashari and Mohammadi (2009, 2010a, 2010b, 2011a, 2011b, 2012).

## 7.2  Fracture Mechanics for Bimaterial Interface Cracks

A traction-free interface crack between two dissimilar orthotropic materials is considered, as depicted in Figure 7.1. The crack is located in the $(x, y)$ plane and $(r, \theta)$ are the local polar coordinates defined on the crack tip. The elastic principal directions of each material are assumed to be normal to or parallel with the crack direction.

The general form of the asymptotic solution for a bimaterial interface crack can be written as,

$$u_i = \sum_j \{K_I A_j F_j(r, \theta) + K_{II} B_j F_j(r, \theta)\}, \quad i = x, y \tag{7.1}$$

where $K_I$ and $K_{II}$ are the mode 1 and 2 stress intensity factors, respectively, $A_j$ and $B_j$ are spatially-independent functions of the elasticity coefficients of the two materials, and $F_j$ are functions of local crack tip coordinates $(r, \theta)$, mainly representing the radial/angular nature of the displacement field.

In the following, two cases of isotropic and orthotropic bimaterials are comprehensively examined.

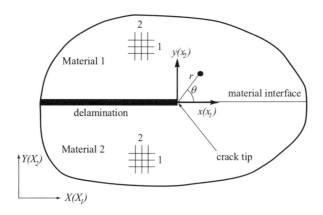

**Figure 7.1**   Interface crack between two orthotropic materials.

### 7.2.1  Isotropic Bimaterial Interfaces

Deng (1995) developed the basic asymptotic solution for stationary and moving cracks with frictional contact along bimaterial interfaces in homogeneous solids. Neglecting the effects of crack propagation and contact, Sukumar *et al.* (2004) adopted the methods of Rice (1988), Suo (1990), Rice, Hawk and Asaro (1990) and Shih (1991) to present the asymptotic solution for isotropic bimaterial cracks. For the material above the interface ($k = 1$),

$$
\begin{aligned}
u_x = {} & \frac{e^{-\varepsilon(\pi-\theta)}\operatorname{Re}\left(\bar{K}r\right)^{i\varepsilon}}{2\mu(1+4\varepsilon^2)\cosh(\pi\varepsilon)}\sqrt{\frac{r}{2\pi}}\left[-e^{2\varepsilon(\pi-\theta)}\left(\cos\frac{\theta}{2}+2\varepsilon\sin\frac{\theta}{2}\right)\right.\\
& \left. + \kappa\left(\cos\frac{\theta}{2}-2\varepsilon\sin\frac{\theta}{2}\right)+(1+4\varepsilon^2)\sin\frac{\theta}{2}\cos\theta\right]\\
& + \frac{e^{-\varepsilon(\pi-\theta)}\operatorname{Im}\left(\bar{K}r\right)^{i\varepsilon}}{2\mu(1+4\varepsilon^2)\cosh(\pi\varepsilon)}\sqrt{\frac{r}{2\pi}}\left[e^{2\varepsilon(\pi-\theta)}\left(\sin\frac{\theta}{2}-2\varepsilon\cos\frac{\theta}{2}\right)\right.\\
& \left. + \kappa\left(\sin\frac{\theta}{2}+2\varepsilon\cos\frac{\theta}{2}\right)+(1+4\varepsilon^2)\cos\frac{\theta}{2}\sin\theta\right]
\end{aligned}
\tag{7.2}
$$

$$
\begin{aligned}
u_y = {} & \frac{e^{-\varepsilon(\pi-\theta)}\operatorname{Re}\left(\bar{K}r\right)^{i\varepsilon}}{2\mu(1+4\varepsilon^2)\cosh(\pi\varepsilon)}\sqrt{\frac{r}{2\pi}}\left[e^{2\varepsilon(\pi-\theta)}\left(\sin\frac{\theta}{2}-2\varepsilon\cos\frac{\theta}{2}\right)\right.\\
& \left. + \kappa\left(\sin\frac{\theta}{2}+2\varepsilon\cos\frac{\theta}{2}\right)-(1+4\varepsilon^2)\cos\frac{\theta}{2}\sin\theta\right]\\
& + \frac{e^{-\varepsilon(\pi-\theta)}\operatorname{Im}\left(\bar{K}r\right)^{i\varepsilon}}{2\mu(1+4\varepsilon^2)\cosh(\pi\varepsilon)}\sqrt{\frac{r}{2\pi}}\left[e^{2\varepsilon(\pi-\theta)}\left(\cos\frac{\theta}{2}+2\varepsilon\sin\frac{\theta}{2}\right)\right.\\
& \left. - \kappa\left(\cos\frac{\theta}{2}-2\varepsilon\sin\frac{\theta}{2}\right)+(1+4\varepsilon^2)\sin\frac{\theta}{2}\sin\theta\right]
\end{aligned}
\tag{7.3}
$$

where $\kappa$ is defined in (2.68) and $\bar{K}$ is the complex stress intensity factor,

$$
\bar{K} = K_{\mathrm{I}} + iK_{\mathrm{II}}
\tag{7.4}
$$

and can be defined in a similar way to classical stress intensity factors,

$$
[\sigma_{yy} + i\sigma_{xy}]_{\substack{r\to 0 \\ \theta=0}} = \frac{\bar{K}r^{i\varepsilon}}{\sqrt{2\pi r}}
\tag{7.5}
$$

$\bar{K}$ can be associated with the energy release rate by,

$$
G = \frac{1}{E^{12}}\frac{|\bar{K}|^2}{\cosh(\pi\varepsilon)} = \frac{1}{E^{12}}\frac{K_{\mathrm{I}}^2+K_{\mathrm{II}}^2}{\cosh(\pi\varepsilon)}
\tag{7.6}
$$

where $E^{12}$ is the equivalent bimaterial elastic modulus defined in terms of the effective Young's modulus $E_k'(k = 1, 2)$ of each material (see (2.113))

$$
\frac{2}{E^{12}} = \frac{1}{E_1'} + \frac{1}{E_2'}
\tag{7.7}
$$

$\varepsilon$ is the index of oscillation defined as

$$\varepsilon = \frac{1}{2\pi} \ln\left(\frac{1 - \beta}{1 + \beta}\right) \tag{7.8}$$

where $\beta$ is the second Dundurs parameter for isotropic materials (Dundurs, 1969)

$$\beta = \frac{\mu_1(\kappa_2 - 1) - \mu_2(\kappa_1 - 1)}{\mu_1(\kappa_2 + 1) + \mu_2(\kappa_1 + 1)} \tag{7.9}$$

The asymptotic solutions (7.2) and (7.3) can be used for points on the lower half plane by replacing $\varepsilon\pi$ with $-\varepsilon\pi$.

Another important parameter in the characterization of interfacial fracture is the phase angle $\psi$, defined at a distance $r$ from the crack tip,

$$\psi = \tan^{-1}\left[\frac{\mathrm{Im}\left(\bar{K}r^{i\varepsilon}\right)}{\mathrm{Re}\left(\bar{K}r^{i\varepsilon}\right)}\right] \tag{7.10}$$

The phase angle $\psi$ is a measure of the relative proportion of shear to normal tractions at a distance $r$ from the crack tip (Sukumar $et$ $al.$, 2004).

## 7.2.2 Orthotropic Bimaterial Interface Cracks

Lee, Hawong and Choi (1996) derived the governing equilibrium, elastic constitutive law and compatibility equations for orthotropic bimaterial interface cracks and combined them into a partial differential equation with the following characteristic equation in terms of a complex variable $z = x + my$,

$$m^4 + 2B_{12}m^2 + B_{66} = 0 \tag{7.11}$$

where

$$B_{12} = \frac{1}{2}\frac{(2c_{12} + c_{66})}{c_{11}} \tag{7.12}$$

$$B_{66} = \frac{c_{22}}{c_{11}} \tag{7.13}$$

The roots $m_l$ and $m_s$ of the characteristic equation (7.11) are purely imaginary when $\sqrt{B_{66}} < B_{12}$; $B_{66} > 0$ or complex when $\sqrt{B_{66}} > |B_{12}|$; $B_{66} > 0$, known as Type 1 and Type 2, respectively. Most orthotropic materials are categorized as Type 1 and their conjugated roots are:

$$m_l = ip, \quad m_s = iq \tag{7.14}$$

where

$$p = \sqrt{B_{12} - \sqrt{B_{12}^2 - B_{66}}} \qquad (7.15)$$

$$q = \sqrt{B_{12} + \sqrt{B_{12}^2 - B_{66}}} \qquad (7.16)$$

The analytical solutions for asymptotic displacement and stress fields around a two-dimensional traction-free interface crack tip were previously obtained by Lee (2000) and reported by Esna Ashari and Mohammadi (2011a). The general crack-tip displacement fields under the mixed mode loading in the upper layer ($k = 1$) of a cracked bimaterial specimen are (Lee, 2000):

$$u_x = \frac{K_{\mathrm{I}}}{\sqrt{2\pi}\,(1 + 4\varepsilon^2)D\cosh(\varepsilon\pi)}$$

$$\times \left\{ e^{\varepsilon(\pi-\theta_l)} p_l \bar{A} \left[ \cos\left(\varepsilon\ln(r_l) - \frac{\theta_l}{2}\right) + 2\varepsilon\sin\left(\varepsilon\ln(r_l) + \frac{\theta_l}{2}\right) \right] (r_l)^{\frac{1}{2}} \right.$$

$$+ e^{-\varepsilon(\pi-\theta_l)} p_l A \left[ \cos\left(\varepsilon\ln(r_l) - \frac{\theta_l}{2}\right) + 2\varepsilon\sin\left(\varepsilon\ln(r_l) - \frac{\theta_l}{2}\right) \right] (r_l)^{\frac{1}{2}}$$

$$- e^{\varepsilon(\pi-\theta_s)} p_s \bar{B} \left[ \cos\left(\varepsilon\ln(r_s) + \frac{\theta_s}{2}\right) + 2\varepsilon\sin\left(\varepsilon\ln(r_s) + \frac{\theta_s}{2}\right) \right] (r_s)^{\frac{1}{2}}$$

$$\left. - e^{-\varepsilon(\pi-\theta_s)} p_s B \left[ \cos\left(\varepsilon\ln(r_s) - \frac{\theta_s}{2}\right) + 2\varepsilon\sin\left(\varepsilon\ln(r_s) - \frac{\theta_s}{2}\right) \right] (r_s)^{\frac{1}{2}} \right\}$$

$$+ \frac{K_{\mathrm{II}}}{\sqrt{2\pi}\,(1 + 4\varepsilon^2)\,D\cosh(\varepsilon\pi)}$$

$$\times \left\{ - e^{\varepsilon(\pi-\theta_l)} p_l \bar{A} \left[ \sin\left(\varepsilon\ln(r_l) + \frac{\theta_l}{2}\right) - 2\varepsilon\cos\left(\varepsilon\ln(r_l) + \frac{\theta_l}{2}\right) \right] (r_l)^{\frac{1}{2}} \right.$$

$$- e^{-\varepsilon(\pi-\theta_l)} p_l A \left[ \sin\left(\varepsilon\ln(r_l) - \frac{\theta_l}{2}\right) - 2\varepsilon\cos\left(\varepsilon\ln(r_l) - \frac{\theta_l}{2}\right) \right] (r_l)^{\frac{1}{2}}$$

$$+ e^{\varepsilon(\pi-\theta_s)} p_s \bar{B} \left[ \sin\left(\varepsilon\ln(r_s) + \frac{\theta_s}{2}\right) - 2\varepsilon\cos\left(\varepsilon\ln(r_s) + \frac{\theta_s}{2}\right) \right] (r_s)^{\frac{1}{2}}$$

$$\left. + e^{-\varepsilon(\pi-\theta_s)} p_s B \left[ \sin\left(\varepsilon\ln(r_s) - \frac{\theta_s}{2}\right) - 2\varepsilon\cos\left(\varepsilon\ln(r_s) - \frac{\theta_s}{2}\right) \right] (r_s)^{\frac{1}{2}} \right\} \quad (7.17)$$

$$u_y = \frac{K_{\mathrm{I}}}{\sqrt{2\pi}\,(1+4\varepsilon^2)D\cosh(\varepsilon\pi)}$$

$$\times \left\{ e^{\varepsilon(\pi-\theta_l)} q_l\bar{A}\left[\sin\left(\varepsilon\ln(r_l)+\frac{\theta_l}{2}\right) - 2\varepsilon\cos\left(\varepsilon\ln(r_l)+\frac{\theta_l}{2}\right)\right](r_l)^{\frac{1}{2}} \right.$$

$$- e^{-\varepsilon(\pi-\theta_l)} q_l A\left[\sin\left(\varepsilon\ln(r_l)-\frac{\theta_l}{2}\right) - 2\varepsilon\cos\left(\varepsilon\ln(r_l)-\frac{\theta_l}{2}\right)\right](r_l)^{\frac{1}{2}}$$

$$- e^{\varepsilon(\pi-\theta_s)} q_s\bar{B}\left[\sin\left(\varepsilon\ln(r_s)+\frac{\theta_s}{2}\right) - 2\varepsilon\cos\left(\varepsilon\ln(r_s)+\frac{\theta_s}{2}\right)\right](r_s)^{\frac{1}{2}}$$

$$\left. + e^{-\varepsilon(\pi-\theta_s)} q_s B\left[\sin\left(\varepsilon\ln(r_s)-\frac{\theta_s}{2}\right) - 2\varepsilon\cos\left(\varepsilon\ln(r_s)-\frac{\theta_s}{2}\right)\right](r_s)^{\frac{1}{2}} \right\}$$

$$+ \frac{K_{\mathrm{II}}}{\sqrt{2\pi}\,(1+4\varepsilon^2)D\cosh(\varepsilon\pi)}$$

$$\times \left\{ e^{\varepsilon(\pi-\theta_l)} q_l\bar{A}\left[\cos\left(\varepsilon\ln(r_l)+\frac{\theta_l}{2}\right) + 2\varepsilon\sin\left(\varepsilon\ln(r_l)+\frac{\theta_l}{2}\right)\right](r_l)^{\frac{1}{2}} \right.$$

$$- e^{-\varepsilon(\pi-\theta_l)} q_l A\left[\cos\left(\varepsilon\ln(r_l)-\frac{\theta_l}{2}\right) + 2\varepsilon\sin\left(\varepsilon\ln(r_l)-\frac{\theta_l}{2}\right)\right](r_l)^{\frac{1}{2}}$$

$$- e^{\varepsilon(\pi-\theta_s)} q_s\bar{B}\left[\cos\left(\varepsilon\ln(r_s)+\frac{\theta_s}{2}\right) + 2\varepsilon\sin\left(\varepsilon\ln(r_s)+\frac{\theta_s}{2}\right)\right](r_s)^{\frac{1}{2}}$$

$$\left. + e^{-\varepsilon(\pi-\theta_s)} q_s B\left[\cos\left(\varepsilon\ln(r_s)-\frac{\theta_s}{2}\right) + 2\varepsilon\sin\left(\varepsilon\ln(r_s)-\frac{\theta_s}{2}\right)\right](r_s)^{\frac{1}{2}} \right\} \quad (7.18)$$

and the associated stress fields are:

$$\sigma_x = \frac{K_{\mathrm{I}}}{2\sqrt{2\pi}D\cosh(\varepsilon\pi)}$$

$$\times \left[ -p^2\left\{ e^{\varepsilon(\pi-\theta_l)}\bar{A}\cos\left(\varepsilon\ln(r_l)-\frac{\theta_l}{2}\right) + e^{-\varepsilon(\pi-\theta_l)}A\cos\left(\varepsilon\ln(r_l)+\frac{\theta_l}{2}\right)\right\}(r_l)^{\frac{-1}{2}} \right.$$

$$\left. + q^2\left\{ e^{\varepsilon(\pi-\theta_s)}\bar{B}\cos\left(\varepsilon\ln(r_s)-\frac{\theta_s}{2}\right) + e^{-\varepsilon(\pi-\theta_s)}B\cos\left(\varepsilon\ln(r_s)+\frac{\theta_s}{2}\right)\right\}(r_s)^{\frac{-1}{2}} \right]$$

$$+ \frac{K_{\mathrm{II}}}{2\sqrt{2\pi}D\cosh(\varepsilon\pi)}$$

$$\times \left[ p^2\left\{ e^{\varepsilon(\pi-\theta_l)}\bar{A}\sin\left(\varepsilon\ln(r_l)-\frac{\theta_l}{2}\right) + e^{-\varepsilon(\pi-\theta_l)}A\sin\left(\varepsilon\ln(r_l)+\frac{\theta_l}{2}\right)\right\}(r_l)^{\frac{-1}{2}} \right.$$

$$\left. - q^2\left\{ e^{\varepsilon(\pi-\theta_s)}\bar{B}\sin\left(\varepsilon\ln(r_s)-\frac{\theta_s}{2}\right) + e^{-\varepsilon(\pi-\theta_s)}B\sin\left(\varepsilon\ln(r_s)+\frac{\theta_s}{2}\right)\right\}(r_s)^{\frac{-1}{2}} \right]$$

$$(7.19)$$

$$\sigma_y = \frac{K_{\mathrm{I}}}{2\sqrt{2\pi}D\cosh(\varepsilon\pi)}$$

$$\times \left[ \left\{ e^{\varepsilon(\pi-\theta_l)}\bar{A}\cos\left(\varepsilon\ln(r_l)-\frac{\theta_l}{2}\right) + e^{-\varepsilon(\pi-\theta_l)}A\cos\left(\varepsilon\ln(r_l)+\frac{\theta_l}{2}\right)\right\}(r_l)^{\frac{-1}{2}} \right.$$

$$\left. - \left\{ e^{\varepsilon(\pi-\theta_s)}\bar{B}\cos\left(\varepsilon\ln(r_s)-\frac{\theta_s}{2}\right) + e^{-\varepsilon(\pi-\theta_s)}B\cos\left(\varepsilon\ln(r_s)+\frac{\theta_s}{2}\right)\right\}(r_s)^{\frac{-1}{2}} \right]$$

$$+ \frac{K_{\mathrm{II}}}{2\sqrt{2\pi}D\cosh(\varepsilon\pi)}$$

$$\times \left[ \left\{ -e^{\varepsilon(\pi-\theta_l)}\bar{A}\sin\left(\varepsilon\ln(r_l)-\frac{\theta_l}{2}\right) - e^{-\varepsilon(\pi-\theta_l)}A\sin\left(\varepsilon\ln(r_l)+\frac{\theta_l}{2}\right)\right\}(r_l)^{\frac{-1}{2}} \right.$$

$$\left. + \left\{ e^{\varepsilon(\pi-\theta_s)}\bar{B}\sin\left(\varepsilon\ln(r_s)-\frac{\theta_s}{2}\right) + e^{-\varepsilon(\pi-\theta_s)}B\sin\left(\varepsilon\ln(r_s)+\frac{\theta_s}{2}\right)\right\}(r_s)^{\frac{-1}{2}} \right]$$

$$(7.20)$$

$$\sigma_{xy} = \frac{K_{\mathrm{I}}}{2\sqrt{2\pi}D\cosh(\varepsilon\pi)}$$

$$\times \left[ \alpha_l \left\{ e^{\varepsilon(\pi-\theta_l)}\bar{A}\sin\left(\varepsilon\ln(r_l)-\frac{\theta_l}{2}\right) - e^{-\varepsilon(\pi-\theta_l)}A\sin\left(\varepsilon\ln(r_l)+\frac{\theta_l}{2}\right)\right\}(r_l)^{\frac{-1}{2}} \right.$$

$$\left. + \alpha_s \left\{ -e^{\varepsilon(\pi-\theta_s)}\bar{B}\sin\left(\varepsilon\ln(r_s)-\frac{\theta_s}{2}\right) + e^{-\varepsilon(\pi-\theta_s)}B\sin\left(\varepsilon\ln(r_s)+\frac{\theta_s}{2}\right)\right\}(r_s)^{\frac{-1}{2}} \right]$$

$$+ \frac{K_{\mathrm{II}}}{2\sqrt{2\pi}D\cosh(\varepsilon\pi)}$$

$$\times \left[ \alpha_l \left\{ e^{\varepsilon(\pi-\theta_l)}\bar{A}\cos\left(\varepsilon\ln(r_l)-\frac{\theta_l}{2}\right) - e^{-\varepsilon(\pi-\theta_l)}A\cos\left(\varepsilon\ln(r_l)+\frac{\theta_l}{2}\right)\right\}(r_l)^{\frac{-1}{2}} \right.$$

$$\left. + \alpha_s \left\{ -e^{\varepsilon(\pi-\theta_s)}\bar{B}\cos\left(\varepsilon\ln(r_s)-\frac{\theta_s}{2}\right) + e^{-\varepsilon(\pi-\theta_s)}B\cos\left(\varepsilon\ln(r_s)+\frac{\theta_s}{2}\right)\right\}(r_s)^{\frac{-1}{2}} \right]$$

$$(7.21)$$

where $\varepsilon$ is the index of oscillation defined in (7.8) but now in terms of the second Dundurs parameter $\beta$ for orthotropic materials (Dundurs, 1969)

$$\beta = \frac{h_{11}}{\sqrt{h_{12}h_{21}}} \tag{7.22}$$

$$\begin{cases} h_{11} = (l_{11})_1 - (l_{11})_2 \\ h_{21} = (l_{21})_1 + (l_{21})_2 \\ h_{12} = (l_{12})_1 + (l_{12})_2 \end{cases} \tag{7.23}$$

and

$$
\begin{cases}
(l_{11})_k = \left\{ \dfrac{p_s p - p_l q}{q - p} \right\}_k = \left\{ \dfrac{q_s - q_l}{q - p} \right\}_k \\[4mm]
(l_{21})_k = \left\{ \dfrac{q q_l - p q_s}{q - p} \right\}_k \qquad\qquad , \quad k = 1, 2 \\[4mm]
(l_{12})_k = \left\{ \dfrac{p_l - p_s}{q - p} \right\}_k
\end{cases}
\tag{7.24}
$$

Subscripts $k = 1, 2$ denote the upper and lower materials, respectively. Also,

$$
p_l = -p^2 c_{11} + c_{12}, \quad p_s = -q^2 c_{11} + c_{12}
\tag{7.25}
$$

$$
q_l = \frac{-p^2 c_{12} + c_{22}}{p}, \quad q_s = \frac{-q^2 c_{12} + c_{22}}{q}
\tag{7.26}
$$

$$
A = q + \eta, \quad \bar{A} = q - \eta, \quad B = p + \eta, \quad \bar{B} = p - \eta, \quad D = q - p
\tag{7.27}
$$

and

$$
\eta = \left( \frac{h_{21}}{h_{12}} \right)^{\frac{1}{2}}
\tag{7.28}
$$

Variables $r_j, \theta_j (j = l, s)$ are related to the polar coordinate system $(r, \theta)$ as

$$
r_j = r \sqrt{\cos^2 \theta + Z_j^2 \sin^2 \theta}, \quad j = l, s, \quad
\begin{cases}
Z_l = p \\
Z_s = q
\end{cases}
\tag{7.29}
$$

$$
\theta_j = \tan^{-1}(Z_j \tan \theta), \quad j = l, s, \quad
\begin{cases}
Z_l = p \\
Z_s = q
\end{cases}
\tag{7.30}
$$

In order to obtain the displacement and stress fields in the material below the interface ($k = 2$), parameter $\varepsilon \pi$ is changed to $-\varepsilon \pi$.

Equations (7.17) and (7.18) will be used to derive the XFEM enrichment functions for reproduction of the singular orthotropic bimaterial interlaminar crack-tip stress fields.

### 7.2.3 Stress Contours for a Crack between Two Dissimilar Orthotropic Materials

In order to further examine the analytical solutions for typical bimaterials, a sample plane stress problem of a crack between dissimilar orthotropic layers is assumed. The materials axes of orthotropy are presumed to be parallel to $y$ and $x$ axes in materials 1 and 2, respectively, and the crack is assumed to extend from $(-5, 0)$ to $(0, 0)$ with the crack tip located at $(0, 0)$. The following material properties are used:

$$
E_1 = 11.84 \text{ GPa}, \quad E_2 = 0.81 \text{ GPa}, \quad G_{12} = 0.63 \text{ GPa}, \quad \nu_{12} = 0.38
\tag{7.31}
$$

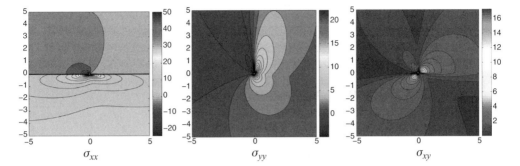

**Figure 7.2** Analytical stress fields for an edge crack between two dissimilar orthotropic materials (Esna Ashari and Mohammadi, 2011a). (Reproduced by permission of John Wiley & Sons, Ltd.)

where the 1 and 2 subscripts denote the local material coordinates. In order to provide a graphical representation of the stress fields, the stress intensity factors $K_I$ and $K_{II}$ are assumed to be

$$K_I = 10, \quad K_{II} = 10 \tag{7.32}$$

Figure 7.2 illustrates the contours of the stress field $\sigma_{xx}$, $\sigma_{yy}$ and $\sigma_{xy}$ based on equations (7.19)–(7.21). The closest data point to the crack tip for drawing the contours was at 0.025 cm. It is clearly observed that while $\sigma_{yy}$ and $\sigma_{xy}$ remain zero along the traction-free crack length, they remain continuous across the interface of two materials. As expected, $\sigma_{xx}$ is not continuous along the crack length nor across the bimaterial interface.

## 7.3 Stress Intensity Factors for Interlaminar Cracks

The domain integral method can be utilized to evaluate mixed-mode stress intensity factors for interfacial cracks between orthotropic materials (Chow, Boem and Alturi, 1995; Li, Shih and Needleman, 1985):

$$J = \int_A (\sigma_{ij} u_{i,1} - w_s \delta_{1j}) q_{,j} \, dA \tag{7.33}$$

where $w_s$ is the strain energy density, $A$ is the interior area of the $\Gamma$ contour surrounding the crack tip (Figure 7.3), and $q$ is a function smoothly varying from $q = 1$ at the crack tip to $q = 0$ at the exterior boundary $\Gamma$.

In the interaction integral method, the auxiliary fields which should satisfy the boundary value problem (equilibrium equation and traction-free boundary condition on the crack surfaces) are superimposed onto the actual fields to extract the mixed-mode stress intensity factors. One of the choices for the auxiliary state is the displacement and stress fields in the vicinity of the interfacial crack tip (7.17)–(7.21).

The $J^S$ integral for the sum of the real and auxiliary states can be defined as:

$$J^S = J + J^{aux} + M \tag{7.34}$$

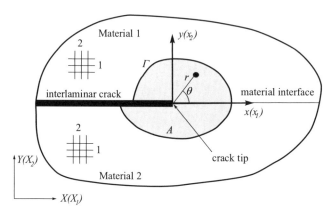

**Figure 7.3**   The contour $\Gamma$ and its interior area, $A$.

where $J$ and $J^{\text{aux}}$ are associated with the actual and auxiliary states, respectively, and $M$ is the interaction integral, defined as

$$M = \int_A \left[ \sigma_{ij} u_{i,1}^{\text{aux}} + \sigma_{ij}^{\text{aux}} u_{i,1} - w^M \delta_{1j} \right] q_{,j} \mathrm{d}A \tag{7.35}$$

Since the $M$ integral shares the same path-independency of the original $J$ integral, it can then be calculated away from the crack tip where the finite element solution is more accurate.

In order to determine the stress intensity factors of the present orthotropic bimaterial problem from the $M$ integral, Chow, Boem and Alturi (1995) proposed the following relation to relate the stress intensity factors to the $M$ integral:

$$K_i = 2 \sum_{m=1}^{2} U_{im} M\left(u, u^{\text{aux}(m)}\right), \quad i = 1, 2 \tag{7.36}$$

where $K_{\text{I}}$ and $K_{\text{II}}$ are obtained for the auxiliary states of $(K_{\text{I}}^{\text{aux}} = 1, K_{\text{II}}^{\text{aux}} = 0)$ and $(K_{\text{I}}^{\text{aux}} = 0, K_{\text{II}}^{\text{aux}} = 1)$, respectively, and

$$U = \left[ \left( L_1^{-1} + L_2^{-1} \right)^{-1} \left( I + \beta^2 \right) \right]^{-1} \tag{7.37}$$

$$\beta = \left( L_1^{-1} + L_2^{-1} \right) \left( S_1 L_1^{-1} - S_2 L_2^{-1} \right) \tag{7.38}$$

The subscripts 1 and 2 denote the upper and lower materials and $S$ and $L$ are $2 \times 2$ real Barnett and Lothe tensors, respectively (Barnett and Lothe, 1973). Explicit expressions of non-zero components of $S$ and $L$ for an orthotropic material have been derived by Dongye and Ting (1989) and Deng (1995),

$$S_{21} = \left[ \frac{d_{66} \left( \sqrt{d_{11} d_{22}} - d_{12} \right)}{d_{22} \left( d_{12} + 2 d_{66} + \sqrt{d_{11} d_{22}} \right)} \right]^{\frac{1}{2}} \tag{7.39}$$

$$S_{12} = -\sqrt{\frac{d_{22}}{d_{11}}} S_{21} \qquad (7.40)$$

$$L_{11} = (d_{12} + \sqrt{d_{11}d_{22}}) S_{21} \qquad (7.41)$$

$$L_{22} = \sqrt{\frac{d_{22}}{d_{11}}} L_{11} \qquad (7.42)$$

where $d_{ij}$ are the components of the contracted form of the fourth order elastic constant tensor **D**, as defined in Eq. (4.10). Components of $S$ and $L$ are computed for each individual layer.

## 7.4  Delamination Propagation

### 7.4.1  Fracture Energy-Based Criteria

The state of stability of a 2D interlaminar crack can be examined by a mixed mode criterion in terms of the fracture energy release rate $G$ and the corresponding critical value $G_c$,

$$\left(\frac{G_\mathrm{I}}{G_{\mathrm{Ic}}}\right)^{m/2} + \left(\frac{G_{\mathrm{II}}}{G_{\mathrm{IIc}}}\right)^{n/2} = 1 \qquad (7.43)$$

Various values for the parameters $m$ and $n$ have been proposed, based on test data curve fitting techniques. For instance, Borovkov *et al.* (1999) have proposed $m = 2.5, n = 3$, and Mi and Crisfield (1996) used $m = n = 4$, however, values $m = n = 2$ are generally adopted (Liu, 1994; Mi and Crisfield, 1996; Mohammadi, Owen and Peric, 1998).

There is usually no need for any specific crack propagation criteria to determine the angle of crack propagation for 2D interlaminar cracks; predefined interface cracks are generally assumed to propagate along the interface. Therefore, once the stability of an interface crack is violated, the interlaminar crack can be assumed to self-similarly propagate in a quasi-static manner. For dynamic problems, however, the extent of delamination can be determined from the crack-tip velocity and the analysis time-step. Asymptotic solutions for a moving orthotropic bimaterial crack have already been developed by Lee, Hawong and Choi (1996), which can be used to analyse dynamic delamination propagation in composites.

### 7.4.2  Stress-Based Criteria

The simple Chang–Springer stress-based criterion has been frequently used for prediction of the initiation or propagation of delamination in 3D problems (Chang and Springer, 1986),

$$\left(\frac{\sigma_n^2}{\bar{S}_n^2}\right) + \left(\frac{\sigma_{t1}^2 + \sigma_{t2}^2}{\bar{S}_t^2}\right) = 1 \qquad (7.44)$$

where $\bar{S}_n$ and $\bar{S}_t$ are the unidirectional normal and tangential bonding strengths, respectively, $\sigma_n$ represents the stress components normal to the interface and $\sigma_{t1}$ and $\sigma_{t2}$ are the two orthogonal tangential stress components on the tangential interface plane.

The Hashin delamination criterion is based on the Chang–Springer approach combined with a linear softening law. The delamination function can be defined as,

$$F(\sigma, \alpha) = f(\sigma) - \bar{S}_n(\eta) \leq 1 \tag{7.45}$$

where

$$f(\sigma) = \sqrt{\sigma^T A \sigma} \tag{7.46}$$

$$\bar{S}_n(\eta) = \bar{S}_{n0}(1 - \mu\eta) \tag{7.47}$$

with

$$A = \text{diag}\left[1, \frac{\bar{S}_n^2}{\bar{S}_t^2}, \frac{\bar{S}_n^2}{\bar{S}_t^2}\right] \tag{7.48}$$

$\eta$ can be assumed as the equivalent inelastic strain and $\mu$ describes the slope of the softening curve $\bar{S}_n(\eta)$, which can be determined from the critical energy release rate $G_c$, the initial tensile strength $\bar{S}_{n0}$ and the characteristic thickness of the bonding layer $h_t$,

$$\mu = \frac{\bar{S}_{n0} h_t}{2G_c} \tag{7.49}$$

Then, the rate of the internal variable can be determined from an evolution law,

$$\dot{\eta} = -\dot{\lambda} \frac{\partial F(\sigma, \eta)}{\partial \bar{S}_n} \tag{7.50}$$

where $\dot{\lambda}$ is the proportionality coefficient. For further details of the stress update procedure see Mohammadi, Forouzan-Sepehr and Asadollahi (2002) and Mohammadi and Moosavi (2007).

### 7.4.3 Contact-Based Criteria

A number of delamination criteria have been proposed, based on the use of a contact interaction approach for simulation of interlaminar behaviour (Mohammadi, Owen and Peric, 1998, 1999; Mohammadi, Forouzan-Sepehr and Asadollahi, 2002; Mohammadi and Forouzan-Sepehr, 2003). The Chang–Springer criterion can also be regarded in this category if $\sigma_n$ and $\sigma_t$ stress components are assumed as the contact interlaminar stresses. Assumptions for interaction between the normal and tangential fracture modes, especially in the softening regime, have resulted in a number of delamination criteria. Such an interaction is usually quantified by a variable $k$ in terms of the ratios of the normal and tangential relative displacements of the interface, $g_n$ and $g_t$, respectively, (Tvergaard, 1990; Espinosa, Dwivedi and Lu, 2000)

$$k = \left\{ \left(\frac{g_n}{g_n^{max}}\right)^{\alpha} + \left(\frac{g_t}{g_t^{max}}\right)^{\alpha} \right\}^{\frac{1}{\alpha}} - 1 \tag{7.51}$$

where $g_n^{max}$ and $g_t^{max}$ are the maximum tolerable relative displacements in the normal and tangential directions, respectively. $\alpha$ is a constant variable between 2 and 4 (usually 2).

The general methodology is usually designed on the basis of defining normal and tangential stress components of the interface for loading and unloading conditions based on the existing relative displacements $g_n$ and $g_t$, maximum tolerable values of $g_n^{max}$ and $g_t^{max}$ and an assumption for the dependence of the energy release rate $G$, the variable $k$ and the strength $\bar{S}_n$. For further details see Tvergaard (1990), Espinosa, Dwivedi and Lu (2000), Hashagen and de Borst (2000), Mi and Crisfield (1996), Mi et al. (1998), Mohammadi, Owen and Peric (1998, 1999), Mohammadi, Forouzan-Sepehr and Asadollahi (2002), Mohammadi and Forouzan-Sepehr (2003) and Mohammadi and Moosavi (2007).

## 7.5 Bimaterial XFEM

The general methodology of orthotropic extended finite element can be similarly extended to include bimaterial problems, if bimaterial enrichment functions are embedded into an existing extended finite element framework. In addition to generalized Heaviside and bimaterial near crack-tip enrichment functions, weak discontinuity enrichments are also required to guarantee continuity of the displacement field across the material interface. A variety of coupling terms between classical, Heaviside, crack tip and weak discontinuity degrees of freedom will form the final discretized stiffness matrix and the nodal force vector.

### 7.5.1 Governing Equation

Consider a bimaterial body in the state of equilibrium with the boundary conditions in the form of traction and displacement that also includes an interlaminar crack between the two parts of the constituents' orthotropic materials (Figure 7.4).

The strong form of the equilibrium equation can be written as:

$$\nabla \cdot \boldsymbol{\sigma} + \mathbf{f}^b = 0 \quad \text{in } \Omega \tag{7.52}$$

with the following boundary conditions:

$$\boldsymbol{\sigma} \cdot \mathbf{n} = \mathbf{f}^t \quad \text{on } \Gamma_t : \quad \text{external traction} \tag{7.53}$$

$$\mathbf{u} = \bar{\mathbf{u}} \quad \text{on } \Gamma_u : \quad \text{prescribed displacement} \tag{7.54}$$

$$\boldsymbol{\sigma} \cdot \mathbf{n} = 0 \quad \text{on } \Gamma_c : \quad \text{traction free crack} \tag{7.55}$$

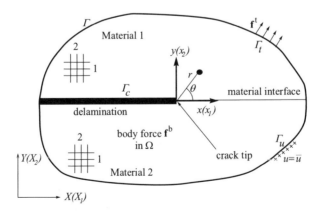

**Figure 7.4** A body with an interface crack between two orthotropic materials.

where $\mathbf{n}$ is normal to the boundary, $\Gamma_t$, $\Gamma_u$ and $\Gamma_c$ are the traction, displacement and crack boundaries, respectively, $\sigma$ is the stress tensor and $\mathbf{f}^b$ and $\mathbf{f}^t$ are the body force and external traction vectors, respectively.

The variational formulation of the boundary value problem can be defined as:

$$\int_\Omega \sigma \cdot \delta\varepsilon \, d\Omega = \int_\Omega \mathbf{f}^b \cdot \delta\mathbf{u} \, d\Omega + \int_\Gamma \mathbf{f}^t \cdot \delta\mathbf{u} \, d\Gamma \tag{7.56}$$

## 7.5.2  XFEM Discretization

In order to model crack surfaces and crack tips in the extended finite element method, the approximate displacement function $\mathbf{u}^h$ can be expressed in terms of the classical displacement $\mathbf{u}$, the crack-split $\mathbf{u}^H$, the crack tip $\mathbf{u}^{\text{tip}}$ and the material interface $\mathbf{u}^{\text{mat}}$ components as

$$\mathbf{u}^h(\mathbf{x}) = \mathbf{u}(\mathbf{x}) + \mathbf{u}^H(\mathbf{x}) + \mathbf{u}^{\text{tip}}(\mathbf{x}) + \mathbf{u}^{\text{mat}}(\mathbf{x}) \tag{7.57}$$

or more explicitly

$$\mathbf{u}^h(\mathbf{x}) = \left[\sum_{j=1}^{n} N_j(\mathbf{x})\mathbf{u}_j\right] + \left[\sum_{h=1}^{mh} N_h(\mathbf{x})H(\mathbf{x})\mathbf{a}_h\right]$$
$$+ \left[\sum_{k=1}^{mt} N_k(\mathbf{x})\left(\sum_{l=1}^{mf} F_l(\mathbf{x})\mathbf{b}_k^l\right)\right] + \left[\sum_{m=1}^{mm} N_m(\mathbf{x})\chi_m(\mathbf{x})\mathbf{c}_m\right] \tag{7.58}$$

and in a simple shifted form for $H$ and $F$ functions to guarantee the interpolation:

$$\mathbf{u}^h(\mathbf{x}) = \left[\sum_{j=1}^{n} N_j(\mathbf{x})\mathbf{u}_j\right] + \left[\sum_{h=1}^{mh} N_h(\mathbf{x})\left(H(\mathbf{x}) - H(\mathbf{x}_h)\right)\mathbf{a}_h\right]$$
$$+ \left[\sum_{k=1}^{mt} N_k(\mathbf{x})\left(\sum_{l=1}^{mf} [F_l(\mathbf{x}) - F_l(\mathbf{x}_k)]\mathbf{b}_k^l\right)\right] + \left[\sum_{m=1}^{mm} N_m(\mathbf{x})\chi_m(\mathbf{x})\mathbf{c}_m\right] \tag{7.59}$$

In Eqs. (7.58) and (7.59) $n$ is the number of nodes of each finite element with classical degrees of freedom $\mathbf{u}_j$ and shape functions $N_j(\mathbf{x})$, $mh$ is the number of nodes that have crack face (but not crack tip) in their support domain, $\mathbf{a}_h$ is the vector of additional degrees of nodal freedom for modelling crack faces (not crack tips) by the Heaviside function $H(\mathbf{x})$, $mt$ is the number of nodes associated with the crack tip in its influence domain, $\mathbf{b}_k^l$ is the vector of additional degrees of nodal freedom for modelling crack tips, $F_l(\mathbf{x})$ are $mf$ crack-tip enrichment functions, $mm$ is the number of nodes that are directly affected by the weak discontinuity, $\mathbf{c}_m$ is the vector of additional degrees of nodal freedom for modelling weak discontinuity interfaces and $\chi_m(\mathbf{x})$ is the enrichment function (3.32) used for modelling weak discontinuities.

The first term in Eqs. (7.58) and (7.59) is the classical finite element approximation, the second and third terms are the enriched approximation related to the crack surface and crack tip, respectively, while the last part is the enriched approximation used for modelling weak discontinuities. Figure 7.5 illustrates these three types of enriched nodes in a finite element modelling of an interface crack. Other nodes and their associated finite element degrees of freedom are not affected by the presence of the crack.

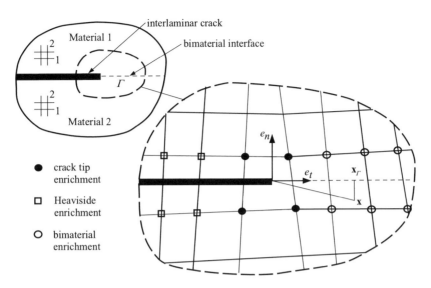

**Figure 7.5** Node selections for enrichment; nodes enriched with crack tip, Heaviside and weak discontinuity functions are marked by solid circles, squares and circles, respectively.

The discretized form of (7.56) based on XFEM approximation (7.59) can be written as a system of linear equations,

$$\mathbf{K}\mathbf{u}^h = \mathbf{f} \tag{7.60}$$

where $\mathbf{u}^h$ is the vector of nodal degrees of freedom for classical and enriched DOFs,

$$\mathbf{u}^h = \{\mathbf{u} \quad \mathbf{a} \quad \mathbf{b} \quad \mathbf{c}\}^T \tag{7.61}$$

$\mathbf{K}$ is the stiffness matrix and $\mathbf{f}$ is the vector of nodal force. The global stiffness matrix and force vector are calculated by assembling the element stiffness matrix and force vector. $\mathbf{K}_{ij}$ for an element $e$ is assembled from the following components:

$$\begin{cases} \mathbf{K}_{ij}^{rs} = \int_{\Omega^e} (\mathbf{B}_i^r)^T \mathbf{D} \mathbf{B}_j^s \, d\Omega & (r, s = u, a, b, c) \\ \mathbf{K}_{ij}^{ac} = \mathbf{K}_{ij}^{ca} = 0 \end{cases} \tag{7.62}$$

where $\mathbf{B}_i^u$, $\mathbf{B}_i^a$ and $\mathbf{B}_i^b$ are derivatives of shape functions defined in Eqs. (3.49)–(3.55), and no shifting procedure is required for derivatives of weak enrichment functions,

$$\mathbf{B}_i^c = \begin{bmatrix} (N_i \chi)_{,x} & 0 \\ 0 & (N_i \chi)_{,y} \\ (N_i \chi)_{,y} & (N_i \chi)_{,x} \end{bmatrix} \tag{7.63}$$

The vector of external forces $\mathbf{f}$ is defined as,

$$\mathbf{f}_i^e = \{\mathbf{f}_i^u \quad \mathbf{f}_i^a \quad \mathbf{f}_i^b \quad \mathbf{f}_i^c\}^T \tag{7.64}$$

and

$$\mathbf{f}_i^{\mathbf{u}} = \int_{\Gamma_t} N_i \mathbf{f}^t \, d\Gamma + \int_{\Omega^e} N_i \mathbf{f}^b \, d\Omega \tag{7.65}$$

$$\mathbf{f}_i^{\mathbf{a}} = \int_{\Gamma_t} N_i H \mathbf{f}^t \, d\Gamma + \int_{\Omega^e} N_i H \mathbf{f}^b \, d\Omega \tag{7.66}$$

$$\mathbf{f}_i^{\mathbf{b}\alpha} = \int_{\Gamma_t} N_i F_\alpha \mathbf{f}^t \, d\Gamma + \int_{\Omega^e} N_i F_\alpha \mathbf{f}^b \, d\Omega \, (\alpha = 1, \dots, mf) \tag{7.67}$$

$$\mathbf{f}_i^{\mathbf{c}} = \int_{\Gamma_t} N_i \chi \mathbf{f}^t \, d\Gamma + \int_{\Omega^e} N_i \chi \mathbf{f}^b \, d\Omega \tag{7.68}$$

Evaluation of (7.68) becomes important when a $\chi$-enriched element (weak discontinuity or material interface) is subjected to a traction boundary condition.

## 7.5.3   XFEM Enrichment Functions for Bimaterial Problems

### 7.5.3.1   Orthotropic Bimaterial Enrichments

The analytical solutions (7.17) and (7.18) can be used to derive the necessary enrichment functions for an orthotropic bimaterial interface crack tip: (Esna Ashari and Mohammadi, 2011a)

$$\{F_l(r, \theta)\}_{l=1}^8 = \left[ e^{-\varepsilon\theta_l} \cos\left(\varepsilon \ln(r_l) + \frac{\theta_l}{2}\right) \sqrt{r_l}, \quad e^{-\varepsilon\theta_l} \sin\left(\varepsilon \ln(r_l) + \frac{\theta_l}{2}\right) \sqrt{r_l}, \right.$$

$$e^{\varepsilon\theta_l} \cos\left(\varepsilon \ln(r_l) - \frac{\theta_l}{2}\right) \sqrt{r_l}, \quad e^{\varepsilon\theta_l} \sin\left(\varepsilon \ln(r_l) - \frac{\theta_l}{2}\right) \sqrt{r_l},$$

$$e^{-\varepsilon\theta_s} \cos\left(\varepsilon \ln(r_s) + \frac{\theta_s}{2}\right) \sqrt{r_s}, \quad e^{-\varepsilon\theta_s} \sin\left(\varepsilon \ln(r_s) + \frac{\theta_s}{2}\right) \sqrt{r_s},$$

$$\left. e^{\varepsilon\theta_s} \cos\left(\varepsilon \ln(r_s) - \frac{\theta_s}{2}\right) \sqrt{r_s}, \quad e^{\varepsilon\theta_s} \sin\left(\varepsilon \ln(r_s) - \frac{\theta_s}{2}\right) \sqrt{r_s} \right] \tag{7.69}$$

with

$$r_j = r\sqrt{\cos^2 \theta + Z_j^2 \sin^2 \theta}, \quad j = l, s, \begin{cases} Z_l = p \\ Z_s = q \end{cases} \tag{7.70}$$

$$\theta_j = \tan^{-1}\left(Z_j \tan \theta\right), \quad j = l, s, \begin{cases} Z_l = p \\ Z_s = q \end{cases} \tag{7.71}$$

where all parameters have been defined in Section 7.2.2. These enrichment functions span the analytical asymptotic displacement fields for a traction-free interfacial crack.

Nevertheless, it is important to note that the enrichment functions (7.69) correspond only to Type I of each composite layer, as defined by the complex roots of the characteristic Eq. (7.11). This is similar to the orthotropic enrichment functions (4.112), (4.115) and (4.116), previously proposed by Asadpoure, Mohammadi and Vafai (2007) for crack analysis in single orthotropic materials.

### 7.5.3.2 Isotropic Bimaterial Enrichments

The orthotropic bimaterial enrichment functions (7.69) can be considered as an orthotropic extension to existing isotropic bimaterial enrichment functions, developed by Sukumar *et al.* (2004),

$$
\begin{aligned}
\{F_l(r,\theta)\}_{l=1}^{12} = \Big\{ & \sqrt{r}\cos(\varepsilon\ln r)e^{-\varepsilon\theta}\sin\frac{\theta}{2}, \ \sqrt{r}\cos(\varepsilon\ln r)e^{-\varepsilon\theta}\cos\frac{\theta}{2}, \\
& \sqrt{r}\cos(\varepsilon\ln r)e^{\varepsilon\theta}\sin\frac{\theta}{2}, \ \sqrt{r}\cos(\varepsilon\ln r)e^{\varepsilon\theta}\cos\frac{\theta}{2}, \\
& \sqrt{r}\sin(\varepsilon\ln r)e^{-\varepsilon\theta}\sin\frac{\theta}{2}, \ \sqrt{r}\sin(\varepsilon\ln r)e^{-\varepsilon\theta}\cos\frac{\theta}{2}, \\
& \sqrt{r}\sin(\varepsilon\ln r)e^{\varepsilon\theta}\sin\frac{\theta}{2}, \ \sqrt{r}\sin(\varepsilon\ln r)e^{\varepsilon\theta}\cos\frac{\theta}{2}, \\
& \sqrt{r}\cos(\varepsilon\ln r)e^{\varepsilon\theta}\sin\frac{\theta}{2}\sin\theta, \ \sqrt{r}\cos(\varepsilon\ln r)e^{\varepsilon\theta}\cos\frac{\theta}{2}\sin\theta, \\
& \sqrt{r}\sin(\varepsilon\ln r)e^{\varepsilon\theta}\sin\frac{\theta}{2}\sin\theta, \ \sqrt{r}\sin(\varepsilon\ln r)e^{\varepsilon\theta}\cos\frac{\theta}{2}\sin\theta \Big\} \quad (7.72)
\end{aligned}
$$

It is worth noting that functions (7.72) cannot be directly used for orthotropic bimaterial problems, because the oscillatory index $\varepsilon$ is computed from isotropic material properties, which are not known for general orthotropic bimaterials, unless approximate average values are employed.

Finally, it should be emphasized that the general set of eight orthotropic bimaterial enrichment functions (7.69) contain all combined typical terms of $\sqrt{r_j}\cos(\theta_j/2)$ or $\sqrt{r}(\cos^2\theta_j + Z^2\sin^2\theta_j)^{0.25}\cos(\theta_j/2)$. As a result, they also include all the twelve terms of $\sin\theta_j$, $\sin\theta_j\sin(\theta_j/2)$, $\sin\theta_j\cos(\theta_j/2)$, ... which are present in the set of isotropic enrichment functions (7.72). Therefore, the enrichment functions (7.69) generally span all the various components of orthotropic and isotropic bimaterial enrichment functions.

### 7.5.3.3 Weak Discontinuity Enrichment

XFEM modelling of weak discontinuities has been comprehensively discussed in Section 3.6. They are used in places where the displacement field remains continuous while the strain field is discontinuous. The weak discontinuity enrichment function $\chi_m(\mathbf{x})$ can be defined as:

$$
\chi_m(\mathbf{x}) = |\xi(\mathbf{x})| - |\xi(\mathbf{x}_m)| \quad (7.73)
$$

where $\xi(\mathbf{x})$ is the signed distance function,

$$
\xi(\mathbf{x}) = \min|\mathbf{x} - \mathbf{x}_\Gamma\| \ \text{sign}[e_n.(\mathbf{x} - \mathbf{x}_\Gamma)] \quad (7.74)
$$

and $\mathbf{x}$, $\mathbf{x}_\Gamma$ ($\mathbf{x}_\Gamma \in \Gamma$) and $\mathbf{e_n}$ are defined in Figure 7.5.

It should be noted that the weak discontinuity enrichment function is used only for the elements that are cut by the interface, avoiding generation of excessive errors in surrounding elements. Alternative functions are also available; for example, Moës *et al.* (2003) have proposed a ridge enrichment function which is centred on the interface and has zero value on the elements not crossed by the interface.

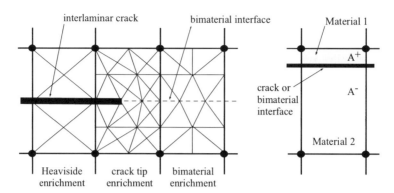

**Figure 7.6**   Cracked and bimaterial elements are subdivided into sub-triangles, and the definition of $A^+$ and $A^-$.

### 7.5.4   Discretization and Integration

It is well known that the ordinary Gaussian rules do not accurately calculate the integration over the enriched elements cut by a crack or an interface. In order to improve the numerical integration procedure, a partitioning technique is adopted, which is based on subdividing such elements at both sides of the crack or interface into sufficient sub-triangles whose edges are adapted to crack faces or the interface, as illustrated in Figure 7.6. In this partitioning approach, if the discontinuity is located very close to an edge, it might be better to avoid enriching that element to avoid numerical ill-conditioning and potential instabilities. For this purpose, if values of $A^-/(A^+ + A^-)$ or $A^+/(A^+ + A^-)$, where $A^+$ and $A^-$ are the area of the influence domain of a node above and below the crack, respectively (see Figure 7.6), are smaller than an allowable tolerance value, the node must not be enriched. The allowable tolerance value depends on each specific problem, but an average value of 0.01%, proposed by Dolbow, Moës and Belytschko (2001), can be used.

## 7.6   Numerical Examples

### 7.6.1   Central Crack in an Infinite Bimaterial Plate

An interfacial central crack in an infinite bimaterial block subjected to a remote unit tensile loading, $\sigma_{22}^0 = 1$ Mpa is considered, as depicted in Figure 7.7. This example was previously studied by Chow, Boem and Atluri (1995), Qu and Bassani (1993) and Chow and Atluri (1998) with slightly different material properties, as defined in Table 7.1. The model is a $[90°/0°]$ orthotropic bimaterial block in a plane strain condition, where the fibre direction in the $90°$ lamina (material 1) is along the 3 axis (out-of-plane $x_3$ direction) while for the $0°$ lamina (material 2), the fibre direction lies along the 1 axis ($x_1$).

The dimensions of the problem for numerical simulation are $a = 1$ m, $b = 20a$, $h = 20a$. Two different methods have been proposed to resemble the infinite domain by a finite model, as depicted in Figure 7.7. In the first approach, a sufficiently large width of the plate is modelled with a simple rolling boundary condition, as depicted in Figure 7.7a. In the second approach, as proposed by Qian and Sun (1998), the following condition must be held in terms of the boundary tractions $\sigma_{11}^1$ and $\sigma_{11}^2$ (Figure 7.7b) and the components of the compliance tensor $\mathbf{C}_{ijkl}$:

$$\mathbf{C}_{1111}^1 \sigma_{11}^1 - \mathbf{C}_{1111}^2 \sigma_{11}^2 = (\mathbf{C}_{1122}^2 - \mathbf{C}_{1122}^1)\sigma_{22}^0 \tag{7.75}$$

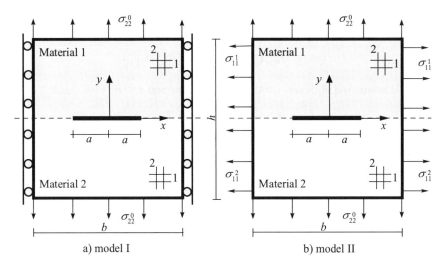

a) model I                                   b) model II

**Figure 7.7** Geometry and various boundary conditions of a bimaterial interface crack between two orthotropic materials (Chow, Boem and Atluri, 1995; Chow and Atluri, 1998).

which results in the following boundary tractions, as derived by Chow, Boem and Atluri (1995)

$$\sigma_{11}^1 = 0 \tag{7.76}$$

$$\sigma_{11}^2 = \sigma_{22}^\circ \left( \frac{[E_1]_{\text{material 2}}}{[E_1]_{\text{material 1}}} \right) \left[ \frac{\nu_{12} + \nu_{13}\nu_{32}}{1 - \nu_{13}\nu_{31}} - \frac{\nu_{12} + \nu_{13}\nu_{32}}{1 - \nu_{13}\nu_{31}} \right]_{\text{material 2}} \tag{7.77}$$

Evaluation of Eq. (7.68) for the classical enrichment interaction components of the force vector is important in this case, because the element placed on the material interface (a weak discontinuity) and next to the right edge of the plate should correctly evaluate Eq. (7.68) in order to take into account the effect of tractions on enriched degrees of freedom. For material setting 1, the computed value of traction $\sigma_{11}^2 = 4.02$ Mpa is about four times the value of the applied tensile loading $\sigma_{22}^\circ = 1$ Mpa, further highlighting the importance of this term.

**Table 7.1** Material properties of the bimaterial orthotropic plate.

| Properties | Chow, Boem and Atluri (1995) | | Qu and Bassani (1993) Chow and Atluri (1998) | |
|---|---|---|---|---|
| | Setting 1 | | Setting 2 | |
| | Material 1 | Material 2 | Material 1 | Material 2 |
| $E_1$ (GPa) | 9.81 | 142 | 10.8 | 137 |
| $E_2$ (GPa) | 9.81 | 9.81 | 10.8 | 10.8 |
| $E_3$ (GPa) | 142 | 9.81 | 137 | 10.8 |
| $G_{12}$ (GPa) | 3.81 | 6.01 | 3.36 | 5.65 |
| $G_{23}$ (GPa) | 6.01 | 3.81 | 5.65 | 3.36 |
| $G_{31}$ (GPa) | 6.01 | 6.01 | 5.65 | 5.65 |
| $\nu_{12}$ | 0.3 | 0.3 | 0.49 | 0.238 |
| $\nu_{23}$ | 0.3 | 0.3 | 0.238 | 0.49 |
| $\nu_{31}$ | 0.3 | 0.3 | 0.238 | 0.238 |

Qu and Bassani (1993) derived the analytical stress intensity factors for a crack between two anisotropic materials subjected to remote tractions $\sigma^0$:

$$k^\infty = \sqrt{\pi a} \mathbf{Y}(\varepsilon) \left[ (1 + 2i\varepsilon)(2a)^{-i\varepsilon} \right] \sigma^0 \tag{7.78}$$

where $\mathbf{Y}(\varepsilon)$ is a function of the oscillatory field $\varepsilon$ (Chow and Atluri, 1998), and

$$k^\infty = \left\{ K_{II}^\infty, K_I^\infty, K_{III}^\infty \right\} \tag{7.79}$$

$$\sigma^0 = \left\{ \sigma_{12}^0, \sigma_{22}^0, \sigma_{23}^0 \right\}^T \tag{7.80}$$

Therefore, the analytical stress intensity factors for the two material settings under unidirectional traction ($\sigma_{12}^\circ = \sigma_{23}^\circ = 0$) are determined as:

$$\begin{cases} \text{Setting 1} : K_I^\infty = 1.778, & K_{II}^\infty = 0.146 \\ \text{Setting 2} : K_I^\infty = 1.775, & K_{II}^\infty = 0.085 \end{cases} \tag{7.81}$$

Due to symmetry along the $x_2$ axis, only one half of the problem is modelled. Different finite element meshes are considered, which include uniform and adaptive structured meshes and unstructured meshes (Figure 7.8).

A partitioning technique, described in Section 7.5.4, is adopted for accurate numerical integration over enriched elements. Elements enriched with the Heaviside or weak discontinuity functions are partitioned into four sub-triangles, whereas crack-tip enriched elements are partitioned into six sub-triangles. A unified $2 \times 2$ Gauss quadrature rule is applied to other (non-enriched) finite elements.

Numerical evaluation of the stress intensity factors is performed by means of the interaction integral. The radius of the circular integral domain $r_J$ is assumed to be a third of the crack length, unless stated otherwise.

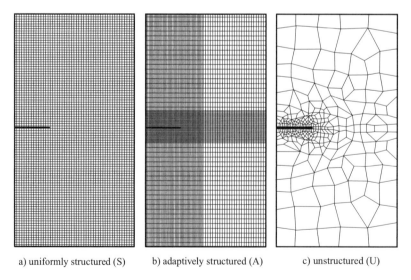

a) uniformly structured (S)          b) adaptively structured (A)          c) unstructured (U)

**Figure 7.8**  Typical FEM discretizations for a single edge crack in a bimaterial rectangular plate including uniformly and adaptively structured grids and unstructured meshes.

**Table 7.2** Values of SIFs relative errors (%) for the bimaterial plate with material settings 1 and 2 (Esna Ashari and Mohammadi, 2011a). (Reproduced by permission of John Wiley & Sons, Ltd.)

| | | | Number of elements and nodes | | Stress intensity factors | | Errors | |
|---|---|---|---|---|---|---|---|---|
| | | Method | $n_e$ | $n_n$ | $K_I$ | $K_{II}$ | $K_I$ | $K_{II}$ |
| Material setting | 1 | Analytical Qu and Bassani, 1993 | — | — | 1.778 | 0.146 | — | — |
| | | Chow, Boem and Atluri, 1995 — Hybrid Element | 56 | 205 | 3.183 | −0.05 | 0.79 | 1.37 |
| | | Chow, Boem and Atluri, 1995 — Mutual Integral | 216 | 679 | 2.969 | 0.047 | 0.67 | 0.68 |
| | | Chow, Boem and Atluri, 1995 — Mutual Integral | 72 | 237 | 2.774 | 1.246 | 0.56 | 7.53 |
| | | Chow, Boem and Atluri, 1995 — Extrapol. | 72 | 237 | 14.38 | −3.05 | 7.09 | 21.92 |
| | | Chow, Boem and Atluri, 1995 — Extrapol. | 216 | 679 | 16.48 | −1.35 | 8.27 | 10.27 |
| | | XFEM with 42 Gauss points in the tip element | 241 | 260 | 1.778 | 0.148 | 0.01 | 1.16 |
| Material setting | 2 | Analytical Qu and Bassani, 1993 | — | — | 1.775 | 0.085 | — | — |
| | | Chow and Atluri, 1998 — Mutual Integral | 216 | 679 | 2.839 | 0.076 | 0.6 | 0.1 |
| | | Chow and Atluri, 1998 — Mutual Integral | 72 | 237 | 3.017 | 1.241 | 0.7 | 13.6 |
| | | Chow and Atluri, 1998 — Extrapol. | 72 | 237 | 18.63 | −0.50 | 9.5 | 6.9 |
| | | Chow and Atluri, 1998 — Extrapol. | 216 | 679 | 25.02 | −0.11 | 13.1 | 2.3 |
| | | XFEM with 78 Gauss points in the tip element | 250 | 268 | 1.775 | 0.084 | 0.05 | 0.82 |
| | | XFEM with 42 Gauss points in the tip element | 250 | 268 | 1.776 | 0.085 | 0.07 | 0.35 |

Table 7.2 compares the differences between the analytical solution of stress intensity factors for an infinite anisotropic bimaterial block by Qu and Bassani (1993), with the numerical results reported by Chow, Boem and Atluri (1995) and Chow and Atluri (1998), based on the hybrid element method and the mutual integral approach, and the orthotropic bimaterial XFEM for both material settings. The reference numerical solutions were based on conforming eight-noded isoparametric elements for the whole block, combined with hybrid elements or quarter-point singular elements to simulate interlaminar cracks for both the mutual integral method and the extrapolation technique. The XFEM modelling was performed by Esna Ashari and Mohammadi (2011a).

It is clearly observed from Table 7.2 that the accuracy of XFEM is considerably higher than other approaches in the evaluation of both stress intensity factors with almost the same number of nodes. The only exception is the hybrid element method, which may have used slightly fewer nodes, but is much less accurate in evaluation of the mode I stress intensity factor (almost one tenth of XFEM). In contrast to XFEM, the hybrid element technique cannot be used directly for simulation of variable length cracks and propagation problems on a fixed mesh (Esna Ashari and Mohammadi, 2011a).

In order to examine the level of accuracy provided by the orthotropic bimaterial enrichment functions, a number of XFEM analyses are performed; with and without the crack-tip orthotropic bimaterial enrichment functions, and with the equivalent (single material) isotropic crack-tip enrichment functions in combination with orthotropic and isotropic auxiliary fields. All cases include the Heaviside and weak discontinuity enrichments for the relevant elements. In the special case where no crack-tip enrichment is used, the crack tip is located just off the edge of the tip element to minimize the approximation error by using a Heaviside enrichment function for the whole crack length. The computed stress intensity factors and the relative errors are compared in Table 7.3. Clearly, inclusion of the orthotropic bimaterial tip enrichments has substantially improved the results. It is also observed that the effect of crack-tip enrichments is higher for evaluation of mode II stress intensity factors. Far lower levels of accuracy are expected for local solutions, such as the stress field in the crack-tip element if the crack tip is positioned well inside the tip element and no tip enrichment is adopted.

The way crack-tip enrichments influence the local solution can be better demonstrated by comparing the variations of stress components over the crack-tip element for material setting 2, as depicted in Figure 7.9. In contrast to the tip-enriched solution, only low levels of stress components are obtained without the tip enrichments, a clear contradiction with the fundamental understandings of LEFM theory for the singular nature of a stress field near a crack tip. Similar results have been reported by Esna Ashari and Mohammadi, 2011a)

Figures 7.10 and 7.11 illustrate the displacement and stress contours of the present example for the material setting 1, based on an adaptively structured mesh with 28 elements along the crack length. The singular nature of the singular stress field is clearly observed, while the displacement field remains continuous along the material interface.

**Table 7.3** Comparison of SIFs with various crack-tip enrichment functions (all cases include Heaviside enrichments) (Esna Ashari and Mohammadi, 2011a). (Reproduced by permission of John Wiley & Sons, Ltd.)

|  |  |  | Bimaterial crack-tip enrichments | | | |
|---|---|---|---|---|---|---|
| Tip enrichments |  |  | Orthotropic | Isotropic | Isotropic | — |
| Auxiliary fields |  |  | Orthotropic | Orthotropic | Isotropic | — |
| Material setting | 1 | $K_I$ | 1.778 | 1.552 | 0.642 | 1.576 |
|  |  | $K_{II}$ | 0.148 | 0.188 | 0.102 | 0.163 |
|  |  | Error $K_I$ | 0.011 | 12.73 | 63.87 | 11.372 |
|  |  | Error $K_{II}$ | 1.164 | 28.56 | 169.93 | 11.712 |
|  | 2 | $K_I$ | 1.776 | 1.776 | 0.826 | 1.778 |
|  |  | $K_{II}$ | 0.085 | 0.096 | 0.169 | 0.075 |
|  |  | Error $K_I$ | 0.068 | 0.096 | 53.480 | 0.197 |
|  |  | Error $K_{II}$ | 0.353 | 13.294 | 298.588 | 11.882 |

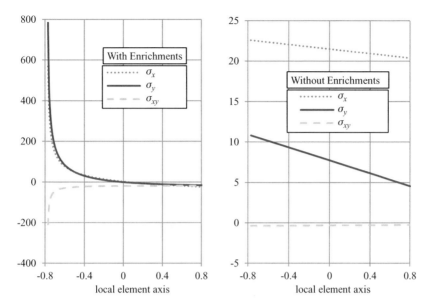

**Figure 7.9** Comparison of stress fields around the crack tip with and without crack tip enrichments for material setting 2. The crack tip is located at $(-0.8,0)$ in the local coordinates of the tip element (see Figure 7.12).

In order to examine the effect of the size of the tip enrichment zone on the solution, two different enrichment zones, depicted in Figure 7.12, are considered: the conventional one which includes only one tip element (indicated by solid circles) and the larger zone which consists of two layers of elements around the crack tip (designated by squares). The results presented in Table 7.4 show that the larger tip enrichment zone has not resulted in better evaluation of stress intensity factors in this example. A number of reasons may explain this

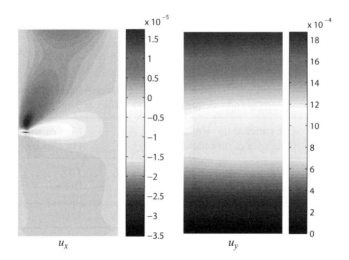

**Figure 7.10** $u_x$ and $u_y$ displacement contours for material setting 1.

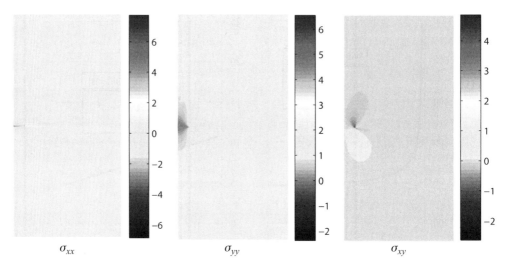

**Figure 7.11**   $\sigma_{xx}$, $\sigma_{yy}$ and $\sigma_{xy}$ contours for material setting 1.

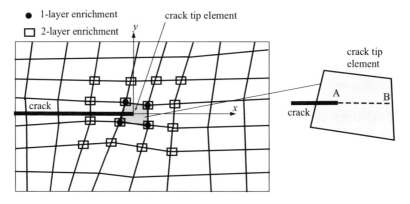

**Figure 7.12**   Different tip enrichment zones.

**Table 7.4**   Stress intensity factors with different tip enrichment zone (material setting 2) (Esna Ashari and Mohammadi, 2011a). (Reproduced by permission of John Wiley & Sons, Ltd.)

|  | One enriched tip element (One layer of tip enrichment) | Nine enriched tip elements (Two layers of tip enrichment) |
|---|---|---|
| $K_{\mathrm{I}}$ | 1.776 | 1.777 |
| $K_{\mathrm{II}}$ | 0.085 | 0.086 |
| Error $(K_{\mathrm{I}})$ | 0.068 | 0.113 |
| Error $(K_{\mathrm{II}})$ | 0.353 | 1.059 |

**Table 7.5**  Comparison of stress intensity factors with different number of Gauss points in the crack tip element for material setting 2 (Esna Ashari and Mohammadi, 2011a). (Reproduced by permission of John Wiley & Sons, Ltd.)

|  | Number of Gauss points in the crack-tip element | | | |
| --- | --- | --- | --- | --- |
|  | 6 | 18 | 42 | 78 |
| $K_I$ | 1.778 | 1.777 | 1.776 | 1.775 |
| $K_{II}$ | 0.088 | 0.084 | 0.085 | 0.084 |
| Error ($K_I$) | 0.192 | 0.135 | 0.068 | 0.051 |
| Error ($K_{II}$) | 3.529 | 1.412 | 0.353 | 0.824 |

small decreased accuracy, such as the validity of the analytical solution only in the vicinity of the crack tip, or the need for a change in radius of the $J$ integral for the larger zone, and so on. For the present case, fortunately, this reduction is very small and negligible and may further be reduced if finer finite elements are used.

The effect of using different numbers of Gauss points in the crack tip element is illustrated in Table 7.5 (and the last two rows of Table 7.2). A fixed quadrature partitioning approach is used in the tip element for all cases. It is observed that while all the results are in an acceptable range, using more Gauss points in the tip element does not necessarily lead to more accurate results. Such a conclusion, previously discussed by Sukumar *et al.* (2000), is also observed in many meshless-based weak form solutions, and is attributed to the non-polynomial nature of integrands, for which the Gauss quadrature rule is no longer necessarily accurate, and an optimal value for the number of Gauss points should be numerically obtained by sensitivity analyses.

Variations of the normalized stress intensity factors with respect to different radii $r_J$ of the circular $J$ integral domain are illustrated in Figure 7.13. The results are for the material setting 1

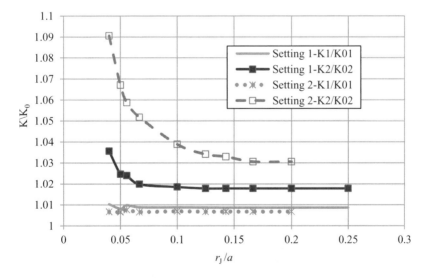

**Figure 7.13**  Variations of the normalized mode I and II stress intensity factors with respect to $r_J$ (Esna Ashari and Mohammadi, 2011a). $K_0(K_{01}, K_{02})$ is the corresponding analytical stress intensity factor, defined in (7.81).

on an adaptively structured mesh, shown in Figure 7.8b, and correspond to radii of about two to ten elements far from the crack tip. Apart from the very small domains, which remain relatively accurate, the results are practically domain independent as $r_J$ reaches up to about $0.15a$.

Also, a large number of uniformly structured (S), adaptively structured (A) and unstructured (U) models with different numbers of elements and nodes, various sizes of elements, especially in the crack-tip region, variable relative position of the crack tip in its corresponding finite element, and a number of $J$ domain radiuses are used to assess the effect of mesh refinement around the crack. According to Table 7.6, except for a very coarse mesh around the crack,

**Table 7.6** Mixed mode stress intensity factors for different meshes (S: uniformly structured, A: adaptively structured, U: unstructured). $n_{ect}$ is the number of elements along the crack length (Esna Ashari and Mohammadi, 2011a). (Reproduced by permission of John Wiley & Sons, Ltd.)

| | Mesh | $n_n$ | $n_e$ | $n_{ect}$ | $\dfrac{r_J}{a}$ | Stress intensity factor $K_I$ | $K_{II}$ | Error % $K_I$ | $K_{II}$ |
|---|---|---|---|---|---|---|---|---|---|
| | | 2904 | 2795 | 20 | 0.33 | 1.792 | 0.148 | 0.79 | 1.16 |
| | A | 17440 | 17172 | 36 | 0.33 | 1.790 | 0.147 | 0.68 | 0.48 |
| | | 23994 | 23680 | 47 | 0.33 | 1.790 | 0.147 | 0.68 | 0.41 |
| 1 | | 36024 | 35639 | 57 | 0.2 | 1.790 | 0.147 | 0.68 | 0.34 |
| | | 154 | 138 | 9 | 0.2 | 1.760 | 0.146 | 1.04 | 0.27 |
| | U | 304 | 288 | 14 | 0.2 | 1.765 | 0.147 | 0.75 | 0.69 |
| | | 260 | 241 | 12 | 0.33 | 1.778 | 0.148 | 0.01 | 1.16 |
| | | 6050 | 5886 | 6 | 0.37 | 1.784 | 0.098 | 0.51 | 14.82 |
| | | 11858 | 11628 | 8 | 0.26 | 1.786 | 0.093 | 0.67 | 9.41 |
| | | 15488 | 15225 | 10 | 0.23 | 1.786 | 0.091 | 0.64 | 6.47 |
| | S | 24200 | 23871 | 12 | 0.2 | 1.786 | 0.090 | 0.65 | 5.41 |
| | | 29282 | 28920 | 13 | 0.2 | 1.786 | 0.088 | 0.65 | 3.88 |
| | | 40898 | 40470 | 15 | 0.2 | 1.788 | 0.089 | 0.76 | 4.82 |
| | | 42050 | 41616 | 15 | 0.2 | 1.788 | 0.088 | 0.77 | 3.65 |
| | | 7252 | 7081 | 26 | 0.2 | 1.789 | 0.088 | 0.80 | 3.18 |
| | | 7252 | 7081 | 28 | 0.2 | 1.787 | 0.086 | 0.72 | 1.29 |
| | | 7742 | 7566 | 28 | 0.16 | 1.787 | 0.086 | 0.69 | 0.82 |
| 2 | | 9006 | 8814 | 31 | 0.16 | 1.789 | 0.086 | 0.81 | 1.29 |
| | | 17440 | 17172 | 32 | 0.16 | 1.787 | 0.086 | 0.70 | 1.06 |
| | A | 17440 | 17172 | 38 | 0.14 | 1.787 | 0.086 | 0.72 | 0.94 |
| | | 17440 | 17172 | 38 | 0.2 | 1.787 | 0.086 | 0.72 | 0.59 |
| | | 17440 | 17172 | 38 | 0.25 | 1.787 | 0.086 | 0.72 | 0.59 |
| | | 23994 | 23680 | 48 | 0.2 | 1.787 | 0.085 | 0.72 | 0.35 |
| | | 23994 | 23680 | 48 | 0.16 | 1.787 | 0.085 | 0.72 | 0.47 |
| | | 222 | 201 | 8 | 0.33 | 1.728 | 0.088 | 2.62 | 3.65 |
| | U | 260 | 241 | 12 | 0.33 | 1.766 | 0.089 | 0.48 | 5.18 |
| | | 385 | 369 | 17 | 0.2 | 1.752 | 0.091 | 1.27 | 6.71 |
| | | 268 | 250 | 18 | 0.33 | 1.775 | 0.084 | 0.05 | 0.82 |
| | | 538 | 511 | 19 | 0.33 | 1.761 | 0.087 | 0.77 | 2.82 |

Material setting (leftmost vertical label spanning all rows)

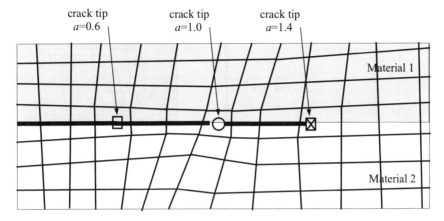

**Figure 7.14** A fixed unstructured finite element mesh for investigating the crack propagation process by analysing three different crack lengths.

the overall error remains well within the acceptable range, indicating the mesh independency of the results. In general, however, evaluation of the mode II stress intensity factor is more sensitive to the quality of the finite element mesh.

Finally, a quasi-static self-similar crack propagation problem for material setting 1 is idealized in this problem to illustrate the efficiency of the present XFEM for simulation of several crack lengths. Delamination is clearly bound to propagate only along the material interface and so no particular criterion is necessary to determine the propagation direction. The propagation process on the fixed finite element mesh is assumed as a set of three different crack lengths of 0.6, 1.0 and 1.4 m, partially shown in Figure 7.14. The results, presented in Table 7.7, show a very good agreement with the analytical solutions for all three cases.

## 7.6.2 Isotropic-Orthotropic Bimaterial Crack

A bimaterial specimen, composed of isotropic and orthotropic layers with an interface crack, and subjected to a remote tensile loading is considered (Figure 7.15).

**Table 7.7** Mixed mode stress intensity factors for different crack lengths on a fixed mesh of 514 elements and 540 nodes (Esna Ashari and Mohammadi, 2011a). (Reproduced by permission of John Wiley & Sons, Ltd.)

| Crack length | Analytical solution | | | XFEM | | Error % | |
| --- | --- | --- | --- | --- | --- | --- | --- |
| | $K_{\mathrm{I}}$ | $K_{\mathrm{II}}$ | $\dfrac{r_J}{a}$ | $K_{\mathrm{I}}$ | $K_{\mathrm{II}}$ | $K_{\mathrm{I}}$ | $K_{\mathrm{II}}$ |
| 0.6 | 1.374 | 0.157 | .33 | 0.376 | 0.155 | 0.102 | 1.271 |
| 1 | 1.778 | 0.146 | .33 | 1.776 | 0.148 | 0.146 | 1.027 |
| 1.4 | 2.106 | 0.128 | .33 | 2.125 | 0.129 | 0.897 | 1.016 |

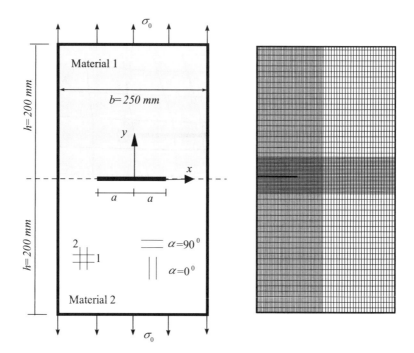

**Figure 7.15** Geometry of a bimaterial plate consisting of isotropic and orthotropic materials, and the typical finite element mesh.

Table 7.8 defines the material properties for the isotropic PSM-1 top layer, and the orthotropic Scotch ply 1002 material (below the interface) in the direction of the fibres and transverse to it, denoted by L and T, respectively. Two fibre orientations, parallel ($\alpha = 0°$) and perpendicular ($\alpha = 90°$) to the interface are considered for material 2.

The analysis is performed for the bimaterial plate in a plane stress state, for the range of crack lengths $a = 0.1b - 0.3b$. Due to the existing symmetry, only one half of the plate is simulated by adaptive structured finite element models, typically shown in Figure 7.15, to ensure sufficiently small element sizes in the vicinity of the crack. Element partitioning is performed to enhance the accuracy of the Gauss quadrature rule, and the complex stress intensity factor $\bar{K}$ (Eq. (7.4)) is determined for each crack length and fibre direction in material 2 ($\alpha = 0°$ or $\alpha = 90°$) by the interaction integral approach based on a $J$ integral domain radius of $r_J = 0.2a$.

Figure 7.16 compares the normalized amplitude of the complex stress intensity factor (NCSIF), $\bar{K}/\sigma_0\sqrt{\pi a}$, for different crack lengths, predicted by orthotropic bimaterial XFEM with the reference experimental and numerical (boundary collocation method, BCM) results of Shukla *et al.* (2003). The experimental results were reported as a band based on the average values with 95% confidence level. It is clear that the normalized complex stress intensity

**Table 7.8** Material properties of the bimaterial constituents.

| PSM-1 | $E = 2.5$ GPa $\quad G = 0.91$ GPa | | |
|---|---|---|---|
| Scotch ply 1002 | $E_{\text{L}} = 39.3$ GPa $\quad E_{\text{T}} = 9.7$ GPa $\quad G_{\text{LT}} = 3.1$ GPa $\quad \nu_{\text{LT}} = 0.25$ | | |

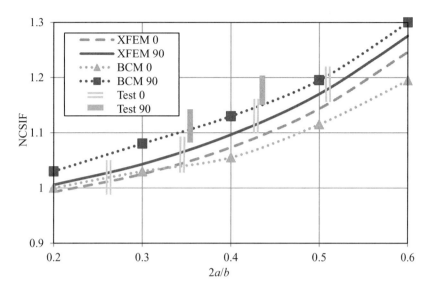

**Figure 7.16** Normalized complex stress intensity factors (NCSIF) as a function of the normalized crack length in comparison with the reference results (Shukla *et al.*, 2003; Esna Ashari and Mohammadi, 2011a). (Reproduced by permission of John Wiley & Sons, Ltd.)

factors increase with the crack length and they are higher for the case where fibres are aligned perpendicular to the interface ($\alpha = 90°$).

## 7.6.3 Orthotropic Double Cantilever Beam

A double cantilever beam (DCB) with orthotropic materials is simulated by orthotropic bi-material XFEM to determine the stress intensity factors and the strain energy release rates for the interface crack. The test specimen, which is composed of two homogeneous elastic layers with thickness $h$ and initial crack length $a$, is subjected to concentrated loads $P$, as depicted in Figure 7.17. The fibre direction in the top layer is along the out-of-plane direction and so the

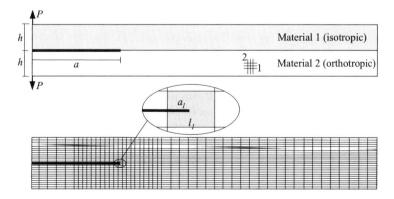

**Figure 7.17** A bilayer orthotropic DCB specimen and the typical finite element mesh.

layer can be treated as an isotropic material with $E_2$ and $v_{12}$ properties. In the lower layer, the fibres are along the horizontal direction and the orthotropic material properties are assumed to be $E_1, E_2, G_{12}$ and $v_{12}$.

Ang, Torrance and Tan (1996) defined two parameters $\eta_1$ and $\eta_2$ in terms of the elastic properties,

$$(\eta_1 \eta_2)^2 = \left( \frac{E_1}{E_2} \right) \tag{7.82}$$

$$(\eta_1 + \eta_2)^2 = 2 \left[ \left( \frac{E_1}{E_2} \right)^{\frac{1}{2}} + \frac{E_1}{2G_{12}} - v_{12} \right] \tag{7.83}$$

and analysed the following three different cases by the boundary element method,

$$\begin{aligned} &\text{Case } 1 : \eta_1 = 0.8, \quad \eta_2 = 2 \\ &\text{Case } 2 : \eta_1 = 1, \quad\ \ \eta_2 = 2 \\ &\text{Case } 3 : \eta_1 = 0.8, \quad \eta_2 = 3 \end{aligned} \tag{7.84}$$

Adaptive structured finite element models, typically depicted in Figure 7.17, with 720 nodes and 663 elements are employed for different crack lengths of $a = 2h, 4h, 6h$. The partitioning technique is used with the Gauss quadrature rule for integration purposes, while different $J$ integral domain radii $r_J$ are examined to obtain the best result for each crack length. The strain energy release rates are computed from the stress intensity factors $K_I$ and $K_{II}$ and the oscillation index $\varepsilon$ (Hemanth *et al.*, 2005):

$$G = \frac{1}{4 \cosh^2 (\pi \varepsilon)} \left[ D_{22} K_I^2 + D_{11} K_{II}^2 + 2 D_{12} K_I^2 K_{II}^2 \right] \tag{7.85}$$

where the components $D_{ij}$ are determined from the Barnett and Lothe tensor $L$ for materials 1 and 2, defined in (7.41) and (7.42),

$$D = L_1^{-1} + L_2^{-1} \tag{7.86}$$

The normalized strain energy release rates, $G/G_0$, are determined from the reference strain energy release rate $G_0$:

$$G_0 = \frac{12 P^2 a^2}{E_1 h^3} \tag{7.87}$$

Table 7.9 and Figure 7.18 compare the results of normalized strain energy release rates obtained by XFEM and the reference BEM values reported by Ang, Torrance and Tan (1996), indicating a very close agreement. Table 7.9 also illustrates the negligible effect of the ratio of crack length in a tip element to the element length ($a_1/l_1$ in Figure 7.17), based on several analyses performed on the case 3 ($\eta_1 = 0.8, \eta_2 = 3$) with $r_J/a = 0.33$.

Finally, the singularity of different stress components around the interlaminar crack tip can be clearly observed from the stress contours for the case of $\eta_1 = 1$ and $\eta_2 = 2$ in Figure 7.19. With the selected contour patterns, high gradient stress fields become constant after a distance almost twice the crack length. Similar results have been reported by Esna Ashari and Mohammadi (2011a) for the case of $\eta_1 = 0.8$ and $\eta_2 = 3$.

**Table 7.9** Comparison of the values of normalized strain energy release rates for different crack lengths, obtained from XFEM (Esna Ashari and Mohammadi, 2011a) and the boundary element method (Ang, Torrance and Tan, 1996). (Reproduced by permission of John Wiley & Sons, Ltd.)

| Case | $\eta_1, \eta_2$ | $\dfrac{a}{h}$ | $\dfrac{a_1}{l_1}$ | $\dfrac{r_J}{a}$ | $\dfrac{G}{G_0}$ XFEM | $\dfrac{G}{G_0}$ BEM |
|---|---|---|---|---|---|---|
| | | 2 | 0.5 | 0.33 | 3.30 | 3.35 |
| 1 | 0.8, 2 | 4 | 0.5 | 0.17 | 2.48 | 2.5 |
| | | 6 | 0.5 | 0.125 | 2.23 | 2.2 |
| | | 2 | 0.5 | 0.33 | 4.52 | 4.6 |
| 2 | 1, 2 | 4 | 0.5 | 0.17 | 3.42 | 3.3 |
| | | 6 | 0.5 | 0.125 | 3.09 | 3.0 |
| | | | 0.052 | | 6.18 | |
| | | | 0.280 | | 6.15 | |
| | | 2 | 0.500 | 0.33 | 6.17 | 6.35 |
| 3 | 0.8, 3 | | 0.556 | | 6.19 | |
| | | | 0.811 | | 6.13 | |
| | | 4 | 0.5 | 0.17 | 4.65 | 4.7 |
| | | 6 | 0.5 | 0.125 | 4.20 | 4.15 |

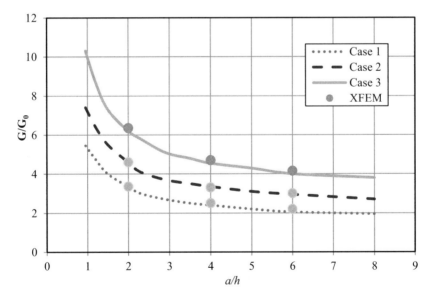

**Figure 7.18** Variations of $G/G_0$ with $a/h$ for a DCB specimen using the boundary element method (Ang, Torrance and Tan, 1996), compared with XFEM results (large circles) (Esna Ashari and Mohammadi, 2011a). (Reproduced by permission of John Wiley & Sons, Ltd.)

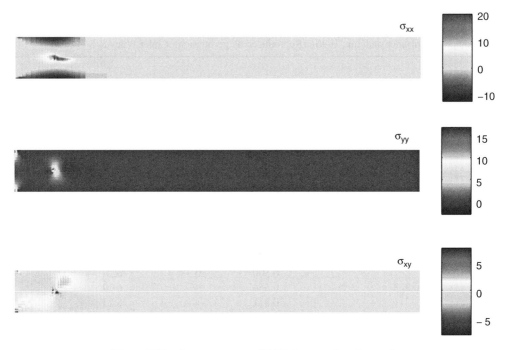

**Figure 7.19**   Stress contours of DCB for $\eta_1 = 1$ and $\eta_2 = 2$.

## 7.6.4   Concrete Beams Strengthened with Fully Bonded GFRP

A reinforced concrete beam, strengthened by epoxy-bonded glass fibre reinforced plate (GFRP), is investigated by the extended finite element method to illustrate the efficiency of XFEM in adopting a uniform mesh of finite elements for simulating the whole beam which includes, in the lowest part, elements simultaneously comprising concrete and thin GFRP, as depicted in Figure 7.20. In fact, XFEM is used to incorporate two different materials and

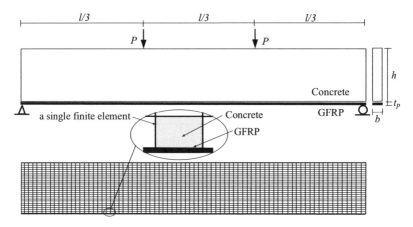

**Figure 7.20**   Geometry of the reinforced concrete beam strengthened by a fully bonded GFRP plate and the typical finite element model, including part of the mesh around the GFRP layer.

**Table 7.10**   Material properties of the bimaterial constituents.

| | |
|---|---|
| Concrete | $E = 21000$ MPa, $v = 0.2$ |
| GFRP | $E_1 = 37230$ MPa, $E_2 = 1861$ MPa, $G_{12} = 1618$ MPa, $v_{12} = 0.27$ |

a material interface within a single element. No debonding is allowed in this test and the XFEM analysis is assumed to remain linearly elastic for the whole range of loading. The reinforced concrete beam is under the four-point bending test and a plane stress state. The beam dimensions are: $b = 0.2$ m, $h = 0.3$ m and $l = 3$ m and two different thicknesses $t_p = 2$ and 4 mm of the GFRP plate are considered. The material properties are defined in Table 7.10.

The model is discretized by a uniform structured finite element mesh of 1500 nodes and 1386 quadrilateral four-noded elements, as typically depicted in Figure 7.20. It re-emphasized that only one row of elements contains the whole GFRP laminate and each element of this row includes the GFRP and part of the concrete. The weak discontinuity enrichment function (7.73) is used for these elements to ensure continuity of the solution.

The same problem was simulated by Hsu (2006) based on nonlinear models of the finite element code ABAQUS and compared with the experimental results. Esna Ashari and Mohammadi (2012) used the XFEM analysis, within the linear elastic part of the response, to simulate the same problem and obtained very close results for the central deflection to the reference numerical and experimental results for different thicknesses of the GFRP plate, as depicted in Figure 7.21.

## 7.6.5   FRP Reinforced Concrete Cantilever Beam Subjected to Edge Loadings

An externally CFRP-strengthened concrete beam subjected to edge bending moment and shear force is considered, as depicted in Figure 7.22. The dimensions of the problem are $l_c = 1500$ mm, $l_p = 1200$ mm, $h = 300$ mm, $t_p = 4$ mm, and a plane stress state is assumed.

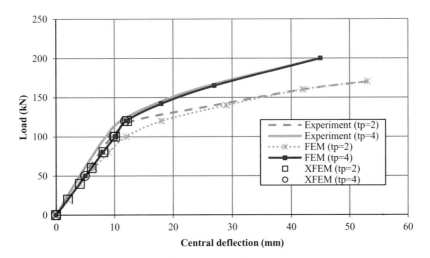

**Figure 7.21**   Central deflection of the GFRP-reinforced beam (Esna Ashari and Mohammadi, 2012). (Reproduced by permissions of SAGE Publications.)

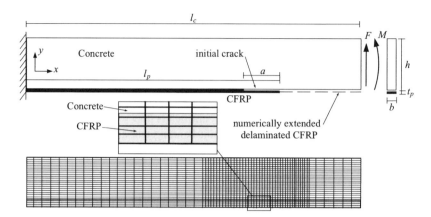

**Figure 7.22**  Geometry and boundary conditions of an interface crack in an FRP-plated cantilever beam under bending/shear loading, and details of typical FEM discretization.

The concrete beam is assumed as a homogeneous isotropic layer, while the CFRP is considered as an orthotropic unidirectional lamina, with material properties defined in Table 7.11.

Greco, Lonetti and Blasi (2007) reported analytical and numerical solutions to the normalized energy release rates in terms of different debonding lengths $a$. The same problem was simulated by Esna Ashari and Mohammadi (2012) based on the orthotropic bimaterial XFEM, and using two different values of $E_2 = 11860$ MPa and $E_2 = 7610$ MPa, since no reference data for $E_2$ was available. Another important assumption which simplifies the numerical simulation is to extend the delaminated CFRP layer to the right end of the beam. This assumption has no effect on the beam response but allows a structured finite element mesh to be used everywhere.

Discretization of the model is performed by an adaptive structured finite element mesh, where element sizes are chosen substantially smaller in the vicinity of the crack (about 4 mm and 1 mm in the $x$ and $y$ directions, respectively). The FRP laminate is discretized by 3.5 elements in the $y$-direction, as typically depicted in Figure 7.22.

The number of nodes and elements for different analyses were kept lower than 8500, which is comparable to the about 40 000 nodes and elements used in the reference study by Greco, Lonetti and Blasi (2007) using the commercial FE code LUSAS based on a crack-tip element size of 1 mm in the $x$-direction. They modelled the adhesive layer and assumed delamination to occur at the interface between the beam and the adhesive layer.

The size of the rectangular $J$ integral domain $r_J$ (the distance between the crack tip and each edge of the integration domain), is chosen in such a way as to ensure sufficient accuracy. Sensitivity analyses have shown that the optimum size for this problem is about the thickness of FRP plate below the crack edge and about 27 mm for the other three edges, as depicted in Figure 7.23 (Esna Ashari and Mohammadi, 2012).

The results of normalized strain energy release rates as a function of the crack length are compared with the reference analytical and numerical results (Greco, Lonetti and Blasi, 2007) in Figure 7.24. Figure 7.24a indicates that the energy release rate for the bending case remains

**Table 7.11**  Material properties of the bimaterial constituents.

| | |
|---|---|
| Concrete | $E = 30000$ MPa, $G_{12} = 12820.53$ MPa |
| CFRP | $E_1 = 160000$ MPa, $G_{12} = 5333.3$ MPa, $\nu_{12} = 0.3$ |

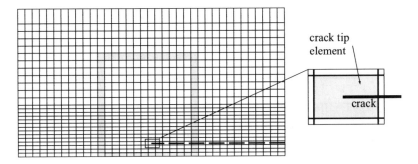

**Figure 7.23**   Rectangular *J* integral domain.

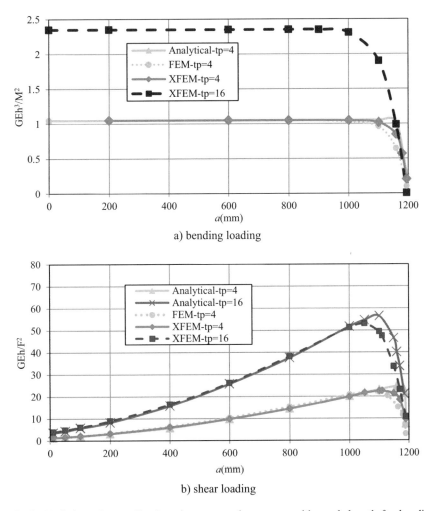

a) bending loading

b) shear loading

**Figure 7.24**   Variation of normalized strain energy release rates with crack length for bending and shear loadings and the influence of the FRP plate thickness ($t_\mathrm{p}$) (Esna Ashari and Mohammadi, 2012). (Reproduced by permission of SAGE Publications.)

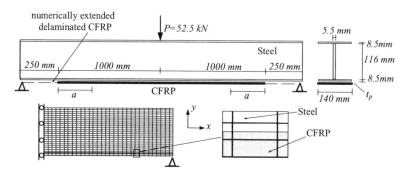

**Figure 7.25**   Geometry, cross-section and the typical FE mesh for half of the metallic beam, including the details of discretization around CFRP.

practically constant for most of the crack lengths and then decreases sharply towards zero in a crack arrest mechanism as the crack reaches the beam end. A similar arrest mechanism is developed for the shear loading with two different CFRP thicknesses $t_p = 4$ and 16 mm (Figure 7.24b). The close agreements indicate the relatively higher accuracy and far higher efficiency of XFEM compared with the conventional numerical approaches. In addition, it can be concluded that an increase in plate thickness leads to a substantial increase in strain energy release rates.

## 7.6.6   Delamination of Metallic I Beams Strengthened by FRP Strips

A simply supported CFRP reinforced I-shaped steel beam subjected to a point loading is considered, as depicted Figure 7.25. The CFRP plate is assumed to be symmetrically debonded at its two ends with the beam. Material properties of the isotropic steel beam and orthotropic CFRP reinforcement are defined in Table 7.12. The original cross-section of CFRP plate was $120 \times 1.4$ mm$^2$, which is changed to an equivalent cross section of $140 \times 1.2$ mm$^2$ to facilitate our 2D numerical modelling.

Due to the symmetry of the problem, only one half of the beam is analysed in a plane stress state for a range of debonding lengths. Esna Ashari and Mohammadi (2012) used an adaptive structured finite element mesh with quadrilateral elements, typically shown in Figure 7.25, to simulate the strengthened beam for different crack lengths ($a = 5 - 160$ mm).

Again, it is assumed that the delaminated CFRP layer is extended to both ends of the beam to simplify the numerical simulation by using a uniform finite element mesh for the whole model. This assumption has no effect on the beam response.

Element sizes are chosen substantially smaller in the vicinity of the crack and the FRP laminate is discretized into 1.5 elements in the $y$-direction. The total number of rows and columns of finite elements are 81 and 134, respectively. The mesh is composed of 266 elements and 402 nodes in the CFRP strip. The distance between the crack tip and each edge of the rectangular $J$ integration domain, $r_J$, is assumed to be about the thickness of the CFRP strip below the crack edge and about 25 mm for the other three edges.

**Table 7.12**   Material properties of the bimaterial constituents.

| Steel | $E = 210$ GPa, $\nu_{12} = 0.3$ |
|---|---|
| CFRP | $E_1 = 197$ GPa, $E_2 = 15.8$ GPa, $G_{12} = 8.18$ GPa, $\nu_{12} = 0.3$ |

**Table 7.13**  Comparison of strain energy release rates $G($ J m$^{-2})$ for different delamination lengths, obtained by the reference results (Colombi, 2006) and XFEM solution (Esna Ashari and Mohammadi, 2012). (Reproduced by permission of SAGE Publications.)

| $a$ (mm) | Transformed Section | Elastic Foundation | FEM | XFEM |
|---|---|---|---|---|
| 0 | 5.00 | 5.28 | 5.0 | — |
| 5 | 5.28 | 5.43 | 5.5 | 5.15 |
| 20 | 6.00 | 6.20 | 6.67 | 5.87 |
| 40 | 6.90 | 7.20 | 7.35 | 6.90 |
| 60 | 8.00 | 8.17 | 8.55 | 8.01 |
| 80 | 9.00 | 9.23 | 9.23 | 9.20 |
| 100 | 10.05 | 10.50 | 10.28 | 10.47 |
| 120 | 11.25 | 11.70 | 11.90 | 11.82 |
| 140 | 12.53 | 12.98 | 12.75 | 12.75 |
| 160 | 14.00 | 14.25 | 14.25 | 14.75 |

Colombi (2006) performed a numerical simulation based on a virtual crack length of 1 mm, a one-dimensional finite element for the beam and a two-dimensional shell finite element model for the composite strip and adhesive layer. According to Table 7.13, XFEM has produced the same level of accuracy with almost one tenth of the reference FEM model which included 3960 elements and 4955 nodes for the CFRP strip.

The effect of tip enrichments on the stress approximation around the crack tip is well demonstrated in Figure 7.26 which shows variations of different stress components on the crack tip element with and without tip enrichments for the special case of $a = 160$ mm.

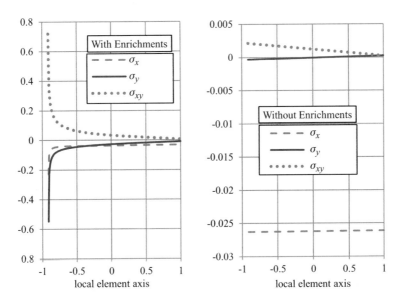

**Figure 7.26**  Variation of stress fields around the crack tip in the local curvilinear coordinates of the tip element. The crack tip is located at $(-0.92,0)$ (see the typical Figure 7.12).

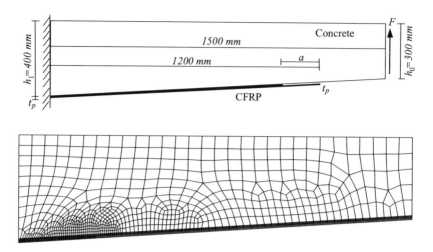

**Figure 7.27**   A cantilever variable section beam subjected to an edge transverse force and the typical finite element mesh for $a = 900$ mm (Esna Ashari and Mohammadi, 2012). (Reproduced by permission of SAGE Publications.)

Clearly, the singular nature of the stress field near the crack tip can only be recovered if the bimaterial crack-tip enrichment functions are included.

### 7.6.7   *Variable Section Beam Reinforced by FRP*

A cantilever, variable section beam was simulated by Esna Ashari and Mohammadi (2012) to illustrate the capability of XFEM in dealing with rather complex delamination problems. The beam is composed of concrete and reinforced with an orthotropic CFRP plate, as depicted in Figure 7.27. The material properties are defined in Table 7.14. The beam is subjected to an edge transverse force $F = 30$ N and the analysis is performed for a range of crack lengths in a plane stress state.

An unstructured finite element model, shown in Figure 7.27, is used for different crack lengths which can represent a quasi-static crack propagation problem. The finite element mesh is composed of 1152 quadrilateral elements and 1232 nodes, and adapted in such a way as to be refined around the crack tip. The delaminated CFRP layer is extended to the right end of the beam to facilitate the finite element discretization without changing the beam response.

**Table 7.14**   Material properties of the bimaterial variable section beam.

| | |
|---|---|
| Concrete | $E = 30$ GPa, $G_{12} = 12.82$ GPa |
| CFRP | $E_1 = 160$ GPa, $E_2 = 7.61$ GPa, $G_{12} = 5.33$ GPa, $\nu_{12} = 0.3$ |

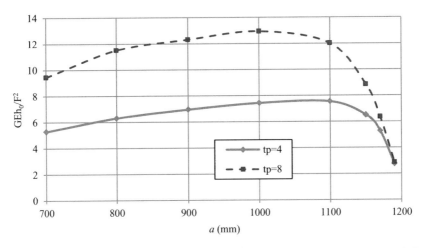

**Figure 7.28** Variation of strain energy release rate as a function of crack length (Esna Ashari and Mohammadi, 2012). (Reproduced by permission of SAGE Publications.)

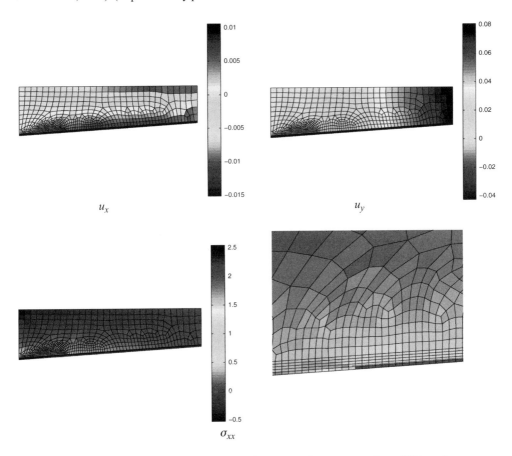

**Figure 7.29** $X$ and $Y$ displacement and $\sigma_{xx}$, $\sigma_{yy}$ and $\sigma_{xy}$ contours ($a = 1000$ mm).

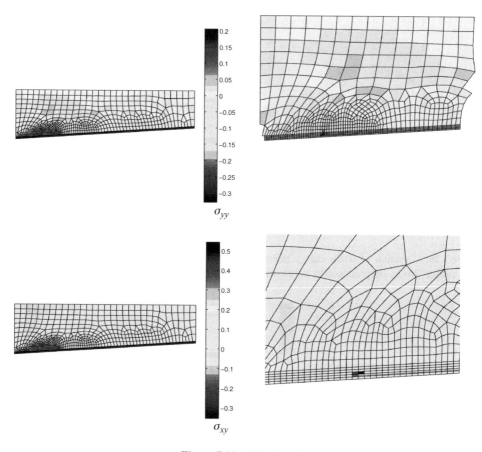

**Figure 7.29**   (*Continued*)

Figure 7.28 depicts the results of $GEh_0/F^2$ obtained by XFEM analysis for different crack lengths. A rapid drop in $G$ is predicted for both cases at about $a = 1100$ mm, where the interlaminar crack extends to the vicinity of the clamped edge.

Displacement and stress contours of the present example for $a = 1000$ mm are plotted in Figure 7.29. Similar results have been reported by Esna Ashari and Mohammadi (2012) for the total delamination length of $a = 900$ mm.

# 8

# New Orthotropic Frontiers

## 8.1 Introduction

Despite the enormous simplifications that isotropy brings to theoretical and numerical simulations, many real engineering applications are in anisotropic states. As a result, development of new numerical and computational techniques, usually with isotropic assumptions, will soon be followed by their extension to anisotropic regimes.

In this chapter, a number of new frontiers of computational mechanics in anisotropic media are briefly introduced. Two new state of the art subjects, extended isogeometric analysis (XIGA), and anisotropic dislocation dynamics by XFEM are discussed in more detail, and a summarized introduction to crack analysis in orthotropic biomechanical applications and piezoelectric bimaterials is presented.

## 8.2 Orthotropic XIGA

Introduction of the isogeometric analysis (IGA) by Hughes, Cottrell and Bazilevs (2005) has revived the earlier works on B-spline finite elements and opened a new and fast growing chapter in unifying computer aided design (CAD) and numerical solutions.

IGA has since been successfully adopted in several engineering problems, including structural dynamics (Cottrell *et al.*, 2006; Hassani, Moghaddam and Tavakkoli, 2009), simulation of blood flow (Zhang *et al.*, 2007), Navier–Stokes flow (Nielsen *et al.*, 2011), fluid–solid interaction (Bazilevs *et al.*, 2009), shells (Benson *et al.*, 2010a, 2011; Kiendl *et al.*, 2009, 2010), damage mechanics (Verhoosel *et al.*, 2010a), cohesive zone simulations (Verhoosel *et al.*, 2010b), heat transfer (Anders, Weinberg and Reichardt, 2012), turbulent flow (Bazilevs and Akkerman, 2010), large deformation (Benson *et al.*, 2011), electromagnetic (Buffa, Sangalli and Vazquez, 2010), strain localization (Elguedj, Rethore and Buteri, 2011), aerodynamics (Hsu, Akkerman and Bazileves, 2011), impact mechanics (Lu, 2011; Temizer, Wriggers and Hughes, 2011), topology optimization (Hassani, Khanzadi and Tavakkoli, 2012) and crack propagation (Verhoosel *et al.*, 2010b, Benson *et al.*, 2010b; De Luycker *et al.*, 2011; Haasemann *et al.*, 2011; Ghorashi, Valizadeh and Mohammadi, 2012).

The first attempt at enhancing IGA for crack problems was reported by Verhoosel *et al.* (2010b) by applying IGA for analysis of cohesive cracks based on a knot insertion scheme to

*XFEM Fracture Analysis of Composites*, First Edition. Soheil Mohammadi.
© 2012 John Wiley & Sons, Ltd. Published 2012 by John Wiley & Sons, Ltd.

represent discontinuities. They used a re-parameterization process in each step of crack growth and adopted simple and systematic refinement strategies to simulate common and complex engineering shapes.

In a different approach, Benson *et al.* (2010b) and De Luycker *et al.* (2011) proposed a combination of XFEM and IGA as a means of crack propagation analysis without remeshing procedures. They used a weighted enrichment function, proposed by Ventura, Gracie and Belytschko (2009), to discuss the effects of enrichments and boundary conditions for a fixed mode I crack. In another work, Haasemann *et al.* (2011) adopted the combination of quadratic NURBS functions and XFEM for analysis of a bimaterial body with a curved interface. Recently, a full combination of XFEM and IGA methodologies has been developed for general mixed mode crack propagation problems by the introduction of extended isogeometric analysis (XIGA) by Ghorashi, Valizadeh and Mohammadi (2012). XIGA uses the superior concepts of XFEM to extrinsically enrich the versatile IGA control points with Heaviside and crack-tip enrichment functions.

In the following, the basic formulation of XIGA and its extension to orthotropic crack problems are briefly discussed.

## 8.2.1 NURBS Basis Function

A non-uniform rational B-spline (NURBS) curve of order $p$ is defined in terms of the positions of a set of $i = 1, 2, \ldots, n_{cp}$ control points $\mathbf{T}_i = (X_{i_1}, X_{i_2})$, as

$$\mathbf{C}(\xi) = \sum_{i=1}^{n_{cp}} R_i^p(\xi)\mathbf{T}_i \quad 0 \leq \xi \leq 1 \tag{8.1}$$

with the NURBS functions $R_i^p(\xi)$

$$R_i^p(\xi) = \frac{B_{i,p}(\xi)W_i}{\sum\limits_{i=1}^{n_{cp}} B_{i,p}(\xi)W_i} \tag{8.2}$$

where $W_i$ are the weights associated with each control point $i$ and $B_{i,p}$ are the B-spline basis functions of order $p$ which are defined in a parametric space of the so-called knot vector $\Xi$,

$$\Xi = \left\{\xi_1, \xi_2, \ldots, \xi_{n_{cp}+p+1}\right\} \quad \xi_i \leq \xi_{i+1}, \quad i = 1, 2, \ldots, n_{cp} + p \tag{8.3}$$

where the knots $\xi_i, i = 1, 2, \ldots, n_{cp} + p + 1$ are real numbers representing the coordinates in the parametric space [0, 1].

B-spline basis functions can be defined in the following recursive form (Piegl and Tiller, 1997)

$$B_{i,p}(\xi) = \frac{\xi - \xi_i}{\xi_{i+p} - \xi_i}B_{i,p-1}(\xi) + \frac{\xi_{i+p+1} - \xi}{\xi_{i+p+1} - \xi_{i+1}}B_{i+1,p-1}(\xi) \quad p = 1, 2, 3, \ldots \tag{8.4}$$

with

$$B_{i,0}(\xi) = \begin{cases} 1 & \xi_i \leq \xi \leq \xi_{i+1} \\ 0 & \text{otherwise} \end{cases} \tag{8.5}$$

Derivatives of the B-spline basis function can then be computed from

$$\frac{dB_{i,p}(\xi)}{d\xi} = \frac{p}{\xi_{i+p} - \xi_i}B_{i,p-1}(\xi) + \frac{p}{\xi_{i+p+1} - \xi_{i+1}}B_{i+1,p-1}(\xi) \tag{8.6}$$

A NURBS surface of order $p$ in the $\xi_1$ direction and order $q$ in the $\xi_2$ direction is defined in the form of

$$\mathbf{S}(\xi_1, \xi_2) = \sum_{i=1}^{n_{cp}}\sum_{j=1}^{m_{cp}} R_{i,j}^{p,q}(\xi_1, \xi_2)\mathbf{T}_{i,j}$$

$$= \sum_{i=1}^{n_{cp}}\sum_{j=1}^{m_{cp}} \frac{B_{i,p}^1(\xi_1)B_{j,q}^2(\xi_2)W_{i,j}}{\sum_{i=1}^{n_{cp}}\sum_{j=1}^{m_{cp}}B_{i,p}^1(\xi_1)B_{j,q}^2(\xi_2)W_{i,j}}\mathbf{T}_{i,j} \quad 0 \le \xi_1, \xi_2 \le 1 \tag{8.7}$$

where $\mathbf{T}_{i,j}$ represent a $n_{cp} \times m_{cp}$ bidirectional set of control points, $W_{i,j}$ are their associate weights, and $B_{i,p}^1(\xi_1)$ and $B_{j,q}^2(\xi_2)$ are the B-spline basis functions defined on the knot vectors $\Xi_1$ and $\Xi_2$, respectively.

For a given 2D knot span (element), the total number of control points per element (also the number of nonzero basis functions) is $n_{cp} = (p+1) \times (q+1)$. Another important aspect of NURBS functions $R_i^p(\xi)$ is the partition of unity: $\sum_{i=1}^{n_{cp}} R_{i,p}(\xi) = 1$. For more details, refer to Piegl and Tiller (1997).

In isogeometric analysis, a particular knot vector, called an open knot vector, where the end knots are repeated $p+1$ times, is usually utilized in order to satisfy the Kronecker delta property at boundary points (Roh and Cho, 2004).

Ghorashi, Valizadeh and Mohammadi (2012) have presented a simple example to discuss various aspects of IGA modelling. The test includes generation of the exact geometry of a circular disk of radius 10 using only one element with the control points and weights given in Figure 8.1, the knot vectors $\Xi_1 = \Xi_2 = \{0, 0, 0, 1, 1, 1\}$ and NURBS function of order 2. The final NURBS basis functions are shown in Figure 8.2. It is observed that the same number of basis functions and control points are used, and each basis function can be uniquely assigned to its corresponding control point. It is also important to note that each basis function is non-zero only within its own support/influence domain, which may not even include its associated control point. The reason may be attributed to the fact that the control points, in general, may not be located in the physical space, and so they are not necessarily in support domains of their basis functions (Ghorashi, Valizadeh and Mohammadi, 2012).

## 8.2.2 Extended Isogeometric Analysis

The general methodology of extended finite element can be similarly extended to isogeometric crack analysis by including the enrichment functions into a conventional isogeometric approximation. The generalized Heaviside and near crack-tip anisotropic enrichment functions are utilized by extra degrees of freedom to model discontinuous and singular fields near a crack.

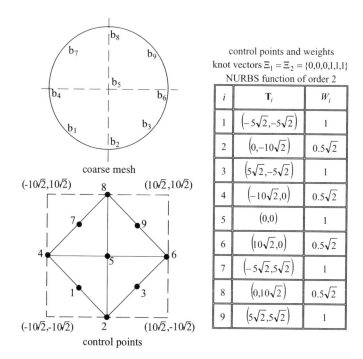

control points and weights
knot vectors $\Xi_1 = \Xi_2 = \{0,0,0,1,1,1\}$
NURBS function of order 2

| $i$ | $\mathbf{T}_i$ | $W_i$ |
|---|---|---|
| 1 | $\left(-5\sqrt{2},-5\sqrt{2}\right)$ | 1 |
| 2 | $\left(0,-10\sqrt{2}\right)$ | $0.5\sqrt{2}$ |
| 3 | $\left(5\sqrt{2},-5\sqrt{2}\right)$ | 1 |
| 4 | $\left(-10\sqrt{2},0\right)$ | $0.5\sqrt{2}$ |
| 5 | $(0,0)$ | 1 |
| 6 | $\left(10\sqrt{2},0\right)$ | $0.5\sqrt{2}$ |
| 7 | $\left(-5\sqrt{2},5\sqrt{2}\right)$ | 1 |
| 8 | $\left(0,10\sqrt{2}\right)$ | $0.5\sqrt{2}$ |
| 9 | $\left(5\sqrt{2},5\sqrt{2}\right)$ | 1 |

**Figure 8.1** Coarse mesh and control net for a disk of radius 10 (Ghorashi, Valizadeh and Mohammadi, 2012). (Reproduced by permission of John Wiley & Sons, Ltd.)

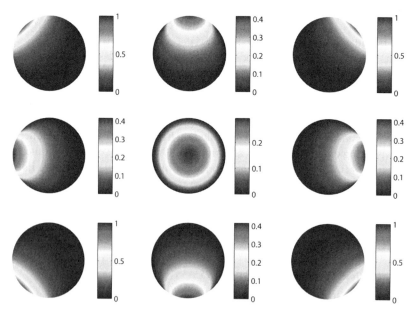

**Figure 8.2** NURBS basis functions for a disk generated by nine control points (Ghorashi, Valizadeh and Mohammadi, 2012). (Reproduced by permission of John Wiley & Sons, Ltd.)

### 8.2.2.1 Governing Equation

Consider a body in the state of equilibrium with the boundary conditions in the form of traction and displacements that also include a crack (Figure 4.1). The strong form of the equilibrium equation can be written as:

$$\nabla.\sigma + \mathbf{f}^b = 0 \quad \text{in } \Omega \tag{8.8}$$

with the following boundary conditions:

$$\sigma \cdot \mathbf{n} = \mathbf{f}^t \quad \text{on } \Gamma_t : \quad \text{external traction} \tag{8.9}$$

$$\mathbf{u} = \bar{\mathbf{u}} \quad \text{on } \Gamma_u : \quad \text{prescribed displacement} \tag{8.10}$$

$$\sigma \cdot \mathbf{n} = 0 \quad \text{on } \Gamma_c : \quad \text{traction free crack} \tag{8.11}$$

where $\Gamma_t$, $\Gamma_u$ and $\Gamma_c$ are traction, displacement and crack boundaries, respectively, $\sigma$ is the stress tensor and $\mathbf{f}^b$ and $\mathbf{f}^t$ are the body force and external traction vectors, respectively (Figure 3.4).

The variational formulation of the boundary value problem can be defined as:

$$\int_\Omega \sigma \cdot \delta\varepsilon \, d\Omega = \int_\Omega \mathbf{f}^b \cdot \delta\mathbf{u} \, d\Omega + \int_\Gamma \mathbf{f}^t \cdot \delta\mathbf{u} \, d\Gamma \tag{8.12}$$

### 8.2.2.2 Extended Isogeometric Discretization

In the extended isogeometric analysis (XIGA), the isogeometric approximation is locally enriched to enhance representation of discontinuities and singular fields near a crack. According to the location of a crack, a few degrees of freedom are added to the selected control points of the original IGA model near to the crack, and contribute to the overall approximation through the use of enrichment functions.

While the NURBS basic definition (8.1) is used for parametric approximation of the geometry $\mathbf{X}(\xi) = \sum_{i=1}^{n_{cp}} R_i^p(\xi)\mathbf{T}_i$, the displacement approximation $\mathbf{u}^h(\xi)$ for a typical point $\xi = (\xi_1, \xi_2)$ (parametric coordinates) can be derived from the partition of unity based enrichment of classical IGA,

$$\mathbf{u}^h(\xi) = \sum_{i=1}^{n_{cp}} R_i(\xi)\mathbf{u}_i + \sum_{h=1}^{mh} R_i(\xi)H(\xi)\mathbf{a}_h + \sum_{k=1}^{mt} R_k(\xi) \left( \sum_{l=1}^{mf} F_l(\xi)\mathbf{b}_k^l \right) \tag{8.13}$$

where $\mathbf{u}_i$ are the classical displacement vector, $\mathbf{a}_h$, $\mathbf{b}_k^l$ are vectors of additional degrees of freedom for modelling crack faces and the crack tip, respectively, $n_{cp}$ is the number of nonzero basis functions for a given knot span (element), $mh$ is the number of basis functions that have crack face (but not crack tip) in their support domain and $mt$ is the number of basis functions associated with the crack tip in their influence domain. $H(\xi)$ is the Heaviside function, which becomes $+1$ if $(X, Y)$(physical coordinates corresponding to the parametric coordinates $\xi = (\xi, \eta)$) is above the crack and $-1$, otherwise (see Section 8.2.2.4) and $F_l(\xi)$ are crack-tip enrichment functions which are defined for orthotropic materials in Section 8.2.2.3.

Strain and stress components can be computed from the approximated displacements $\mathbf{u}^h$,

$$\varepsilon(\xi) = L\mathbf{u}^h(\xi) \tag{8.14}$$

$$\sigma(\xi) = \mathbf{D}\varepsilon(\xi) \tag{8.15}$$

where $L$ is the differential operator and the elasticity tensor $\mathbf{D}$ is defined in Section 4.2.1. Then, discretization of Eq. (8.12) using the XIGA procedure (8.13) results in a discrete system of linear equilibrium equations:

$$\mathbf{K}\mathbf{u}^h = \mathbf{f} \tag{8.16}$$

where $\mathbf{u}^h$ is the global vector of DOFs that collects the displacement control variables $\mathbf{u}$ and additional enrichment degrees of freedom $\mathbf{a}$ and $\mathbf{b}$:

$$\mathbf{u}^h = \{\mathbf{u} \quad \mathbf{a} \quad \mathbf{b}\}^T \tag{8.17}$$

$\mathbf{K}$ and $\mathbf{f}$ are assembled from the element stiffness matrix and element force vector, respectively. The global stiffness matrix $\mathbf{K}$ is assembled from the stiffness of each element $\Omega^e$,

$$\mathbf{K}_{ij}^e = \begin{bmatrix} \mathbf{K}_{ij}^{uu} & \mathbf{K}_{ij}^{ua} & \mathbf{K}_{ij}^{ub} \\ \mathbf{K}_{ij}^{au} & \mathbf{K}_{ij}^{aa} & \mathbf{K}_{ij}^{ab} \\ \mathbf{K}_{ij}^{bu} & \mathbf{K}_{ij}^{ba} & \mathbf{K}_{ij}^{bb} \end{bmatrix}, \quad (i, j = 1, 2, \ldots, n_{cp}) \tag{8.18}$$

with

$$\mathbf{K}_{ij}^{rs} = \int_{\Omega^e} \left(\mathbf{B}_i^r\right)^T \mathbf{D}\mathbf{B}_j^s \, d\Omega \quad (r, s = \mathbf{u}, \mathbf{a}, \mathbf{b}) \tag{8.19}$$

where $\mathbf{B}_i^u$, $\mathbf{B}_i^a$ and $\mathbf{B}_i^b$ are derivatives of basis functions,

$$\mathbf{B}_i^u = \begin{bmatrix} R_{i,X_1} & 0 \\ 0 & R_{i,X_2} \\ R_{i,X_2} & R_{i,X_1} \end{bmatrix} \tag{8.20}$$

$$\mathbf{B}_i^a = \begin{bmatrix} (R_iH)_{,X_1} & 0 \\ 0 & (R_iH)_{,X_2} \\ (R_iH)_{,X_2} & (R_iH)_{,X_1} \end{bmatrix} \tag{8.21}$$

$$\mathbf{B}_i^b = \begin{bmatrix} (R_iF_\alpha)_{,X_1} & 0 \\ 0 & (R_iF_\alpha)_{,X_2} \\ (R_iF_\alpha)_{,X_2} & (R_iF_\alpha)_{,X_1} \end{bmatrix} \quad (\alpha = 1, 2, 3 \text{ and } 4) \tag{8.22}$$

The derivative of the basis functions $R_i$ with respect to the physical coordinates $X_i (i = 1, 2)$ in Eqs. (8.20)–(8.22) can be calculated from the derivatives with respect to the parametric coordinates $\xi_i (i = 1, 2)$,

$$\begin{Bmatrix} R_{i,X_1} \\ R_{i,X_2} \end{Bmatrix} = \mathbf{J}^{-1} \begin{Bmatrix} R_{i,\xi_1} \\ R_{i,\xi_2} \end{Bmatrix} \tag{8.23}$$

where $\mathbf{J}$ is the Jacobian matrix for transformation between physical and parametric spaces,

$$\mathbf{J} = \begin{bmatrix} X_{1,\xi_1} & X_{2,\xi_1} \\ X_{1,\xi_2} & X_{2,\xi_2} \end{bmatrix} \tag{8.24}$$

$\mathbf{f}$ is the vector of external forces,

$$\mathbf{f}_i^e = \{\mathbf{f}_i^u \quad \mathbf{f}_i^a \quad \mathbf{f}_i^{b1} \quad \mathbf{f}_i^{b2} \quad \mathbf{f}_i^{b3} \quad \mathbf{f}_i^{b4}\}^T \tag{8.25}$$

and

$$\mathbf{f}_i^{\mathbf{u}} = \int_{\Gamma_t} R_i \mathbf{f}^t \, d\Gamma + \int_{\Omega^e} R_i \mathbf{f}^b \, d\Omega \tag{8.26}$$

$$\mathbf{f}_i^{\mathbf{a}} = \int_{\Gamma_t} R_i H \mathbf{f}^t \, d\Gamma + \int_{\Omega^e} R_i H \mathbf{f}^b \, d\Omega \tag{8.27}$$

$$\mathbf{f}_i^{\mathbf{b}\alpha} = \int_{\Gamma_t} R_i F_\alpha \mathbf{f}^t \, d\Gamma + \int_{\Omega^e} R_i F_\alpha \mathbf{f}^b \, d\Omega \quad (\alpha = 1, 2, 3 \text{ and } 4) \tag{8.28}$$

### 8.2.2.3 Orthotropic Enrichment Functions

The first introduction of XIGA by Ghorashi, Valizadeh and Mohammadi (2012) was based on isotropic crack-tip enrichment functions (4.119). Alternatively, orthotropic crack-tip enrichment functions (4.112) can be used for enhancing the IGA approximation in the vicinity of the crack tip in orthotropic media,

$$\{F_l(r, \theta)\}_{l=1}^4 = \left\{ \sqrt{r} \cos \frac{\theta_1}{2} \sqrt{g_1(\theta)}, \ \sqrt{r} \cos \frac{\theta_2}{2} \sqrt{g_2(\theta)}, \ \sqrt{r} \sin \frac{\theta_1}{2} \sqrt{g_1(\theta)}, \ \sqrt{r} \sin \frac{\theta_2}{2} \sqrt{g_2(\theta)} \right\} \tag{8.29}$$

where functions $g_k(\theta)$ and $\theta_k (k = 1, 2)$ are defined in (4.113) and (4.114).

### 8.2.2.4 Crack Face Modelling

In order to illustrate the effect of the generalized Heaviside function $H(\xi)$ on the basis functions in Eq. (8.13), Ghorashi, Valizadeh and Mohammadi (2012) presented an example of cubic B-spline basis functions for an open, uniform knot vector $\Xi = \{0, 0, 0, 0, 0.2, 0.4, 0.6, 0.8, 1, 1, 1, 1\}$, affected by a discontinuity at $\xi = 0.5$, as depicted in Figure 8.3. Accordingly, the cubic basis functions $B$ are modified to enriched $BH$ functions. Except for the position of discontinuity, the functions remain $C^2$ continuous everywhere.

### 8.2.2.5 Selection of Enriched Control Points

The conclusion on the one-to-one dependence of control points and basis functions, made at the end of Section 8.2.1, is now used to select the crack-face and crack-tip enriched control points.

The Heaviside enrichment is applied on the control points which support domains of their corresponding basis functions that intersect with the crack face (not crack tip). For example in Figure 8.4, the typical control points $\mathbf{T}_j$ are enriched for modelling crack face discontinuity.

Similarly, crack-tip enrichments are adopted for the control points where their corresponding basis functions have influence domains that include the crack tip (typical control points $\mathbf{T}_k$ in Figure 8.4).

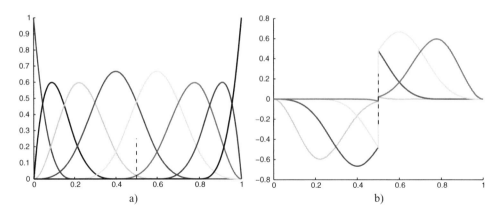

a)                                                    b)

**Figure 8.3**   (a) The cubic B-spline basis functions $B$, (b) Enriched $BH$ functions by the effect of discontinuity at $\xi = 0.5$ (Ghorashi, Valizadeh and Mohammadi, 2012). (Reproduced by permission of John Wiley & Sons, Ltd.)

The selection procedure can now be explained in more detail as:

1. The parametric coordinates of the crack tip ($\xi_{\text{tip}}$) are determined.
2. For all non-zero NURBS values ($R_k(\xi_{\text{tip}}) \neq 0$), their associated control points $\mathbf{T}_k$ are selected for crack-tip enrichment.
3. For all typical points $\mathbf{X}_a$ from the predefined set of points on the crack face ($\mathbf{X}_A$, $\mathbf{X}_B$, ... in Figure 8.4), check for all non-zero NURBS values ($R_j(\xi(\mathbf{X}_a)) \neq 0$).

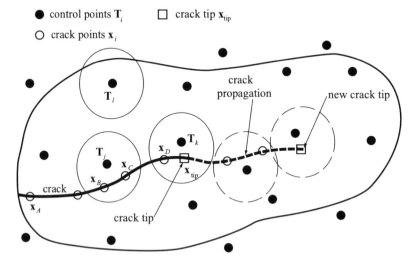

**Figure 8.4**   Selection of enriched control points: ordinary control points $\mathbf{T}_l(R_l(\xi_{\text{tip}}) = R_l(\xi_a) = 0, a =$ A, B, ...), crack-tip enriched control points $\mathbf{T}_k(R_k(\xi_{\text{tip}}), R_k(\xi_D) \neq 0)$ and Heaviside-enriched control points $\mathbf{T}_j$ $(R_j(\xi_B), R_j(\xi_C) \neq 0)$. The same procedure is followed for selection of new enriched crack face and tip control points in a crack propagation analysis. The large circles represent the influence domain of the basis function associated with each control point.

4. Apply the Heaviside enrichment to control points $\mathbf{T}_j$.
5. A control point can be selected for only one type of enrichment. For example, while the control point $\mathbf{T}_k$ in Figure 8.4 is candidate for both enrichments: that is, $R_k(\xi_{\text{tip}}) \neq 0$, $R_k(\xi(\mathbf{X}_D)) \neq 0$, it is enriched only by the crack-tip enrichment functions.

Similar to XFEM simulations the so-called topological and geometrical enrichments can be used. In the topological enrichment, the size of the enrichment domain depends on the order of NURBS or the sizes of the support/influence domains, as described in the above procedure. In contrast, the geometrical enrichment predefines a constant area of enrichment for different XIGA simulations, to limit the enrichment to only those control points where the influence domains of corresponding basis functions completely lie in that area.

### 8.2.2.6   Numerical Integration

It is well known that the accuracy of the conventional Gauss quadrature rule is substantially decreased if a discontinuity is present within the integration domain. Therefore, the sub-triangulation technique, originally developed in XFEM, is adopted to integrate over the XIGA elements. The method has been successfully adopted by Ghorashi, Valizadeh and Mohammadi (2012) to analyse several crack problems.

The first step is to subdivide the elements, cut by a crack or including a crack tip, into a set of sub-triangles on all sides of the crack, as depicted in Figures 8.5 and 8.6. Then, the Gauss rule is adopted for integration over each triangle. The method, however, is more complicated because of the difference between the geometric and parametric spaces, and additional necessary transformations.

Other ordinary IGA elements which are not cut by the crack, only require conventional transformations $T_1$ and $T_2$. In contrast, triangulation of a cracked element, which is performed on the parent parametric element, defined in the $[-1, 1] \times [-1, 1]$ domain (see Figure 8.6), requires an extra transformation $T_3$.

The intersection points and the point of the crack tip in the parent parametric element are obtained from coordinates of the element vertices and the crack tip in the physical space, using a numerical procedure for solving nonlinear algebraic equations.

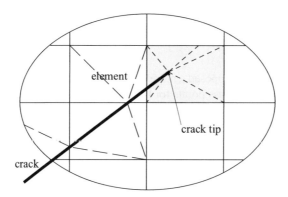

**Figure 8.5**   The sub-triangles technique for partitioning the cracked elements.

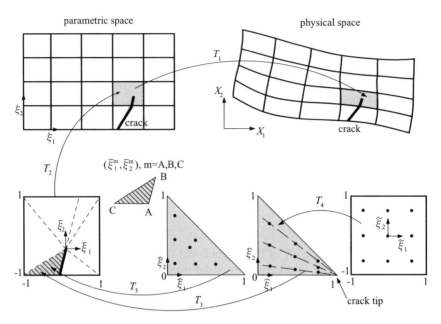

**Figure 8.6** Necessary mappings and transformations from the physical space to sub-triangle integration (Ghorashi, Valizadeh and Mohammadi, 2012; Laborde *et al.*, 2005). (Reproduced by permission of John Wiley & Sons, Ltd.)

Then, the transformation $T_3$ can be defined as (Figure 8.6),

$$T_3 : \begin{cases} \bar{\xi}_1 = \bar{\xi}_1^A(1 - \widehat{\xi}_1 - \widehat{\xi}_2) + \bar{\xi}_1^B(\widehat{\xi}_1) + \bar{\xi}_1^C(\widehat{\xi}_2) \\ \bar{\xi}_2 = \bar{\xi}_2^A(1 - \widehat{\xi}_1 - \widehat{\xi}_2) + \bar{\xi}_2^B(\widehat{\xi}_1) + \bar{\xi}_2^C(\widehat{\xi}_2) \end{cases} \tag{8.30}$$

with the Jacobian of transformation $\mathbf{J}_3$,

$$\mathbf{J}_3 = \begin{bmatrix} \bar{\xi}_{1,\widehat{\xi}_1} & \bar{\xi}_{2,\widehat{\xi}_1} \\ \bar{\xi}_{1,\widehat{\xi}_2} & \bar{\xi}_{2,\widehat{\xi}_2} \end{bmatrix} = \begin{bmatrix} -\bar{\xi}_1^A + \bar{\xi}_1^B & -\bar{\xi}_2^A + \bar{\xi}_2^B \\ -\bar{\xi}_1^A + \bar{\xi}_1^C & -\bar{\xi}_2^A + \bar{\xi}_2^C \end{bmatrix} \tag{8.31}$$

Laborde *et al.* (2005) have proposed a more accurate integration procedure in the form of an almost-polar distribution of Gauss points for the crack-tip element, as depicted in Figure 8.6. Accordingly, application of an extra transformation $T_4$ is required to relate the sub-triangle to an equivalent square integration domain,

$$T_4 : \begin{cases} \widehat{\xi}_1 = \dfrac{1}{4}(1 + \tilde{\xi}_1 - \tilde{\xi}_2 - \tilde{\xi}_1\tilde{\xi}_2) \\ \tilde{\xi}_2 = \dfrac{1}{2}(1 + \tilde{\xi}_2) \end{cases} \tag{8.32}$$

with the Jacobian of transformation $\mathbf{J}_4$,

$$\mathbf{J}_4 = \begin{bmatrix} \widehat{\xi}_{1,\tilde{\xi}_1} & \widehat{\xi}_{2,\tilde{\xi}_1} \\ \widehat{\xi}_{1,\tilde{\xi}_2} & \widehat{\xi}_{2,\tilde{\xi}_2} \end{bmatrix} = \begin{bmatrix} \dfrac{1}{4}(1-\tilde{\xi}_2) & 0 \\ \dfrac{1}{4}(-1-\tilde{\xi}_1) & \dfrac{1}{2} \end{bmatrix} \tag{8.33}$$

These transformations remain accurate as long as the crack path is straight in the parametric space and then in the parent element. They also remain reliable provided that the crack is composed of straight segments. When the crack is not straight in the parametric space, alternative integration procedures are necessary (Sevilla, Fernández-Méndez and Huerta, 2008; Haasemann *et al.*, 2011). Further studies are necessary to develop reliable and efficient approaches for curved cracks in the parametric space.

### 8.2.3   XIGA Simulations

Here, a number of 2D static and quasi-static problems are simulated to demonstrate the efficiency of XIGA. A rather simple case of B-splines (unit weights) is adopted for all examples and the order of NURBS functions in both parametric directions $\xi_1$ and $\xi_2$ are considered three ($p = q = 3$), unless stated otherwise. All knot vectors are open and uniform without any interior repetition and the Lagrange multiplier method is adopted for imposition of essential boundary conditions.

Cracks in both physical and parametric spaces are assumed to remain straight or to be composed of some straight segments. As a result, the sub-triangles technique with 13 Gauss points in each sub-triangle is reliably adopted for crack-face and crack-tip elements, while a simple $4 \times 4$ Gauss quadrature rule is used for ordinary elements.

The post-processing of a solution for determining the stress intensity factors is based on the equivalent domain interaction integral approach.

#### 8.2.3.1   Centre Crack Under Far Field Uniform Tensile Loading

The first example is to simulate a plane strain infinite isotropic plate containing a line crack of length $2a$ subjected to uniform tensile traction, which is governed by the analytical solution of pure mode I fracture mechanics. To avoid the numerical difficulties of modelling an infinite domain, the analytical solution is prescribed on the finite area ABCD, as depicted in Figure 8.7.

Figure 8.7 also illustrates the typical non-uniform distribution of conventional and enriched control points for XIGA simulation of the ABCD region. Because of the singular nature of the stress fields at the crack tip, the concentration of control points and the decrease of element sizes around the crack tip are expected to enhance the accuracy of the solution for the same number of DOFs. Accordingly, in addition to uniform models, a variety of non-uniformly distributed control meshes (36 to 1444 control points) are modelled. The same number of nodes has been utilized in XFEM analyses.

This test is appropriate for convergence analysis because of the availability of an analytical solution. Therefore, various measures of error norm can be computed for XIGA and compared

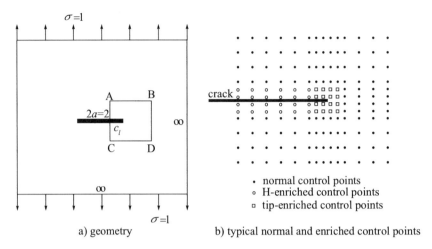

a) geometry

b) typical normal and enriched control points

**Figure 8.7**  Geometry, loading and typical distribution of normal and enriched control points for simulation by XIGA.

with similar results by XFEM. The adopted $L_2$, $H_1$ and $E$ norms are defined as,

$$\|\Theta\|_{L_2} = \left[ \int_\Omega \Theta^T \Theta \, d\Omega \right]^{\frac{1}{2}} \tag{8.34}$$

$$\|\Theta\|_{H_1} = \left\{ \int_\Omega \left[ \Theta^T \Theta + \Theta^T_{,x} \Theta_{,x} + \Theta^T_{,y} \Theta_{,y} \right] d\Omega \right\}^{\frac{1}{2}} \tag{8.35}$$

$$\|\Theta\|_E = \left\{ \int_\Omega \left[ (\nabla \Theta^T)(\mathbf{D}\nabla\Theta) \right] d\Omega \right\}^{\frac{1}{2}} \tag{8.36}$$

where $\Theta$ can be used for the displacement field $\Theta = \mathbf{u}$ and the displacement error $\Theta = \mathbf{e} = \mathbf{u} - \mathbf{u}^h$. All XFEM and XIGA analyses are performed with the same number of nodes and control points, respectively.

Figure 8.8 compares the convergence rates of the $L_2$ error norms of XIGA and XFEM. Clearly, XIGA has higher accuracy and convergence rate than XFEM; the error of XFEM with 1444 nodes and 1369 elements (2992 DOFs) is almost 0.13%, while the error of XIGA for a model of 144 control points and 81 elements (516 DOFs) is even lower at about 0.08%.

It should be noted that XIGA has fewer elements and more DOFs than XFEM for the same number of points. Ghorashi, Valizadeh and Mohammadi (2012) have reported that non-uniform models in this problem have resulted in lower $L_2$ error norms than their equivalent uniform elements for the same number of degrees of freedom. They have also concluded that suboptimal convergence rates (slopes 4 and 3 for $L_2$ and $H_1$ error norms on the displacement, respectively, for cubic basis functions) are obtained if no transition zone is adopted for the enriched domains.

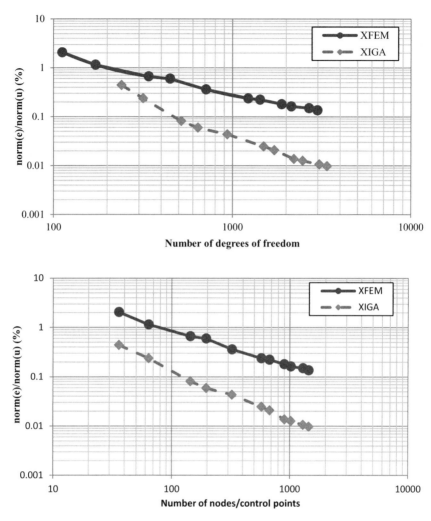

**Figure 8.8**  XFEM and XIGA $L_2$ error norms (%) for different DOFs and number of control points/nodes (Ghorashi, Valizadeh and Mohammadi, 2012). (Reproduced by permission of John Wiley & Sons, Ltd.)

Similar conclusions can be drawn from Table 8.1, where the $L_2$, $H_1$ and $E$ error norms of two XFEM models and several XIGA simulations have been compared for various numbers of nodes, elements, control points and degrees of freedom and different options of inclusion of crack-tip enrichments, and the sub-triangle approach for numerical integration.

As expected, a significant improvement is observed if XIGA is combined with the sub-triangles technique and if the crack-tip enrichment functions are included.

Another important comparison of Table 8.1 is related to the order of NURBS functions. It is observed that the most accurate solution in this example is obtained with an order of four, while higher orders have produced less accurate solutions. The reason can be attributed to the way a constant number of Gauss points has been used in each element ($4 \times 4$) and

**Table 8.1** Comparison of error norms of various XFEM and XIGA models (Ghorashi, Valizadeh and Mohammadi, 2012). (Reproduced by permission of John Wiley & Sons, Ltd.)

| Method | $n_e$ | $n_n$ or $n_{cp}$ | Order | $n_{DOF}$ | Tip-enrichment | Sub-triangle | Norm of error (%) | | |
|--------|-------|------------------|-------|-----------|----------------|--------------|-------|-------|--------|
| | | | | | | | $L_2$ | $H_1$ | Energy |
| | 121 | 144 | — | 340 | ✓ | ✓ | 0.66 | 2.52 | 15.28 |
| XFEM | 1369 | 1444 | — | 2992 | ✓ | ✓ | 0.14 | 1.19 | 7.31 |
| | 81 | 144 | 3 | 448 | ✓ | — | 1.10 | 3.47 | 18.26 |
| | 81 | | 3 | 448 | ✓ | ✓ | 0.46 | 0.64 | 3.33 |
| | 81 | | 3 | 404 | — | ✓ | 2.30 | 4.45 | 24.19 |
| | 81 | | 3 | 516 | ✓ | ✓ | 0.08 | 0.40 | 2.97 |
| | 121 | | 1 | 408 | ✓ | ✓ | 0.65 | 2.52 | 15.53 |
| | 100 | | 2 | 458 | ✓ | ✓ | 0.12 | 0.65 | 4.46 |
| | 81 | | 3 | 516 | ✓ | ✓ | 0.08 | 0.40 | 2.97 |
| | 64 | | 4 | 596 | ✓ | ✓ | 0.07 | 0.41 | 3.39 |
| XIGA | 49 | | 5 | 680 | ✓ | ✓ | 0.09 | 0.49 | 3.69 |
| | 529 | | 1 | 1368 | ✓ | ✓ | 0.23 | 1.56 | 9.59 |
| | 484 | | 2 | 1430 | ✓ | ✓ | 0.04 | 0.42 | 2.87 |
| | 441 | | 3 | 1500 | ✓ | ✓ | 0.02 | 0.29 | 2.12 |
| | 400 | 576 | 4 | 1592 | ✓ | ✓ | 0.03 | 0.27 | 2.26 |
| | 361 | | 5 | 1688 | ✓ | ✓ | 0.02 | 0.29 | 2.36 |
| | 324 | | 6 | 1810 | ✓ | ✓ | 0.04 | 0.35 | 2.91 |
| | 289 | | 7 | 1932 | ✓ | ✓ | 0.03 | 0.36 | 2.93 |

each sub-triangle (13), regardless of the order of simulation. In other words, for higher order of NURBS, higher degree Gauss rules are required for elements and sub-triangles to achieve more accurate results.

Finally, in order to demonstrate the effect of crack-tip enrichments in the present study, Figure 8.9 compares the expanded ($\times 30$) illustration of exact and XIGA deformations of the plate around the crack tip for a model of 144 control points with and without the crack-tip enrichment functions. Without the inclusion of crack-tip enrichments, even the accuracy of the displacement field around the crack tip has been substantially reduced. This is further highlighted for stress components, where only a finite stress value can be obtained at the crack tip if XIGA is used without any crack-tip enrichments.

#### 8.2.3.2   Crack Propagation in a Mixed-Mode Double Cantilever Beam

The second example is the simulation of crack propagation in a pre-cracked double cantilever beam (DCB test) in a quasi-static state (Figure 8.10). The initial crack of length $a = 20\,\text{mm}$ is located slightly off the mid-plane ($\Delta h$), so it propagates in a mixed-mode curved path, as depicted schematically in Figure 8.10c.

The beam is analysed in a plane stress state, with elastic properties of $E = 3 \times 10^7\,\text{N}\,\text{mm}^{-2}$ and $\nu = 0.3$ is considered.

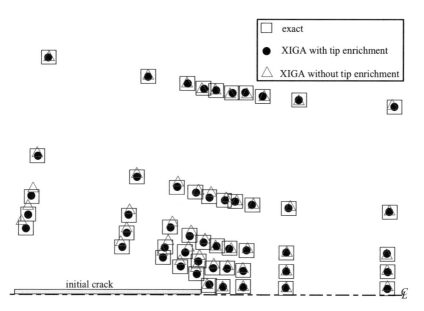

**Figure 8.9** Comparison of exact and XIGA deformations (with a magnification factor of 30) of element vertices around the crack tip with and without using crack-tip enrichment functions (Ghorashi, Valizadeh and Mohammadi, 2012). (Reproduced by permission of John Wiley & Sons, Ltd.)

Several XIGA models are utilized to simulate the crack propagation for different crack offsets and crack length increment $\Delta a$. Mixed mode SIFs are computed using the interaction integral approach with a circular domain of radius $r_J = 0.15a$. The maximum hoop stress criterion is employed to predict the crack propagation angle. The adopted control points and elements remain unchanged in the whole crack propagation process. Ghorashi, Valizadeh and Mohammadi (2012) have reported a similar propagation path to the experimental observations

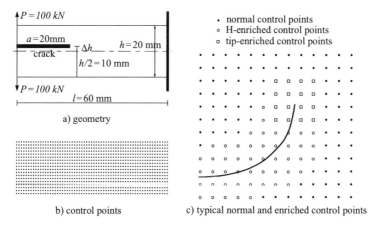

**Figure 8.10** The DCB test, (a) geometry and loading conditions, (b) uniform distribution of control points, (c) Heaviside and crack-tip enriched control points near a propagated crack tip.

**Table 8.2**   Mixed mode SIFs $K_I$ and $K_{II}$ and the crack inclination angle in local crack tip ($\theta_c$) and global coordinate systems ($\alpha$) for each step of crack propagation (Ghorashi, Valizadeh and Mohammadi, 2012). (Reproduced by permission of John Wiley & Sons, Ltd.)

| Step | $K_I$ | $K_{II}$ | $\theta_c(°)$ | $\alpha(°)$ |
|------|-------|----------|---------------|-------------|
| 1 | 293204.77 | −5166.34 | 2.0177 | 2.0177 |
| 2 | 300539.44 | −7827.16 | 2.9797 | 4.9974 |
| 3 | 312732.60 | −9010.08 | 3.2951 | 8.2925 |
| 4 | 328311.78 | −15221.38 | 5.2864 | 13.5788 |
| 5 | 344832.20 | −26711.72 | 8.7555 | 22.3343 |
| 6 | 369742.88 | −30384.36 | 9.2726 | 31.6069 |
| 7 | 408041.16 | −46464.16 | 12.6748 | 44.2818 |
| 8 | 491816.86 | −61558.34 | 13.8517 | 58.1335 |
| 9 | 563412.67 | −47398.64 | 9.4859 | 67.6194 |
| 10 | 760009.40 | −51844.11 | 7.7337 | 75.3531 |
| 11 | 939907.63 | −36301.03 | 4.4104 | 79.7635 |

of Sumi and Kagohashi (1983) and the numerical predictions reported in Sukumar and Prevost (2003) using XFEM and a fixed mesh of 120 × 41 elements. Table 8.2 shows the estimated mixed mode SIFs and the local and global crack angles for each step of the quasi-static crack propagation on a mesh of 63 × 24 control points and 60 × 21 elements using a crack increment of 1 mm.

Figure 8.11 compares the predicted crack growth paths for different eccentricities of $\Delta h$ ($\Delta h = 0, 0.015.0.035, 0.07, 0.14, 0.3, 0.5, 1$ mm) on the same mesh and with a similar crack increment of 1 mm. For the case where the initial crack locates in the mid-plane ($\Delta h = 0$), it propagates horizontally along the mid-plane ($Y = 10$ mm). In contrast, the crack tends to propagate towards a vertical path for higher eccentricity values.

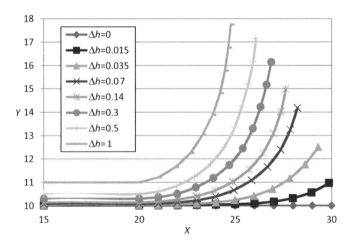

**Figure 8.11**   The effect of initial crack eccentricity ($\Delta h$) on the crack growth path (Ghorashi, Valizadeh and Mohammadi, 2012). (Reproduced by permission of John Wiley & Sons, Ltd.)

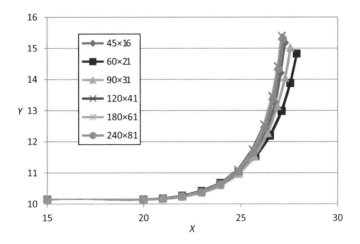

**Figure 8.12** The effect of XIGA elements on the predicted crack path (Ghorashi, Valizadeh and Mohammadi, 2012). (Reproduced by permission of John Wiley & Sons, Ltd.)

Another important aspect of a reliable numerical solution is its mesh insensitivity. To investigate this characteristic for XIGA, several different uniformly distributed control points within a degree 4 XIGA have been analysed by Ghorashi, Valizadeh and Mohammadi (2012), and resulted in more or less similar crack propagation paths (Figure 8.12). The existing differences may be attributed to the insufficiency of numerical discretization for some of the coarse models. They have also illustrated that the results of XFEM and various 2–4 orders XIGA converge to a very close path of crack propagation.

### 8.2.3.3 Central Crack in a Circular Orthotropic Disk

An orthotropic disk with an inclined central crack subjected to double point loads (Figure 8.13a) is solved. The material properties and material orthotropy axes are displayed in Figure 8.13a and b. 841 control points and 441 elements are used for modelling the problem, as illustrated in Figure 8.13c and d, respectively. The order of NURBS functions in both parametric directions $\xi$ and $\eta$ are considered three ($p = q = 3$), all knot vectors are open and uniform without any interior repetition and a $2 \times 2$ Gauss quadrature and sub-triangles technique with 13 Gauss points in each sub-triangle are used for integration. The Lagrange multiplier method has been adopted for imposition of essential boundary conditions. To determine the fracture properties, stress intensity factors (SIFs) are obtained by means of the interaction integral method which developed by Kim and Paulino (2002c).

This problem has previously been solved by Kim and Paulino (2002c), Asadpoure and Mohammadi (2007), Ghorashi, Mohammadi and Sabbagh-Yazdi (2011) and Ghorashi, Valizadeh and Mohammadi (2011). Table 8.3 compares the predicted stress intensity factors with the reference values for the case of $\alpha = 30°$.

XIGA is used for analysis of this problem considering different inclination angles of the crack. The computed values of mixed mode SIFs compared with Asadpoure and Mohammadi (2007) and Ghorashi, Mohammadi and Sabbagh-Yazdi (2011) are illustrated in Figure 8.14

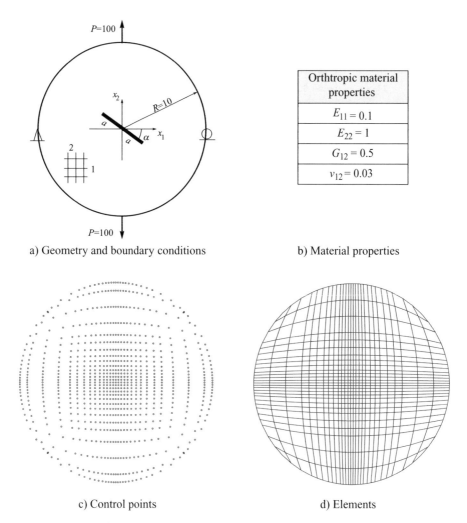

a) Geometry and boundary conditions                    b) Material properties

c) Control points                                       d) Elements

**Figure 8.13**  An orthotropic disk with inclined central crack: (a) geometry and boundary conditions, (b) material properties, (c) control points and (d) elements (Ghorashi, Valizadeh and Mohammadi, 2011).

**Table 8.3**  Stress intensity factors for an inclined central crack in an orthotropic disk subjected to point loads ($\alpha = 30°$).

| Method | DOFs | Elements | Cells | $K_I$ | $K_{II}$ |
|---|---|---|---|---|---|
| Kim and Paulino (2002c) (MCC) | 5424 | 999 | — | 16.73 | 11.33 |
| Kim and Paulino (2002c) (M- integral) | 5424 | 999 | — | 16.75 | 11.38 |
| Asadpoure and Mohammadi (2007) XFEM | 1960 | 920 | — | 17.08 | 11.65 |
| Ghorashi, Mohammadi and Sabbagh-Yazdi (2011) XEFG | 1507 | — | 641 | 16.98 | 11.95 |
| Ghorashi, Valizadeh and Mohammadi (2011) XIGA | 1223 | 441 | — | 16.68 | 11.53 |

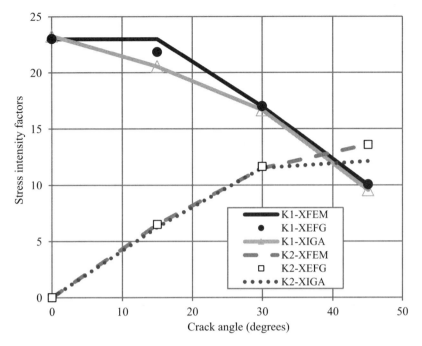

**Figure 8.14** Mode I and II SIFs corresponding to different central crack angles in the orthotropic disk (Ghorashi, Valizadeh and Mohammadi, 2011).

show good agreement. Note that, similar to XFEM and EFG methods, XIGA is capable of analysing the problem of all crack inclinations on a single model.

## 8.3 Orthotropic Dislocation Dynamics

Despite the enormous simplifications that isotropy brings to theoretical and numerical simulations, many real crystals behave anisotropically and a reliable solution is only obtained if an anisotropic model is adopted.

### 8.3.1 Straight Dislocations in Anisotropic Materials

Eshelby, Read and Shockley (1953) were the first to develop the theory of anisotropic elasticity for straight dislocations, followed by Foreman (1955) and Stroh (1958) to include anisotropy in energy calculations (Hirth and Lothe, 1982). The method consists of the following steps for a dislocation in the $z(x_3)$ direction.

First, three sets of complex conjugate roots ($p_1$, $p_2$ and $p_3$ and their conjugates) of the following 6th order characteristic equation are determined:

$$\det\{a_{ik}\} = 0 \tag{8.37}$$

$$a_{ik} = D_{i1k1} + (D_{i1k2} + D_{i2k1})p + D_{i2k2}p^2 \tag{8.38}$$

where $D_{ijkl}$ is the material constitutive modulus (see Section 4.2.1). Then, $A_k(n)$ is solved for three values of $n$ from

$$a_{ik}(n)A_k(n) = 0 \qquad (8.39)$$

The next step is to solve the following simultaneous equations for $Q(n)$

$$\text{Re}\left[\sum_{n=1}^{3} \pm A_k(n)Q(n)\right] = b_k \quad k = 1, 2, 3 \qquad (8.40)$$

$$\text{Re}\left[\sum_{n=1}^{3} \pm G_{i2k}(n)A_k(n)Q(n)\right] = 0 \quad i = 1, 2, 3 \qquad (8.41)$$

where

$$G_{ijk} = D_{ijk1} + D_{ijk2}p_n \qquad (8.42)$$

Positive and negative signs are used for positive and negative imaginary parts of $p_n$, respectively. The final displacement and stress fields are then computed from,

$$u_k = \text{Re}\left[\frac{-1}{2\pi i}\sum_{n=1}^{3} G_k(n)Q(n)\ln \eta_n\right] \quad k = 1, 2, 3 \qquad (8.43)$$

$$\sigma_{ij} = \text{Re}\left[\frac{-1}{2\pi i}\sum_{n=1}^{3} G_{ijk}(n)A_k(n)Q(n)\ln \eta_n^{-1}\right] \quad k = 1, 2, 3 \qquad (8.44)$$

where $i^2 = -1$ and

$$\eta_n = x_1 + p_n x_2 \qquad (8.45)$$

## 8.3.2 Edge Dislocations in Anisotropic Materials

The displacement field for an edge dislocation with the Burgers vector $\mathbf{b} = (b_x, b_y, 0)$ can be written in terms of the components $d'_{ij}$ of the rotated constitutive tensor (see Section 4.2.1) as (Hirth and Lothe, 1982),

$$
u_x = -\frac{b_x}{4\pi}\left[\tan^{-1}\left(\frac{2xy\sin\phi}{x^2 - \lambda^2 y^2}\right) + \frac{\bar{d}'^2 - d'^2_{12}}{2\bar{d}'d'_{66}\sin 2\phi}\ln\frac{q}{t}\right]
$$
$$
- \frac{b_y}{4\pi\bar{d}'\sin 2\phi}\left[(\bar{d}' - d'_{12})\cos\phi\ln qt - (\bar{d}' + d'_{12})\sin\phi\tan^{-1}\left(\frac{x^2\sin 2\phi}{\lambda^2 y^2 - x^2\cos 2\phi}\right)\right]
$$

$$(8.46)$$

$$u_y = -\frac{b_y}{4\pi}\left[\tan^{-1}\left(\frac{2xy\sin\phi}{x^2 - \lambda^2 y^2}\right) - \frac{\bar{d}'^2 - d'_{12}{}^2}{2\bar{d}'d'_{66}\sin 2\phi}\ln\frac{q}{t}\right]$$

$$+ \frac{\lambda b_x}{4\pi\bar{c}'\sin 2\phi}\left[(\bar{d}' - d'_{12})\cos\phi\ln qt - (\bar{d}' + d'_{12})\sin\phi\tan^{-1}\left(\frac{\lambda^2 y^2\sin 2\phi}{x^2 - \lambda^2 y^2\cos 2\phi}\right)\right]$$

$$(8.47)$$

where

$$\lambda^4 = \frac{d'_{11}}{d'_{22}} \tag{8.48}$$

$$\bar{d}' = \sqrt{d'_{11}d'_{22}} \tag{8.49}$$

$$\phi = \frac{1}{2}\cos^{-1}\frac{d'^2_{12} + 2d'_{12}d'_{66} - \bar{d}'^2}{2\bar{d}'d'_{66}} \tag{8.50}$$

$$q^2 = x^2 + 2xy\lambda\cos\phi + y^2\lambda^2 \tag{8.51}$$

$$t^2 = x^2 - 2xy\lambda\cos\phi + y^2\lambda^2 \tag{8.52}$$

and for stress components,

$$\sigma_{xx} = -\frac{Mb_x}{2\pi s^4 d'_{22}}\left\{\left[(\bar{d}' - d'_{12})(\bar{d}' + d'_{12} + 2d'_{66}) - \bar{d}'d'_{66}\right]x^2 y + \frac{\bar{d}'^2 d'_{66}}{d'_{22}}y^3\right\}$$

$$+ \frac{Mb_y d'_{66}}{2\pi s^4}\left\{\frac{\bar{d}'}{d'_{22}}xy^2 - x^3\right\} \tag{8.53}$$

$$\sigma_{yy} = -\frac{Mb_y}{2\pi s^4\bar{d}'}\left\{\left[(\bar{d}' - d'_{12})(\bar{d}' + d'_{12} + 2d'_{66}) - \bar{d}'d'_{66}\right]xy^2 + d'_{22}d'_{66}x^3\right\}$$

$$+ \frac{Mb_x d'_{66}}{2\pi s^4}\left\{\frac{\bar{d}'}{d'_{22}}y^3 - x^2 y\right\} \tag{8.54}$$

$$\sigma_{zz} = \frac{Mb_x}{2\pi s^4}\left\{\frac{\bar{d}'}{d'_{22}}xy^2 - x^3\right\} + \frac{Mb_y d'_{66}}{2\pi s^4}\left\{\frac{\bar{d}'}{d'_{22}}y^3 - x^2 y\right\} \tag{8.55}$$

where

$$M = (\bar{d}' + d'_{12})\left[\frac{(\bar{d}' - d'_{12})}{d'_{22}d'_{66}(\bar{d}' + d'_{12} + 2d'_{66})}\right] \tag{8.56}$$

$$s = \left(x^2 + \frac{\bar{d}'}{d'_{22}}y^2\right)^2 + \left[\frac{(\bar{d}' + d'_{12})(\bar{d}' - d'_{12} - 2d'_{66})}{d'_{22}d'_{66}}\right]x^2 y^2 \tag{8.57}$$

### 8.3.3 Curve Dislocations in Anisotropic Materials

The displacement field for a pure screw dislocation with the Burgers vector $\mathbf{b} = (0, 0, b_z)$ can be expressed as (Hirth and Lothe, 1982),

$$u_z = -\frac{b_z}{2\pi} \tan^{-1} \left[ \frac{y\sqrt{d'_{44}d'_{55} - d'^2_{45}}}{d'_{44}x - d'_{45}y} \right] \tag{8.58}$$

and the stress components,

$$\sigma_{xz} = -\frac{b_z}{2\pi}\sqrt{d'_{44}d'_{55} - d'^2_{45}} \, \frac{d'_{45}x - d'_{55}y}{d'_{44}x^2 - 2d'_{45}xy + d'_{55}y^2} \tag{8.59}$$

$$\sigma_{yz} = -\frac{b_z}{2\pi}\sqrt{d'_{44}d'_{55} - d'^2_{45}} \, \frac{d'_{44}x - d'_{45}y}{d'_{44}x^2 - 2d'_{45}xy + d'_{55}y^2} \tag{8.60}$$

$$\sigma_{xz} = -\frac{b_z}{2\pi}\sqrt{d'_{44}d'_{55} - d'^2_{45}} \, \frac{d'_{34}x - d'_{35}y}{d'_{44}x^2 - 2d'_{45}xy + d'_{55}y^2} \tag{8.61}$$

### 8.3.4 Anisotropic Dislocation XFEM

The general methodology of the orthotropic extended finite element of Section 4.5 can be extended to solve anisotropic dislocation problems. A body with $nd$ internal dislocations $\Gamma_d^\alpha$, which is subjected to body and traction forces, $\mathbf{f}^b$ and $\mathbf{f}^t$, respectively, is considered, as depicted in Figure 8.15.

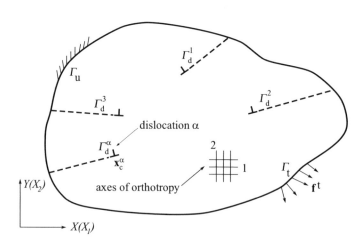

**Figure 8.15**   An anisotropic body which includes dislocations.

### 8.3.4.1　Governing Equation

The strong form of the equilibrium equation in terms of the stress tensor $\sigma$ can be written as:

$$\nabla \cdot \sigma + \mathbf{f}^{b} = 0 \quad \text{in } \Omega \tag{8.62}$$

with the following boundary conditions:

$$\sigma \cdot \mathbf{n} = \mathbf{f}^{t} \quad \text{on } \Gamma_{t}: \quad \text{external traction} \tag{8.63}$$

$$\mathbf{u} = \bar{\mathbf{u}} \quad \text{on } \Gamma_{u}: \quad \text{prescribed displacement} \tag{8.64}$$

where $\Gamma_{t}$ and $\Gamma_{u}$ are traction and displacement boundaries, respectively.

The variational formulation of the boundary value problem can be defined as:

$$\int_{\Omega} \sigma \cdot \delta\varepsilon \, d\Omega = \int_{\Omega} \mathbf{f}^{b} \cdot \delta\mathbf{u} \, d\Omega + \int_{\Gamma} \mathbf{f}^{t} \cdot \delta\mathbf{u} \, d\Gamma \tag{8.65}$$

### 8.3.4.2　XFEM Approximation

According to Belytschko and Gracie (2007) and Gracie, Ventura and Belytschko (2007), the XFEM approximation of the displacement field $\mathbf{u}^{h}(\mathbf{x})$ for a point $\mathbf{x}$ located within a domain which includes anisotropic dislocations can be expressed in terms of the classical FEM approximation $\mathbf{u}(\mathbf{x})$, and the dislocation discontinuous and core parts of the approximation $\mathbf{u}_{H}^{D}(\mathbf{x})$ and $\mathbf{u}_{c}^{D}(\mathbf{x})$, respectively,

$$\mathbf{u}^{h}(\mathbf{x}) = \mathbf{u}(\mathbf{x}) + \mathbf{u}_{H}^{D}(\mathbf{x}) + \mathbf{u}_{c}^{D}(\mathbf{x}) \tag{8.66}$$

and more explicitly,

$$\mathbf{u}^{h}(\mathbf{x}) = \left[\sum_{j=1}^{n_{cp}} N_{j}(\mathbf{x})\mathbf{u}_{j}\right] + \left[\sum_{\alpha=1}^{n_{d}} \Psi_{H}^{\alpha}\right] + \left[\sum_{\alpha=1}^{n_{d}} \Psi_{c}^{\alpha}\right] \tag{8.67}$$

where $\mathbf{u}_{j}$ is the displacement vector at node $j$, $N_{j}$ are the standard finite element shape functions, and $n_{d}$ is the number of dislocations. $\Psi_{c}^{\alpha}$ is the analytical displacement field associated with the core of an edge dislocation $\alpha$, defined in its local coordinate systems by (8.46) and (8.47).

A number of different approximations for dislocation discontinuity enrichments $\Psi_{H}^{\alpha}$ have been proposed. The incompatible formulation (Figure 8.16a) by Belytschko and Gracie (2007) can be written as,

$$\Psi_{H}^{\alpha} = \mathbf{b}^{\alpha} \sum_{h=1}^{mh} N_{h}(\mathbf{x}) \left[H(f^{\alpha}(\mathbf{x})) - H(f^{\alpha}(\mathbf{x}_{I})) + H(f^{\alpha}(\mathbf{x}_{I}))\delta_{Ic}^{\alpha}\right] H(g^{\alpha}(\mathbf{x})) \tag{8.68}$$

where $\delta_{Ic}^{\alpha}$ is the Kronecker delta at the core of dislocation $\alpha$, $H$ is the Heaviside function, $\mathbf{b}^{\alpha}$ is the Burgers vector with magnitude $b_{\alpha}$ for dislocation $\alpha$ on $\Gamma_{d}^{\alpha}$ and $mh$ is the set of Heaviside enriched nodes (nodes belonging to the elements with at least one edge crossed by the glide plane).

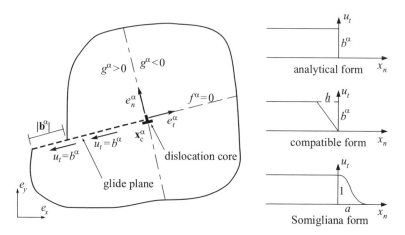

**Figure 8.16** Definition of functions $f(\mathbf{x})$ and $g(\mathbf{x})$ for an edge dislocation, and the incompatible, linearly regularized and tapered Somigliana profiles for a dislocation glide plane.

Functions $f^{\alpha}(\mathbf{x})$ and $g^{\alpha}(\mathbf{x})$, which define the glide plane and the position of thecore for an edge dislocation, respectively, can be written in the general forms of (with summation rule for repeated indices):

$$f^{\alpha}(\mathbf{x}) = \alpha_0 + \alpha_j x_j \tag{8.69}$$

$$g^{\alpha}(\mathbf{x}) = \beta_0 + \beta_j x_j \tag{8.70}$$

Other options are the compatible form (Figure 8.16),

$$\boldsymbol{\Psi}_H^{\alpha} = \mathbf{b}^{\alpha} \sum_{h=1}^{mh} N_h(\mathbf{x}) \left[ H(f^{\alpha}(\mathbf{x})) - H(f^{\alpha}(\mathbf{x}_I)) \right] H(g^{\alpha}(\mathbf{x})) \tag{8.71}$$

and the tapered Somigliana dislocation proposed by Gracie, Ventura and Belytschko (2007) (Figure 8.16),

$$\boldsymbol{\Psi}_H^{\alpha} = \mathbf{b}^{\alpha} \sum_{h=1}^{mh} N_h(\mathbf{x}) \left[ H(f^{\alpha}(\mathbf{x})) - H(f^{\alpha}(\mathbf{x}_I)) \right] \psi_H(g^{\alpha}(\mathbf{x})) \tag{8.72}$$

$$\psi_H(g(\mathbf{x})) = \begin{cases} 1 & g(\mathbf{x}) > 0 \\ \dfrac{(a^m - g(\mathbf{x})^m)^n}{a^{mn}} & -a < g(\mathbf{x}) < 0 \\ 0 & g(\mathbf{x}) < -a \end{cases} \tag{8.73}$$

where parameters $n$, $m$ and $a$ are material properties which have to be obtained from experimental results or atomistic simulations.

It is important to note that, dissimilar to conventional XFEM solutions, no additional degrees of freedom are introduced in approximation (8.67) due to the fact that the problem is solved for known prescribed sliding $\mathbf{b}^{\alpha} = b_{\alpha} \mathbf{e}_t^{\alpha}$ along the sliding plane $\Gamma_d^{\alpha}$ (Malekafzali, 2010; Mohammadi and Malekafzali, 2011).

### 8.3.4.3    Discretized XFEM

Discretization of Eq. (8.65) using the XFEM approximation (8.67) results in a discrete system of linear equilibrium equations:

$$\mathbf{K^u u} + \mathbf{K^D b} = \mathbf{f} \tag{8.74}$$

where $\mathbf{f}$ is the nodal force vector, $\mathbf{u}$ is the vector of nodal degrees of freedom and $\mathbf{b}$ is the vector of slidings $\mathbf{b}^\alpha$ along the glide planes,

$$\mathbf{b} = \left\{ \mathbf{b}^1 \mathbf{b}^2 \cdots \mathbf{b}^{n_d} \right\}^{\mathrm{T}} \tag{8.75}$$

$\mathbf{K^u}$ is the classical finite element stiffness matrix, and $\mathbf{K^D}$ is the enrichment stiffness matrix,

$$\mathbf{K^u} = \int_\Omega (\mathbf{B^u})^{\mathrm{T}} \mathbf{D} \mathbf{B^u} \, d\Omega \tag{8.76}$$

$$\mathbf{K}^D_{H\alpha} = \int_\Omega (\mathbf{B^u})^{\mathrm{T}} \mathbf{D} \mathbf{B}^{DH}_\alpha \, d\Omega \tag{8.77}$$

$$\mathbf{K}^D_{c\alpha} = \int_\Omega (\mathbf{B^u})^{\mathrm{T}} \mathbf{D} \mathbf{B}^{Dc}_\alpha \, d\Omega \tag{8.78}$$

with

$$\mathbf{B^u} = \begin{bmatrix} N_{,x} & 0 \\ 0 & N_{,y} \\ N_{,y} & N_{,x} \end{bmatrix} \tag{8.79}$$

$$\mathbf{B}^{DH}_\alpha = \begin{bmatrix} \mathbf{e}_x \cdot \mathbf{\Psi}^\alpha_{H,x} \\ \mathbf{e}_y \cdot \mathbf{\Psi}^\alpha_{H,y} \\ \mathbf{e}_x \cdot \mathbf{\Psi}^\alpha_{H,y} + \mathbf{e}_y \cdot \mathbf{\Psi}^\alpha_{H,x} \end{bmatrix} \tag{8.80}$$

$$\mathbf{B}^{Dc}_\alpha = \begin{bmatrix} \mathbf{e}_x \cdot \mathbf{\Psi}^\alpha_{c,x} \\ \mathbf{e}_y \cdot \mathbf{\Psi}^\alpha_{c,y} \\ \mathbf{e}_x \cdot \mathbf{\Psi}^\alpha_{c,y} + \mathbf{e}_y \cdot \mathbf{\Psi}^\alpha_{c,x} \end{bmatrix} \tag{8.81}$$

It is important to note that the core enrichment functions are defined in the dislocation local coordinate system $(x', y')$, which requires a set of chain rules for evaluation of their derivatives with respect to global coordinates and $(x, y)$.

### 8.3.4.4    Transition Domain

Malekafzali (2010) and Mohammadi and Malekafzali (2011), among others, have discussed the necessity for a transition domain (Figure 8.17) for the XFEM simulation of dislocations mainly due to the prescribed enrichment part of the approximation (8.67). XFEM approximation can then be extended to,

$$\mathbf{u}^h(\mathbf{x}) = \left[ \sum_{j=1}^{n_{cp}} N_j(\mathbf{x})\mathbf{u}_j \right] + \left[ \sum_{\alpha=1}^{n_d} \mathbf{\Psi}^\alpha_H \varphi_H(\mathbf{x}) \right] + \left[ \sum_{\alpha=1}^{n_d} \mathbf{\Psi}^\alpha_c \varphi_c(\mathbf{x}) \right] \tag{8.82}$$

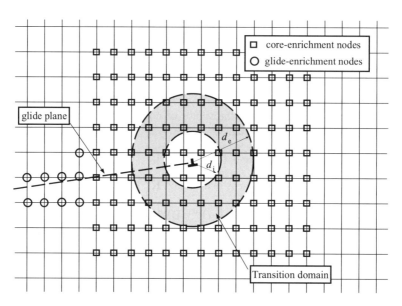

**Figure 8.17**   Transition domain.

$\varphi$ is the ramp function for the transition domain, defined as:

$$\varphi(d) = \begin{cases} 1 & 0 \le |d| \le d_i \\ 1 - R & d_i \le |d| \le d_e \\ 0 & d_e \le |d| \end{cases} \tag{8.83}$$

where $d = d(\mathbf{x})$ is the distance of $\mathbf{x}$ from $\Gamma_d$. Function $R$ can be selected as the linear ramp function,

$$R = \frac{|d| - d_i}{d_e - d_i} \tag{8.84}$$

In order to avoid derivative discontinuity inside the finite elements, the ramp function $\varphi$ is defined as a piecewise continuous function. This is achieved by defining the nodal values of $R_I$,

$$R_I = R(\mathbf{x}_I) = \frac{|d(\mathbf{x}_I)| - d_i}{d_e - d_i} \tag{8.85}$$

and associated nodal values of the ramp function

$$\varphi_I = \begin{cases} 1 & R_I \le 0 \\ 1 - R_I & 0 < R_I < 1 \\ 0 & R_I \ge 1 \end{cases} \tag{8.86}$$

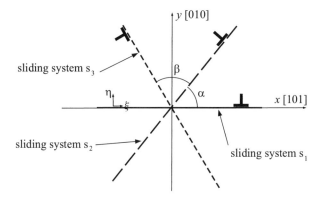

**Figure 8.18**   (a) FCC crystal and simultaneous sliding systems for a plane strain state in the [101]–[010] plane. (b) 2D plane strain system and the sliding systems $s_1$, $s_2$ and $s_3$. The edge dislocations are along the $z$-direction (Shishvan *et al.*, 2011; Malekafzali, 2010).

Then, the finite element shape functions $N_I(\mathbf{x})$ are used to discretize the ramp function $\varphi(\mathbf{x})$,

$$\varphi(\mathbf{x}) = \sum_{I=1}^{m} N_I(\mathbf{x})\varphi_I \tag{8.87}$$

## 8.3.5   *Plane Strain Anisotropic Solution*

Face centred cubic (FCC) crystals only require three elastic constants, and the important dislocations are those in the close-packed {111} planes (Hirth and Lothe, 1982). Therefore, for an FCC crystal, sliding systems {111} <110> take place in 12 planes of higher atom density <110>. Clearly, sliding occurs along the diameters of the cube faces (see Figure 8.18).

Symmetry conditions that are required to simplify the anisotropic elasticity in edge dislocations can be obtained in special plane strain conditions. Rice (1987) explained that by simultaneous application of certain sliding systems, plane strain conditions in FCC can be achieved. This is further discussed by Kysar, Gan and Mendez-Arzuza (2005), Soleymani Shishvan, Mohammadi and Rahimian (2008) and Shishvan *et al.* (2011). To activate a plane strain mode, it is assumed that the shear stress parallel to each sliding plane results in a simultaneous similar sliding along the diameter in the same plane. For instance, on a plane composed of two directions [101] − [010], it is possible to define three independent plane strain sliding systems (see Figure 8.18),

$$\begin{cases} \text{sliding system } (s_1) & (\bar{1}11) \text{ and } (11\bar{1}) \Rightarrow [101] \\ \text{sliding system } (s_2) & (\bar{1}1\bar{1}), [110] \text{ and } (011) \Rightarrow [1\bar{2}1] \\ \text{sliding system } (s_3) & (111), [\bar{1}10] \text{ and } (01\bar{1}) \Rightarrow [\bar{1}2\bar{1}] \end{cases} \tag{8.88}$$

The angles between the sliding systems are,

$$\begin{cases} (s_1),(s_2) & \tan^{-1}(\sqrt{2},1) \approx 54.7° \\ (s_1),(s_3) & \tan^{-1}(\sqrt{2},-1) \approx 125.3° \\ (s_2),(s_3) & 125.3 - 54.7 \approx 70.5° \end{cases} \tag{8.89}$$

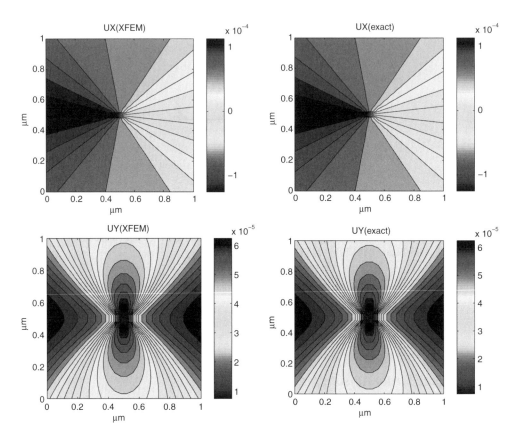

**Figure 8.19**   Comparison of analytical and XFEM displacements for the sliding system $s_1$ (Malekafzali, 2010).

## 8.3.6   Individual Sliding Systems $s_1$ and $s_2$ in an Infinite Domain

The numerical simulation by anisotropic XFEM can be verified by the available analytical solution for the sliding system $s_1$. The simulation is performed for Cu under the plane strain sliding system $s_1$, and the results match the available analytical solution well, as depicted in Figures 8.19 and 8.20, for displacement and stress components, respectively.

Similar conclusions can be drawn for the numerical and analytical solutions for the sliding system $s_2$, as depicted in Figure 8.21 for the displacement components.

## 8.3.7   Simultaneous Sliding Systems in an Infinite Domain

The final example is to demonstrate the flexibility of anisotropic XFEM in dealing with simultaneous plane strain sliding systems in a domain. Three distinct edge dislocations within a $1 \times 1$ area of Cu material are assumed along the three sliding systems (see Figures 8.18 and 8.22). No external loading is present and the generated stress state is a dislocation self-stress.

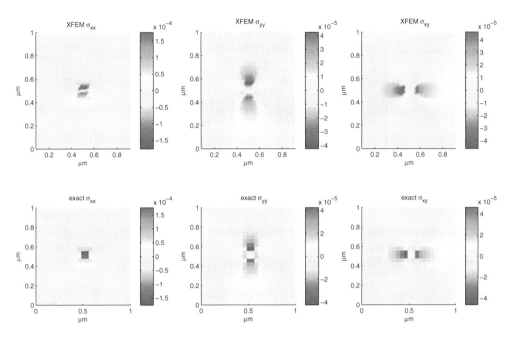

**Figure 8.20** Comparison of analytical and XFEM stress components for the sliding system $s_1$ (Malekafzali, 2010).

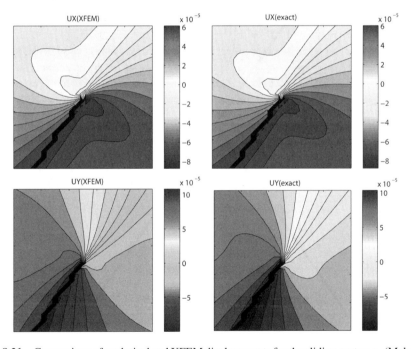

**Figure 8.21** Comparison of analytical and XFEM displacements for the sliding system $s_2$ (Malekafzali, 2010).

For a coordinate origin placed at the lower left corner, the coordinates of the assumed dislocations are:

$$
\begin{cases}
s_1 & \left(0, \dfrac{1}{2}\right) - \left(\dfrac{3}{4}, \dfrac{1}{2}\right) \\[2ex]
s_2 & \left(\dfrac{1}{2}\left(1 - \dfrac{1}{\sqrt{2}}\right), 0\right) - \left(\dfrac{1}{2}\left(1 + \dfrac{1}{\sqrt{2}}\right), \dfrac{3}{4}\right) \\[2ex]
s_3 & \left(\dfrac{1}{2}\left(1 + \dfrac{1}{\sqrt{2}}\right), 0\right) - \left(\dfrac{1}{2}\left(1 - \dfrac{1}{\sqrt{2}}\right), \dfrac{3}{4}\right)
\end{cases}
\tag{8.90}
$$

As a result, each dislocation is placed on its corresponding sliding system, while concurrent at the centre of the square. The two lower corners of the square are restrained, which contradicts the presumptions of the present plane strain theory. Nevertheless, they are included for numerical purposes and to avoid rigid body instabilities.

Figure 8.22 depicts the kernel and Heaviside-enriched nodes for various parts of the model, including the adopted transition zone.

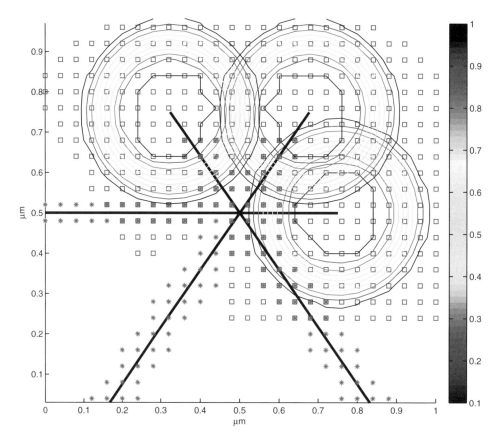

**Figure 8.22** Kernel and Heaviside-enriched nodes and the three simultaneous sliding systems (Mohammadi and Malekafzali, 2011).

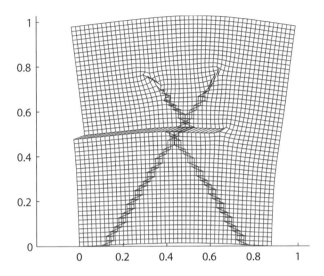

**Figure 8.23**  Deformed shape of the model subjected to dislocation self-stress (Mohammadi and Malekafzali, 2011).

The deformed shape of the model (with a magnification factor of 200) is illustrated in Figure 8.23, which clearly shows the discontinuity of the displacement field along the sliding lines.

Finally, Figures 8.24 and 8.25 depict the predicted contours of the displacement and stress components of the problem, generated by the three simultaneous sliding systems.

## 8.4  Other Anisotropic Applications

### 8.4.1  Biomechanics

Recent advances in multiscale analysis have allowed the development of an emerging science at the interface of engineering and biology; that is, computational biomechanics (Buehler, 2012). Progress in this field and the synthesis of new bio- and bio-inspired materials have become

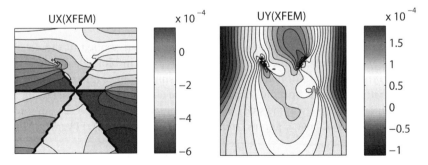

**Figure 8.24**  Displacement contours generated by the three simultaneous sliding systems (Mohammadi and Malekafzali, 2011).

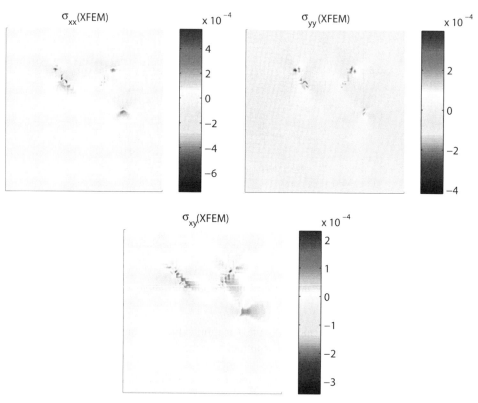

**Figure 8.25** Contours of the stress components generated by the three simultaneous sliding systems (Mohammadi and Malekafzali, 2011).

possible by the introduction of dramatically new experimental, mathematical, analytical and numerical tools and integration of them across the scales.

Development of biomaterials allows one to probe how the structure, material properties and composition of the cell/biomaterial interface affect fundamental cellular processes (Isenberg *et al.*, 2009). Introduction of a 3D electrokinetic assembly of nano-wires and filaments and biomolecules has allowed the simulation and design of revolutionary drug delivery systems to achieve a desired therapeutic effect (Liu *et al.*, 2006; Liu *et al.*, 2007).

In a different field, synthetic biology approaches have been applied to the development of self-assembling DNA nanostructures and devices for use in biomedical applications. In addition to carrying genetic information, DNA is increasingly being explored by programming it to fold in on itself to create specific shapes for its use as a building material (Shih and Lin, 2010). Another diverse topic was the study of the membrane fusion between a virus and a host cell through the model of coarse grained molecular dynamics (CGMD) (Vaidya, Huang and Takagi, 2010).

The study of organ failures has been a major topic of biomechanics. Cimrman *et al.* (2002) developed a finite element macroscopic model of contracting myocardium and used the constitutive model of the tissue based on the theory of mixtures, involving a simple model of

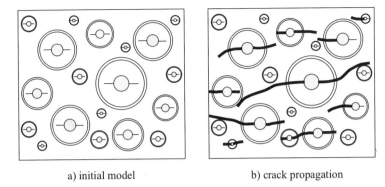

a) initial model                                      b) crack propagation

**Figure 8.26**   Numerical model of fracture in cortical bone microstructure (Budyn and Hoc, 2007).

intra-myocardial blood perfusion. Anisotropy of the tissue was described in terms of preferential directions of the muscle and connective fibres. Sun and Sacks (2005) numerically simulated the anisotropic mechanical properties of soft tissues and tissue-derived biomaterials using a generalized Fung-elastic constitutive model to achieve numerical stability by the convexity of the strain energy function and limiting the condition number of the material stiffness matrix. Göktepe *et al.* (2011) simulated passive myocardial tissues of a generic biventricular heart model by a mixed finite element formulation through a three field Hu–Washizu functional based on a convex, anisotropic hyperelastic model that accounts for the locally orthotropic microstructure of cardiac muscle. In this rather complex model, representation of anisotropy was incorporated through relevant invariants of the Cauchy–Green deformation and structural tensors, combined with additive decoupling of the strain energy function into volumetric and isochoric parts and the multiplicative decomposition of the deformation gradient. Similar methodologies have been adopted by Shahi and Mohammadi (2012) for multiscale simulation of heart valves using triple organ-tissue-cell scales.

There also exist a number of biomechanical problems which are closer to solid mechanics applications. Budyn and Hoc (2007) have simulated cortical bone microstructures by a four-phase model, composed of an interstitial matrix and osteonal fibres (hollow disks) coated by viscous cement lines and hollowed by Haversian canals, representing free boundaries where normally cytoplasmic fluid flows freely (Figure 8.26).

They used a linear elastic isotropic XFEM to study multiple crack growth in cortical bone microstructures. An elastic-damage strain driven criterion was used to initiate cracks and the crack propagation was governed by the maximum hoop criterion.

With the introduction of XFEM within the biomechanical applications, it is expected that it will soon be adopted in various problems that involve inhomogeneity, anisotropic behaviour or multilayer compositions in the macro-level of organ simulations, the micro-model of tissue or even the cell scale itself.

## 8.4.2   Piezoelectric

Piezoelectric materials are widely used in a variety of electronic and electro-mechanical devices, such as actuators, sensors, transducers, multiplayer systems, smart structures, and

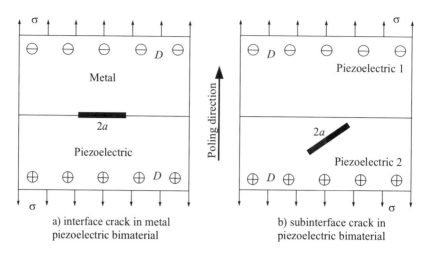

a) interface crack in metal
piezoelectric bimaterial

b) subinterface crack in
piezoelectric bimaterial

**Figure 8.27**  Interface and subinterface cracks in piezoelectric bimaterials (Rojas-Diaz *et al.*, 2011; Sharma *et al.*, 2012).

micro-electromechanical-systems (MEMS) and for microbalancing, optical ultrafine focusing, sound detection and production, and generation of high voltage and electronic frequency (Rojas-Diaz *et al.*, 2011; Sharma *et al.*, 2012).

Piezoelectricity is the charge that is accumulated in certain materials, notably some crystals, ceramics, bone, proteins and even DNA, in response to applied mechanical stresses (Rojas-Diaz *et al.*, 2011; Sharma *et al.*, 2012). The piezoelectric effect can be described as the electromechanical interaction between the mechanical and the electrical states in crystalline materials.

Piezoelectric ceramics are very brittle and susceptible to fracture at all scales, ranging from electric domains to devices, especially under cyclic loading and fatigue conditions (Liu and Hsia, 2003). They are vulnerable to various types of defects, such as grain boundaries, impurities, inclusions, flaws and pores, and so on, which may cause geometric, electric, thermal, and mechanical discontinuities or induce high stress and electric field concentrations, leading to cracking, partial discharge, and dielectric breakdown (Zhang and Gao, 2004).

Metal/piezoelectric bimaterials are used in multiplayer devices to generate high-applied electric fields by placing metal electrodes on the piezoelectric surfaces (Figure 8.27a). In addition to the usual defects in piezoelectric materials, the mismatch between the properties of the two dissimilar materials (metal and piezoelectric ceramic) and the existing interface micro-cracks due to dislocations pile up (Xiao and Zhao, 2004) may result in delamination under thermo-electromechanical loadings. As a result, the study of the anticipated degradation in mechanical and electrical performance of such bimaterials has received considerable attention. Two types of interfacial cracks can be distinguished: impermeable cracks, where the crack surfaces are free of surface charges, and permeable cracks, where the continuity in both the normal electric displacement component and the electric potential across the crack gap are assumed.

Kudryavtsev, Parton and Rakitin (1975a, 1975b) were the first to study the problems of rectilinear and axisymmetric cracks on an interface between a linear transversely isotropic piezoelectric and an isotropic elastic conductor. Several authors have discussed the credibility of impermeable crack assumption (Parton (1976), McMeeking (1989, 2004), Pak and Tobin (1993), Dunn (1994), Sosa and Khutoryansky (1996), Zhang, Qian and Tong (1998), Gao and Fan (1998) and Shindo *et al.* (2002)). Wang and Shen (2002) developed a mixed boundary value formulation for anisotropic piezoelectric bimaterial interfaces. An interfacial crack between a piezoelectric actuator and an elastic substrate under in-plane elastic loading was analysed by Liu and Hsia (2003). Ou and Chen (2004) and Qun and Yiheng (2007) studied interfacial impermeable and permeable cracks in metal/piezoelectric bimaterials and derived expressions for the crack-tip energy release rate (ERR) and the crack-tip generalized stress intensity factors (GSIF) for an interfacial permeable crack in metal/piezoelectric bimaterials. Later, Bakirov and Kim (2009) discussed the oscillating nature and the order of crack tip singularities and proposed a unified approach for determining the crack tip energy release rate for both permeable and impermeable cracks.

Recently, Rojas-Diaz *et al.* (2011) and Sharma *et al.* (2012) have studied the fracture in magnetoelectroelastic materials and the subinterface cracks in piezoelectric bimaterials with the extended finite element method (Figure 8.27b). It is expected that XFEM will soon be used in other studies of fracture in piezoelectric applications, including anisotropic problems.

# References

Abd-Alla, A.M., Mahmoud, S.R., Abo-Dahab, S.M. and Helmy, M.I. (2011) Propagation of s-wave in a non-homogeneous anisotropic incompressible and initially stressed medium under influence of gravity field. *Applied Mathematics and Computation*, **217**, 4321–4332.

Achenbach, J.D. (1974) Dynamic effects in brittle fractures, in *Mechanics Today*, vol. **1** (ed. S. Nemat-Nasser), Pergamon Press.

Achenbach, J.D. and Bazant, Z.P. (1975) Elastodynamic near-tip stress and displacement fields for rapidly propagating crack in orthotropic materials. *Journal of Applied Mechanics*, **42**(1), 183–189.

Adalsteinsson, D. and Sethian, J.A. (2003) Transport and diffusion of material quantities on propagating interfaces via level set methods. *Journal of Computational Physics*, **85**(1), 271–288.

Aliabadi, M.H. and Sollero, P. (1998) Crack growth analysis in homogeneous orthotropic laminates. *Composites Science and Technology*, **58**, 1697–1703.

Anahid, M. and Khoei, A.R. (2008) New development in the extended finite element modelling of large elasto-plastic deformations. *International Journal for Numerical Methods in Engineering*, **75**(10) 1133–1171.

Anders, D. Weinberg, K. and Reichardt, R. (2012) Isogeometric analysis of thermal diffusion in binary blends. *Computational Materials Science*, **52**(1), 182–188.

Anderson, T. (1995) *Fracture Mechanics: Fundamentals and Applications*, CRC Press, USA.

Ang, H.E., Torrance, J.E. and Tan, C.L. (1996) Boundary element analysis of orthotropic delamination specimens with interface cracks. *Engineering Fracture Mechanics*, **54**, 601–615.

Anlas, G., Santare, M. and Lambros, J. (2000) Numerical calculation of stress intensity factors in functionally graded materials. *International Journal of Fracture*, **104**, 131–143.

Arcisz, M. and Sih, G.C. (1984) Effect of orthotropy on crack propagation. *Theoretical Applied Fracture Mechanics*, **1**, 225–238.

Areias, P.M.A. and Belytschko, T. (2005a) Non-linear analysis of shells with arbitrary evolving cracks using XFEM. *International Journal for Numerical Methods in Engineering*, **62**, 384–415.

Areias, P.M.A. and Belytschko, T. (2005b) Analysis of three-dimensional crack initiation and propagation using the extended finite element method. *International Journal for Numerical Methods in Engineering*, **63**(55), 760–788.

Areias, P.M.A. and Belytschko, T. (2006) Two-scale shear band evolution by local partition of unity. *International Journal for Numerical Methods in Engineering*, **66**, 878–910.

Areias, P.M.A., Song, J.H. and Belytschko, T. (2006) Analysis of fracture in thin shells by overlapping paired elements. *Computer Methods in Applied Mechanics and Engineering*, **195**, 5343–5360.

Armero, F. and Linder, C. (2008) New finite elements with embedded strong discontinuities in the finite deformation range. *Computer Methods in Applied Mechanics and Engineering*, **197**(33–40), 3138–3170.

Asadpoure, A. (2006) Analysis of layered composites by the extended finite element method. MSc thesis, Sharif University of Technology, Tehran, Iran.

Asadpoure, A. and Mohammadi, S. (2007) A new approach to simulate the crack with the extended finite element method in orthotropic media. *International Journal for Numerical Methods in Engineering*, **69**, 2150–2172.

Asadpoure, A., Mohammadi, S. and Vafai, A. (2006) Modeling crack in orthotropic media using a coupled finite element and partition of unity methods. *Finite Elements in Analysis and Design*, **42**(13), 1165–1175.

Asadpoure, A., Mohammadi S. and Vafai, A. (2007) Crack analysis in orthotropic media using the extended finite element method. *Thin Walled Structures*, **44**(9), 1031–1038.

Atluri, S.N. (1982) Path-independent integrals in finite elasticity and inelasticity, with body forces, inertiam and arbitrary crack-face conditions. *Engineering Fracture Mechanics*, **16**(3), 341–364.

Atluri, S.N. and Shen, S. (2002) *The Meshless Local Petrov–Galerkin (MLPG) Method*, Tech Science Press, USA.

Atluri, S.N., Kobayashi, A.S. and Nakagaki, M.A. (1975a) Finite element program for fracture mechanics analysis of composite material. *Fracture Mechanics of Composites,* ASTM STP, **593**, 86–98.

Atluri, S.N., Kobayashi, A.S. and Nakagaki, M. (1975b) An assumed displacement hybrid finite element model for linear fracture mechanics. *International Journal of Fracture*, **11**(2), 257–271.

Attigui, M. and Petit, C. (1997) Mixed-mode separation in dynamic fracture mechanics: New path independent integrals. *International Journal of Fracture*, **84**, 19–36.

Ayari, M.L. and Ye, Z. (1995) Maximum strain theory for mixed mode crack propagation in anisotropic solids. *Engineering Fracture Mechanics*, **52**(3), 389–400.

Ayhan, A.O. (2009) Three-dimensional mixed-mode stress intensity factors for cracks in functionally graded materials using enriched finite elements. *International Journal of Solids and Structures*, **46**, 796–810.

Babuska, I. and Miller, A. (1984) The post-processing approach to finite element method – part II. *International Journal for Numerical Methods in Engineering*, **20**, 1111–1129.

Baker, B.R. (1962) Dynamic stresses created by a moving crack. *ASME Journal of Applied Mechanics*, **29**(3), 449–458.

Bakirov, V.F. and Kim, T.W. (2009) Analysis of a crack at the piezoceramic–metal interface and estimates of adhesion fracture energy. *International Journal of Engineering Science*, **47**(7–8), 793–804.

Barber, J.R. (2004) Stresses in a half space due to Newtonian gravitation. *Journal of Elasticity*, **75**, 187–192.

Barenblatt, G.I. (1959a) The formation of equilibrium cracks during brittle fracture: general ideas and hypotheses for axially symmetric cracks. *Journal of Applied Mathematics and Mechanics*, **23**, 622–636.

Barenblatt, G.I. (1959b) Equilibrium cracks formed during brittle fracture rectilinear cracks in plane plates. *Journal of Applied Mathematics and Mechanics*, **23**(4), 1009–1029.

Barnett, D.M. and Asaro, R.J. (1972) The fracture mechanics of slit-like cracks in anisotropic elastic media. *Journal of the Mechanics and Physics of Solids*, **20**, 353–66.

Barnett, D.M. and Lothe, J. (1973) Synthesis of the sextic and the integral formalism for dislocation, Green's function and surface waves in anisotropic elastic solids. *Physica Norvegica*, **7**, 13–19.

Barsom, J. M. and Rolfe, S.T. (1987) *Fracture and Fatigue Control in Structures Application of Fracture Mechanics*, 2nd edn, Prentice Hall, Englewood Cliffs, NJ.

Barsoum, R. (1974) Application of quadratic isoparametric finite elements in linear fracture mechanics. *International Journal of Fracture*, **10**, 603–605.

Barsoum, R. (1975) Further application of quadratic isoparametric finite elements to linear fracture mechanics of plate bending and general shells. *International Journal of Fracture*, **11**, 167–169.

Barsoum, R. (1976a) On the use of isoparametric finite elements in linear fracture mechanics. *International Journal for Numerical Methods in Engineering*, **10**, 25–37.

Barsoum, R. (1976b) A degenerate solid element for linear fracture analysis of plate bending and general shells. *International Journal for Numerical Methods in Engineering*, **10**, 551–564.

Barsoum, R. (1977) Triangular quarter point elements as elastic and perfectly plastic crack tip elements. *International Journal for Numerical Methods in Engineering*, **11**, 85–98.

Barsoum, R. (1981) An assessment of the quarter-point elements in pressure vessel fracture analysis. Sixth International Conference on Structural Mechanics in Reactor Technology, Paris.

Bassani, J.L. and Qu, J. (1989) Finite crack on bimaterial and bicrystals. *Journal of the Mechanics and Physics of Solids*, **37**, 435–453.

Batakis, A. P. and Vogan, W. (1985) Rocket thrust chamber thermal barrier coatings. NASA Contractor Report 1750222.

Bayesteh, H. and Mohammadi, S. (2011) XFEM fracture analysis of shells: The effect of crack tip enrichments. *Computational Materials Science*, **50**, 2793–2813.

Bayesteh, H. and Mohammadi, S. (2012) Fracture analysis of orthotropic functionally graded materials by XFEM. *Journal of Composites, Part B*, Submitted.

Bazant, Z.P. and Planas, J. (1997) *Fracture and Size Effects in Concrete and Other Quasibrittle Materials*. CRC Press, USA.

Bazilevs, Y. and Akkerman, I. (2010) Large eddy simulation of turbulent taylor-couette flow using isogeometric analysis and the residual-based variational multiscale method. *Journal of Computational Physics*, **229**, 3402–3414.

Bazilevs, Y., Gohean, J.R., Hughes, T.J.R. *et al.* (2009) Patient-specific isogeometric fluid–structure interaction analysis of thoracic aortic blood flow due to implantation of the Jarvik 2000 left ventricular assist device. *Computer Methods in Applied Mechanics and Engineering*, **198**(45–46), 3534–3550.

Béchet, E., Minnebo, H., Moës, N. and Burgardt, B. (2005) Improved implementation and robustness study of the X-FEM for stress analysis around cracks. *International Journal for Numerical Methods in Engineering*, **64**, 1033–1056.

Bechet, E., Scherzer, M. and Kuna, M (2009) Application of the X-FEM to the fracture of piezoelectric materials. *International Journal for Numerical Methods in Engineering*, **77**, 1535–1565.

Begley, J.A. and Landes, J.D. (1972) The J integral as a fracture criterion. Special ASTM publication, 1–23.

Bellec, J. and Dolbow, J.E. (2003) A note on enrichment functions for modelling crack nucleation. *Communications in Numerical Methods in Engineering*, **19**, 921–932.

Belytschko, T. and Black, T. (1999) Elastic crack growth in finite elements with minimal remeshing. *International Journal of Fracture Mechanics*, **45**, 601–620.

Belytschko, T. and Chen, H. (2004) Singular enrichment finite element method for elastodynamic crack propagation. *International Journal of Computational Methods*, **1**(1), 1–15.

Belytschko, T., Gracie, R. (2007) On XFEM applications to dislocations and interface. *International Journal of Plasticity*, **23**, 1721–1738.

Belytschko, T. and Tabbara, M. (1996) Dynamic fracture using element- free Galerkin methods. *International Journal for Numerical Methods in Engineering*, **39**, 923–938.

Belytschko, T., Lu, Y.Y. and Gu, L. (1994) Element-free Galerkin methods. *International Journal for Numerical Methods in Engineering*, **37**, 229–256.

Belytschko, T., Organ, D. and Krongauz, Y. (1995) A coupled finite element–Element free Galerkin method. *Computational Mechanics* **17**(3), 186–195.

Belytschko, T., Krongauz, Y., Organ, D., Fleming, M. and Krysl, P. (1996) Meshless methods: An overview and recent developments. *Computer Methods in Applied Mechanics and Engineering*, **139**, 3–47.

Belytschko, T., Guo, Y., Liu, W.K. and Xiao, S.P. (2000) A unified stability analysis of meshless particle methods. *International Journal for Numerical Methods in Engineering*, **48**(9), 1359–1400.

Belytschko, T., Organ, D. and Gerlach, C. (2000) Element-free Galerkin methods for dynamic fracture in concrete. *Computer Methods in Applied Mechanics and Engineering*, **187**(3–4), 385–399.

Belytschko, T., Moës, N., Usui, S. and Parimik, S. (2001) Arbitrary discontinuities in finite elements. *International Journal for Numerical Methods in Engineering*, **50**, 993–1013.

Belytschko, T., Daniel, W.J.T. and Ventura, G. (2002) A monolithic smoothing-gap algorithm for contact-impact based on the signed distance function. *International Journal for Numerical Methods in Engineering*, **55**, 101–125.

Belytschko, T., Xu, J.X., Chessa, J., Moës, N. and Gravouil, A. (2002) Recent developments in meshless and extended finite element methods. Second International Conference on Advanced Computational Methods in Engineering, Eindhoven, The Netherlands.

Belytschko, T., Chen, H., Xu, J. and Zi, G. (2003) Dynamic crack propagation based on loss of hyperbolicity and a new discontinuous enrichment. *International Journal for Numerical Methods in Engineering*, **58**, 1873–1905.

Benson, D.J., Bazilevs, Y., Hsu, M.C. and Hughes, T.J.R. (2010a) Isogeometric shell analysis: the reissner-mindlin shell. *Computer Methods in Applied Mechanics and Engineering*, **199**, 276–289.

Benson, D.J., Bazilevs, Y., De Luycker, E. *et al.* (2010b) A generalized finite element formulation for arbitrary basis functions: From isogeometric analysis to XFEM. *International Journal for Numerical Methods in Engineering*, **83**(6), 765–785.

Benson, D.J., Bazilevs, Y., Hsu, M.C. and Hughes, T.J.R. (2011) A large deformation, rotation-free, isogeometric shell. *Computer Methods in Applied Mechanics and Engineering*, **200**(13–16), 1367–1378.

Benzley, S.E. (1974) Representation of singularities with isoparametric finite elements. *International Journal for Numerical Methods in Engineering*, **8**, 537–545.

Bitaraf, M. and Mohammadi, S. (2010) Large deflection analysis of flexible plates by the finite point method. *Thin-Walled Structures*, **48**, 200–214.

Bogy, D.B. (1972) The plane solution for anisotropic elastic wedges under normal and shear traction. *Journal of Applied Mechanics*, **39**, 1103–9.

Boone, T.J., Wawrzynek, P.A. and Ingraffea A.R. (1987) Finite element modeling of fracture propagation in orthotropic materials. *Engineering Fracture Mechanics*. **26**, 185–201.

Bordas, S. and Legay, A. (2005) X-FEM Mini-Course. EPFL, Lausanne, Switzerland.

Borovkov, A., Kiylo, O., Misnik, Yu and Tripolnikov, T. (1999) Finite element stress analysis of multidirectional laminated composite structures, 2. H-p- refinement and m- adaptive procedures. *Zeitschrift fur Angewandte Mathematik and Mechanik*, **79**(2), S527–S528.

de Borst, R., Remmers, J.J.C. and Needleman, A. (2004) Computational aspects of cohesive-zone models. Fifteenth European Conference of Fracture, Sweden.

de Borst, R., Gutiérrez, M.A., Wells, G.N. *et al.* (2004a) Cohesive-zone models, higher-order continuum theories and reliability methods for computational failure analysis. *International Journal for Numerical Methods in Engineering*, **60**, 289–315.

de Borst, R., Remmers, J.J.C., Needleman, A. and Abellan, M.A. (2004b) Discrete vs smeared crack models for concrete fracture: bridging the gap. *International Journal for Numerical and Analytical Methods in Geomechanics*, **28**, 583–607.

Bouchbinder, E., Livne, A. and Finebergm, J. (2009) The 1/r singularity in weakly nonlinear fracture mechanics. *Journal of the Mechanics and Physics of Solids*, **57**, 1568–1577.

Bouchbinder, E., Livne, A. and Finebergm, J. (2010) Weakly nonlinear fracture mechanics: experiments and theory. *International Journal of Fracture*, **162**, 3–20.

Bower, A.F. (2012) http://solidmechanics.org/text/chapter3_2/chapter3_2.htm (accessed 1 September 2011).

Bowie, O.L. and Freese, C.E. (1972) Central crack in plane orthotropic rectangular sheet. *International Journal of Fracture Mechanics*, **8**, 49–58.

Brindley, W. (1995) Properties of plasma sprayed bond coats, Proceedings of the Thermal Barrier Coating Workshop. NASA-Lewis, Cleveland, OR, 1995.

Broberg, K.B. (1999) Intersonic crack propagation in an orthotropic material. *International Journal of Fracture*, **99**, 1–11.

Brock, L.M., Georgiadis, H.G. and Hanson, M.T. (2001) Rapid indentation of transversely isotropic or orthotropic half-spaces. *Transactions of the ASME*, **68**, 490–495.

Broumand, P. and Khoei, A.R. (2011) Modeling ductile fracture with damage plasticity using XFEM technique, in *International Conference on Extended Finite Element Methods - XFEM 2011* (eds S. Bordas, B. Karihaloo and P. Kerfriden), Cardiff, United Kingdom.

Budyn, E. and Hoc, T. (2007) Multiple scale modelling for cortical bone fracture in tension using X-FEM. RMN-16/2007, 215–238.

Budyn, E., Zi, G., Moës, N. and Belytschko, T. (2004) A method for multiple crack growth in brittle materials without remeshing. *International Journal for Numerical Methods in Engineering*, **61**, 1741–1770.

Buehler, M.J. (2012) Materiomics, merging biology and engineering in multiscale structures and materials. NSF Short Course on Nanomechanics, Nanomaterials and Micro/Nanomanufacturing, Northwestern University.

Bruno, D., Carpino, R. and Greco, F. (2007) Modelling of mixed mode debonding in externally FRP reinforced beams. *Composites Science and Technology*, **67**, 1459–1474.

Buffa, A., Sangalli, G. and Vazquez, R. (2010) Isogeometric analysis in electromagnetics: B-splines approximation. *Computer Methods in Applied Mechanics and Engineering*, **199**(17–20), 1143–1152.

Carloni, C. and Nobile, L. (2002) Crack initiation behaviour of orthotropic solids as predicted by the strain energy density theory. *Theoretical and Applied Fracture Mechanics*, **38**, 109–119.

Carloni, C., Piva, A. and Viola, E. (2003) An alternative complex variable formulation for an inclined crack in an orthotropic medium. *Engineering Fracture Mechanics*, **70**, 2033–2058.

Chahine, E., Laborde, P. and Renard, Y. (2006) A quasi-optimal convergence result for fracture mechanics with XFEM. *Comptes Rendus Mathematique*, **342**(7), 527–532.

Chahine, E. (2009) *Extended Finite Element Method (XFEM): Mathematical and Numerical Study for Cracked Domains Computations*, VDM Verlag Dr Muller.

Chan, S.K., Tuba, I.S. and Wilson, W.K. (1970) On the finite element method in liner fracture mechanics: Finite element method with first order displacement functions to compute crack tip stress intensity factors in various shapes under different loading conditions. *Engineering Fracture Mechanics*, **2**, 1–17.

Chandler, D. (1989) Metal fatigue seen as most likely cause. *Boston Globe*, (February 25).

Chang, F.-K. and Springer, G.S. (1986) The strengths of fiber reinforced composite bends. *Journal of Composite Materials*, **20**(1) 30–45.

Chang, J., Xu, J.-q. and Mutoh, Y. (2006) A general mixed-mode brittle fracture criterion for cracked materials. *Engineering Fracture Mechanics*, **73**, 1249–1263.

Chen, E.P. (1978) Sudden appearance of a crack in a stretched finite strip. *Journal of Applied Mechanics*, **45**, 270–280.

Chen, Y.F. and Erdogan, F. (1996) The interface crack problem for a nonhomogeneous coating bonded to a homogeneous substrate. *Journal of the Mechanics and Physics of Solids*, **44**, 771–787.

Chessa, J. and Belytschko, T. (2003) An enriched finite element method and level sets for axisymmetric two-phase flow with surface tension. *International Journal for Numerical Methods in Engineering*, **58**, 2041–2064.

Chessa, J. and Belytschko, T. (2004) Arbitrary discontinuities in space–time finite elements by level sets and X-FEM. *International Journal for Numerical Methods in Engineering*, **61**, 2595–2614.

Chessa, J. and Belytschko, T. (2006) A local space–time discontinuous finite element method. *Computer Methods in Applied Mechanics and Engineering*, **195**, 1325–1343.

Chessa, J., Wang, H. and Belytschko, T. (2003) On the construction of blending elements for local partition of unity enriched finite elements. *International Journal for Numerical Methods in Engineering*, **57**, 1015–1038.

Chopp, D.L. (2001) Some improvements on the fast marching method. *SIAM Journal of Scientific Computations*, **23**(1), 230–244.

Chopp, D.L. and Sukumar, N. (2003) Fatigue crack propagation of multiple coplanar cracks with the coupled extended finite element/fast marching method. *International Journal of Engineering Science*, **41**, 845–869.

Chow, W.T. and Atluri, S.N. (1998) Stress intensity factors as the fracture parameters for delamination crack growth in composite laminates. *Computational Mechanics*, **21**, 1–10.

Chow, W.T., Boem, H.G. and Atluri, S.N. (1995) Calculation of stress intensity factors for interfacial crack between dissimilar anisotropic media using a hybrid element method and the mutual integral. *Computational Mechanics*, **15**, 546–557.

Cimrman, R., Kroc, J., Rohan, E. *et al.* (2002) On coupling cellular automata based activation and finite element muscle model applied to heart ventricle modelling.

Clarke, K. and Hess, D. (1978) *Communication Circuits: Analysis and Design*, Addison-Wesley Publishing Company.

Clements, D.L. (1971) A crack between dissimilar anisotropic media. *International Journal of Engineering Science*, **9**, 257–265.

Colombi, C. (2006) Reinforcement delamination of metallic beams strengthened by FRP strips: Fracture mechanics based approach. *Engineering Fracture Mechanics*, **73**, 1980–1995.

Combescure, A., Elguedj, T., Haboussa, D. *et al.* (2011) Modeling damage brittle ductile fracture transition with combined X-FEM cohesive zone, in *International Conference on Extended Finite Element Methods - XFEM 2011* (eds S. Bordas, B. Karihaloo and P. Kerfriden), Cardiff, United Kingdom.

Combescure, A., Gravouil, A., Gregoire, D. and Rethore, J. (2008) X-FEM a good candidate for energy conservation in simulation of brittle dynamic crack propagation. *Computational Methods in Applied Mechanics and Engineering*. **197**, 309–318.

Comi, C. and Mariani, S. (2007) Extended finite element simulation of quasi-brittle fracture in functionally graded materials. *Computational Methods in Applied Mechanics and Engineering*, **196**, 4013–4026.

Comninou, M. (1977) The interface crack. *Journal of Applied Mechanics*, **44**, 631–636.

Comninou, M. and Schmuser, D. (1979) The interface crack in a combined tension-compression and shear field. *Journal of Applied Mechanics*, **46**, 345–348.

Cotterell, B. (1964) On the nature of moving cracks. *ASME Journal of Applied Mechanics*, **31**(1), 12–16.

Cottrell, A.H. (1961) Theoretical aspects of radiation damage and brittle fracture in steel pressure vessels. Iron and Steel Institute Special Report No. 69, pp. 281–296. *Iron and Steel Institute*, London, UK.

Cottrell, J.A., Reali, A., Bazilevs, Y. and Hughes, T.J.R. (2006) Isogeometric analysis of structural vibrations. *Computer Methods in Applied Mechanics and Engineering*. **195**, 5257–5296.

Cottrell, J.A., Hughes, T.J.R. and Bazilevs, Y. (2009) *Isogeometric analysis: Towards integration of CAD and FEA*, John Wiley & Sons, Chichester.

Craggs, J.W. (1960) On the propagation of a crack in an elastic-brittle material. *Journal of the Mechanics and Physics of Solids*, **8**, 66.

Cruse, T. (1988) *Boundary Element Analysis in Computational Fracture Mechanics*, Kluwer, Dordrecht, The Netherlands.

Dag, S., Yildirim, B. and Sarikaya, D. (2007) Mixed-mode fracture analysis of orthotropic functionally graded materials under mechanical and thermal loads. *International Journal of Solids and Structures*, **44**, 7816–7840.

Dally, J.W. (1987) Dynamic photoelasticity and its application to stress wave propagation, fracture mechanics and fracture control. in *Static and Dynamic Photoelasticity and Caustics* (eds A. Lagarde), Springer-Verlag, New York, pp. 247–406.

Daneshyar, A.R. (2012) Shear Band Simulation by the Extended Finite Element Method. M.Sc. Thesis, University of Tehran, Iran.

Daneshyar, A.R. and Mohammadi, S. (2011) Simulation of strong tangential discontinuity for XFEM shear band evolution, in *International Conference on Extended Finite Element Methods - XFEM 2011 Cardiff, United Kingdom*, (eds S. Bordas, B. Karihaloo and P. Kerfriden),

Daneshyar, A.R. and Mohammadi, S. (2012a) An extended finite element for simulation of shear band frictional contact. 20th Annual Conference on Mechanical Engineering, ISME 2012, Shiraz, Iran.

Daneshyar, A.R. and Mohammadi, S. (2012b) XFEM simulation of shear band initiation and propagation based on a strong tangential discontinuity formulation, Submitted.

Daux, C., Moës, N., Dolbow, J. *et al.* (2000) Arbitrary branched and intersecting cracks with the extended finite element method. *International Journal for Numerical Methods in Engineering*, **48**, 1741–1760.

De, J. and Patra, B. (1992) Elastodynamic crack problems in an orthotropic medium through complex variable approach. *Engineering Fracture Mechanics* **43**, 895–909.

Delale, F. and Erdogan, F. (1983) The crack problem for a non-homogeneous plane. *ASME Journal of Applied Mechanics*, **50**(3), 609–614.

De Luycker, E., Benson, D.J., Belytschko, T., et al. (2011) X-FEM in isogeometric analysis for linear fracture mechanics. *International Journal for Numerical Methods in Engineering*, **87**(6) 541–565.

DeMasi, T., Sheffler, K D. and Ortiz, M. (1989) Thermal barrier coating life prediction model development. NASA Contractor Report 182230, 1989.

Deng, X. (1995) Mechanics of debonding and delamination in composites: Asymptotic studies. *Composites Engineering*, **5**, 1299–1315.

Destrade, M. (2007) Seismic Rayleigh waves on an exponentially graded, orthotropic half-space. *Proceedings of the Royal Society of London, Series A*, **463**, 495–502.

Dini, D., Barber, J.R., Churchman, C.M. *et al.* (2008) The application of asymptotic solutions to contact problems characterised by logarithmic singularities. *European Journal of Mechanics A-Solids*, **27**, 847–858.

Dolbow, J.E. (1999) An extended finite element method with discontinuous enrichment for applied mechanics. *PhD dissertation, Theoretical and Applied Mechanics*, Northwestern University, USA.

Dolbow, J.E. and Nadeau, J.C. (2002) On the use of effective properties for the fracture analysis of microstructured materials. *Engineering Fracture Mechanics*, **69**, 1607–1634.

Dolbow, J.E. and Gosz, M. (2002) On the computation of mixed-mode stress intensity factors in functionally graded materials. *International Journal of Solids and Structures*, **39**, 2557–2574.

Dolbow, J.E. and Devan, A. (2004) Enrichment of enhanced assumed strain approximations for representing strong discontinuities: addressing volumetric incompressibility and the discontinuous patch test. *International Journal for Numerical Methods in Engineering*, **59**, 47–67.

Dolbow, J., Moës, N. and Belytschko, T. (2000a) Discontinuous enrichment in finite elements with a partition of unity method. *Finite Elements in Analysis and Design*, **36**, 235–260.

Dolbow, J., Moës, N. and Belytschko, T. (2000b) Modeling fracture in Mindlin–Reissner plates with the extended finite element method. *International Journal of Solids and Structures*, **37**, 7161–7183.

Dolbow, J., Moës, N. and Belytschko, T. (2000c) An extended finite element method for modeling crack growth with frictional contact. *Finite Elements in Analysis and Design*, **36**(3) 235–260.

Dolbow, J., Moës, N. and Belytschko, T. (2001) An extended finite element method for modeling crack growth with frictional contact. *Computer Methods in Applied Mechanics and Engineering*, **190**, 6825–6846.

Dongye, C. and Ting, T.C.T. (1989) Explicit expressions of Barnett-Lothe tensors and their associated tensors for orthotropic materials. *Quarterly of Applied Mathematics*, **47**, 723–734.

Duarte, C.A. (2011) Bridging scaleswith a generalized fem, in *International Conference on Extended Finite Element Methods - XFEM 2011* Cardiff, United Kingdom, (eds S. Bordas, B. Karihaloo and P. Kerfriden).

Duarte, C.A. and Oden, J.T. (1995) Hp clouds – a meshless method to solve boundary-value problems. TICAM Report 95-05.

Duarte, C.A. and Oden, J.T. (1996) An H-p adaptive method using clouds. *Computer Methods in Applied Mechanics and Engineering*, **139**, 237–262.

Duarte, C.A., Babuska, I. and Oden J.T. (1998) Generalized finite element methods for three dimensional structural mechanics problems, Proceeding of the International Conference on Computational Science, *Tech.* Science Press, Atlanta, GA, **1**, pp. 53–58.

Dugdale, D. (1960) Yielding of steel sheets containing slits. *Journal of Mechanics and Physics of Solids*, **8**, 100–108.

Dumstorff, P. and Meschke, G. (2003) Finite element modelling of cracks based on the partition of unity method. *Proceedings of Applied Mathematics and Mechanics (PAMM)*, **2**, 226–227.

Dundurs, J. (1969) Edge-bonded dissimilar orthogonal elastic wedges. *Journal of Applied Mechanics*, **36**, 650–652.

Dunn, M. (1994) The effects of crack face boundary conditions on the fracture mechanics of piezoelectric solids. *Engineering Fracture Mechanics*, **48**, 25–39.

Ebrahimi, S.H., Mohammadi, S. and Assadpoure, A. (2008) An extended finite element (XFEM) approach for crack analysis in composite media. *International Journal of Civil Engineering*, **3**, 198–207.

Ebrahimi, S.H., Mohammadi, S. and Kani, I.M. (2012) A local PUFEM modelling of stress singularity in complete sliding contact problems with minimal enrichment, Submitted.

Ebrahimi, S.H. (2012) Development of the extended finite element method in modelling frictional contact problems. PhD. Thesis, University of Tehran, Iran.

Edgerton, H. E. and Barstow, F. E. (1941) Further studies of glass fracture with high-speed photography. *Journal of the American Ceramic Society*, **24**, 131–7.

Eftekhari, M. and Mohammadi, S. (2012) on the buckling analysis of defective carbon nanotubes by molecular structural mechanics approach. *Mechanics of Nano, Micro and Macro Composite Structures*, Politecnico di Torino, Italy.

Eischen, J. (1983) Fracture of nonhomogeneous materials. *International Journal of Fracture*, **34**, 3–22.

Elguedj, T., Gravouil, A. and Combescure, A. (2006) Appropriate extended functions for X-FEM simulation of plastic fracture mechanics. *Computer Methods in Applied Mechanics and Engineering*, **195**, 501–515.

Elguedj, T. , Gravouil, A. and Maigre, H. (2009) An explicit dynamics extended finite element method. Part 1: Mass lumping for arbitrary enrichment functions. *Computer Methods in Applied Mechanics and Engineering*, **198**, 2297–2317.

Elguedj, T., Rethore, J. and Buteri, A. (2011) Isogeometric analysis for strain field measurements. *Computer Methods in Applied Mechanics and Engineering*, **200**(1–4), 40–56.

England, A.H. (1965) A crack between dissimilar media. *Journal of Applied Mechanics*, **32**, 400–402.

Erdogan, F. (1963) Stress distribution in a nonhomogeneous elastic plane with cracks. *Journal of Applied Mechanics*, **30**, 232–237.

Erdogan, F. (1995) Fracture mechanics of functionally graded materials. *Composites Engineering*, **5**(7), 753–770.

Erdogan, F. and Sih, G.C. (1963) On the crack extension in plates under plane loading and transverse shear. *Journal of Basic Engineering*, **85**, 519–527.

Erdogan, E. and Wu, B.H. (1993) Interface crack problems in layered orthotropic materials. *Journal of Mechanics, Physics and Solids*, **41**(5), 889–917.

Eshelby, J.D. (1956) The continuum theory of lattice defects, in *Solid State Physics*, vol. **3** (eds F. Seitz and D. Turnbull), Academic Press, New York, pp. 79–141.

Eshelby, J.D. (1974) Calculation of energy release rate, in *Prospect of Fracture Mechanics*, (eds G.C. Sih, H.C. Van Elst and D. Brock) Nordhoff, UK, pp. 69–84.

Eshelby, J.D., Read, W.T. and Shockley, W. (1953) Anisotropic elasticity with applications to dislocation theory. *Acta Metallurgica*, **1**, 251–259.

Esna Ashari, S. (2009) An Extended Finite Element Method for Delamination Analysis of Layered Composites. M.Sc. Thesis, University of Tehran, Iran.

Esna Ashari, S. and Mohammadi, S. (2009) XFEM delamination analysis of composite laminates by new orthotropic enrichment functions. Proceedings of 1st International Conference on Extended Finite Element Method - Recent Developments and Applications (XFEM2009), Aachen, Germany.

Esna Ashari, S. and Mohammadi, S. (2010a) Modeling interface crack between orthotropic and isotropic materials using extended finite element method. 9th International Congress on Advances in Civil Engineering, 27–30 September 2010, Karadeniz Technical University, Trabzon, Turkey.

Esna Ashari, S. and Mohammadi, S. (2010b) Modeling delamination in composite laminates using XFEM by new orthotropic enrichment functions. WCCM/APCOM 2010; IOP Conference Series: Materials Science and Engineering 10, 012240, 1-8, Sydney, Australia.

Esna Ashari, S. and Mohammadi, S. (2011a) Delamination analysis of composites by new orthotropic bimaterial extended finite element method. *International Journal for Numerical Methods in Engineering*, **86**(13), 1507–1543.

Esna Ashari, S. and Mohammadi, S. (2011b) Debonding propagation analysis of FRP reinforced beams by the extended finite element method, in International Conference on Extended Finite Element Methods - *XFEM 2011* Cardiff, United Kingdom (eds S. Bordas, B. Karihaloo and P. Kerfriden).

Esna Ashari, S. and Mohammadi, S. (2012) Fracture Analysis of FRP Reinforced Beams by Orthotropic XFEM. *Journal of Composite Materials – Part B*, **46**(11), 1367–1389.

Espinosa, H.D., Dwivedi, S. and Lu, H.-C. (2000) Modeling impact induced delamination of woven fiber reinforced composites with contact/cohesive laws. *Computer Methods in Applied Mechanics and Engineering*, **183**(3–4), 259–290.

Esslinger, V., Kieselbach, R., Koller, R. and Weisse, B. (2004) The railway accident of Eschede technical background. *Engineering Failure Analysis*, **11**, 515–535.

Fagerström, M. and Larsson, R. (2006) Theory and numerics for finite deformation fracture modelling using strong discontinuities. International Journal for Numerical Methods in Engineering, **66**, 911–948.

Fawkes, A., Owen, D.R.J. and Luxmoore, A. (1979) An assessment of crack tip singularity models for use with isoparamtreic elements. *Engineering Fracture Mechanics*, **11**, 143–159.

Federici, L., Nobile, L., Piva, A. and Viola, E. (2001) On the intersonic crack propagation in an orthotropic medium. *International Journal of Fracture*, **112**, 69–85.

Ferrie, E., Buffiere, J.Y., Ludwig, W. *et al.* (2006) Fatigue crack propagation: In situ visualization using X-ray microtomography and 3D simulation using the extended finite element method. *Acta Materialia*, **54**, 1111–1122.

Folias, E.S. (1999) Failure correlation between cylindrical pressurized vessels and flat plates. *International Journal of Pressure Vessels and Piping*, **76**(11), 803–811.

Foreman, A.J.E. (1955) Dislocation energies in anisotropic crystals. *Acta Metallurgica*, **3**, 322–330.

Forghani, A. (2005) Application of the extended finite element method for modeling cohesive crack problems. MSc thesis, Sharif University of Technology, Tehran, Iran.

Forschi, R.O. and Barret. J.D. (1976) Stress intensity factors in anisotropic plates using singular isoparametric elements. *International Journal for Numerical Methods in Engineering*, **10**, 1281–1287.

Freund, L.B. (1972) Energy flux into the tip of an extended crack in an elastic solid. *Journal of Elasticity*, **2**, 341–360.

Freund, L.B. (1976) Dynamic crack propagation, in *Mechanics of Fracture* (ed. F. Erdogan), ASME, pp. 105–134.

Freund, L.B. (1990) *Dynamic Fracture Mechanics*, Cambridge University Press, Cambridge.

Freund, L.B. and Clifton, R.J. (1974) On the uniqueness of plane elastodynamic solutions for running cracks. *Journal of Elasticity*, **4**(4), 293–299.

Freund, L.B. and Douglass, A.S. (1982) The influence of inertia on elastic-plastic anti-plane shear crack growth. *Journal of Mechanics of Physics of Solids*, **30**(1/2), 59–74.

Fries, T.P. and Belytschko, T. (2006) The intrinsic XFEM: a method for arbitrary discontinuities without additional unknowns. *International Journal for Numerical Methods in Engineering*, **68**(13) 1358–1385.

Gao, C.F. and Fan, J.X. (1998) The foundation solution for the plane problem of piezoelectric media with an elliptical hole or crack. *Applied Mathematics and Mechanics*, **19**, 965–973.

Gao, H. and Klein, P. (1998) Numerical simulation of crack growth in an isotropic solid with randomized internal cohesive bonds. *Journal of the Mechanics and Physics of Solids*, **42**(6), 187–218.

Gao, H., Abbudi, M. and Barnett, D.M. (1992) Interfacial crack-tip field in anisotropic elastic solids. *Journal of Applied Mechanics*, **40**, 393–416.

García-Sánchez, F., Zhang, C. and Sáez, A. (2008) A two-dimensional time-domain boundary element method for dynamic crack problems in anisotropic solids. *Engineering Fracture Mechanics*, **75**, 1412–1430.

Gentilini, C., Piva, A. and Viola, E. (2004) On crack propagation in orthotropic media for degenerate states. *European Journal of Mechanics A/Solids*, **23**, 247–258.

Ghorashi, S.Sh., Mohammadi, S. and Sabbagh-Yazdi, S.R. (2011) Orthotropic enriched element free Galerkin method for fracture analysis of composites. *Engineering Fracture Mechanics*, **78**, 1906–1927.

Ghorashi, S.S., Valizadeh, N. and Mohammadi, S. (2011) An improved isogeometric analysis using the Lagrange multiplier method. *International Conference on Extended Finite Element Methods – XFEM2011*, Cardiff, UK, eds. S. Bordas, B. Karihaloo and P. Kerfriden.

Ghorashi, S.S., Valizadeh, N. and Mohammadi, S. (2012) Extended isogeometric analysis (XIGA) for analysis of stationary and propagating crack. *International Journal for Numerical Methods in Engineering*, **89**(9), 1069–1101.

Gifford, N.L. and Hilton, P.D. (1978) Stress intensity factors by enriched finite elements. *Engineering Fracture Mechanics*, **20**, 485–296.

Giner, E., Sukumar, N., Fuenmayor, F.J. and Vercher, A. (2008) Singularity enrichment for complete sliding contact using the partition of unity finite element method. *International Journal for Numerical Methods in Engineering*, **76**(9), 1402–1418.

Göktepe, S., Acharya, S.N.S., Wong, J. and Kuhl, E. (2011) Computational modeling of passive myocardium. *International Journal for Numerical Methods in Biomedical Engineering*, **27**, 1–12.

Goodarzi, M., Mohammadi, S. and Jafari, A. (2011) Numerical study of induced gas pressure on controlled blasting. 6th National Congress of Civil Engineering (6NCCE), Semnan, Iran (In Persian).

Gosz, M., Dolbow, J., Moran, B. (1998) Domain integral formulation for stress intensity factor computation along curved threedimensional interface cracks. *International Journal of Solids and Structures*, **35**, 1763–1783.

Gotoh, M. (1967) Some problems of bonded anisotropic plates with crack along the bond. *International Journal of Fracture*, **3**, 253–265.

Gracie, R., Ventura, G. and Belytschko, T. (2007) A new fast finite element method for dislocations based on interior discontinuities. *International Journal for Numerical Methods in Engineering*, **69**, 423–441.

Gravouil, A., Moës, N. and Belytschko, T. (2002) Non-planar 3D crack growth by the extended finite element and level sets–Part II: Level set update. *International Journal for Numerical Methods in Engineering*, **53**, 2569–2586.

Gravouil, A., Elguedj, T. and Maigre, H. (2009a) An explicit dynamics extended finite element method. Part 1: Mass lumping for arbitrary enrichment functions. *Computer Methods in Applied Mechanics and Engineering*, **198**, 2297–2317.

Gravouil, A., Elguedj, T. and Maigre, H. (2009b) An explicit dynamics extended finite element method. Part 2: Element-by-element stable-explicit/explicit dynamic scheme. *Computer Methods in Applied Mechanics and Engineering*, **198**, 2318–2328.

Greco, F., Lonetti, P. and Blasi, N. (2007) An analytical investigation of debonding problems in beams strengthened using composite plates. *Engineering Fracture Mechanics*, **74**(3), 346–372.

Gregoire, D., Maigre, H. and Combescure, A. (2008) On the growth, the arrest and the restart of a crack during a dynamic brittle fracture experiment. *Key Engineering Materials*, **385**, 245–248.

Gregoire, D., Maigre, H., Rethore, J. and Combescure, A. (2007) Dynamic crack propagation under mixed-mode loading – Comparison between experiments and X-FEM simulations. *International Journal of Solids and Structures*, **44**, 6517–6534.

Griffith, A.A. (1921) The phenomena of rupture and flow in solids. *Philosophical Transactions of the Royal Society A: Mathematical, Physical and Engineering*, **221**, 163–197.

Griffith, A.A. (1924) The theory of rupture. Proceedings of International Congress on Applied Mechanics, Delft, pp. 55–62.

Gtoudos, E. (1993) *Fracture Mechanics: An Introduction*, Kluwer Academic Press, The Netherlands.

Gu, P., Dao, M. and Asaro, R. (1999) A simplified method for calculating the crack-tip field of functionally graded materials using the domain integral. *Journal of Applied Mechanics*, **66**, 101–108.

Gu, P. and Asaro, R.J. (1997) Cracks in functionally graded materials. *International Journal of Solids and Structures*, **34**(1), 1–17.

Haasemann, G., Kästner, M., Prüger, S. and Ulbricht, V. (2011) Development of a quadratic finite element formulation based on the XFEM and NURBS. *International Journal for Numerical Methods in Engineering*, 2011; Available online: doi: 10.1002/nme.3120

Hajikarimi, P., Mohammadi, S. and Aflaki, S. (2012) Two dimensional creep analysis of a linear cracked viscoelastic medium using the extended finite element method. 6th European Congress on Computational Methods in Applied Sciences and Engineering (ECCOMAS 2012), Vienna, Austria.

Hallai, J. (2008) Fracture of orthotropic materials under mixed mode loading. *EM 388F Fracture Mechanics*: Term Paper, University of Texas at Austin.

Harrison, R. P., Loosemore, K., Milne, I. and Dowling, A.R. (1980) Assessment of the integrity of structures containing defects. CEGB Report R/H/R6-Rev.2.

Hashagen, F. and de Borst, R. (2000) Numerical assessment of delamination in fibre metal laminates. *Computer Methods in Applied Mechanics and Engineering*, **185**, 141–159.

Hassani, B., Moghaddam, N.Z. and Tavakkoli, S.M. (2009) Isogeometrical solution of Laplace equation. *Asian Journalof Civil Engineering*, **10**, 579–592.

Hassani, B., Khanzadi, M. and Tavakkoli, S.M. (2012) An isogeometric approach to structural topology optimization by optimality criteria. *Structural and Mulitdisciplinary Optimization*, **45**(2), 223–233.

Hatefi, S. and Mohammadi, S. (2012) An XFEM Model for Transition of Micro Damage Mechanics to Macro Crack Analysis, University of Tehran, Report: 8102051.

Hellen, T.K. and Blackburn, W.S. (1975) The calculation of stress intensity factors for combined tensile and shear loading. *International Journal of Fracture*, **11**(4), 605–617.

Hemanth, D., Shivakumar Aradhya, K.S., Rama Murthy, T.S. and Govinda Raju, N. (2005) Strain energy release rates for an interface crack in orthotropic media-a finite element investigation. *Engineering Fracture Mechanics*, **72**, 759–772.

Henshell, R.D. and Shaw, K.G. (1975) Crack tip elements are unnecessary. *International Journal of Numerical Methods in Engineering*, **9**(3), 495–509.

Hettich, T. and Ramm, E. (2006) Interface material failure modeled by the extended finite-element method and level sets. *Computer Methods in Applied Mechanics and Engineering*, **195**, 4753–4767.

Hibbit, H. (1977) Some properties of singular isoparametric elements. *International Journal for Numerical Methods in Engineering*, **11**, 180–184.

Hirth, J.P. and Lothe, J. (1982) *Theory of Dislocations*, 2nd edn, John Wiley & Sons, New York.

Holt, J.B., Koizumi, M., Hirai, T. and Munir, Z.A. (eds) (1993) Proceedings of the Second International Symposium on Functionally Graded Materials, *Ceramic Transactions*, vol. **34**. American Ceramic Society, Westerville, OH.

Houck, D.L. (1987) *Proceedings of the National Thermal Spray Conference*, ASM International.

Hsu, M.-C., Akkerman, I. and Bazileves, Y. (2011) High-performance computing of wind turbine aerodynamics using isogeometric analysis. *Computers and Fluids*, **49**(1), 93–100.

Hsu, M.H. (2006) Concrete beams strengthened with externally bonded glass fiber reinforced plastic plates. *Tamkang Journal of Science and Engineering*, **9**(3) 223–232.

Huang, R., Sukumar, N. and Prevost, J.H. (2003) Modeling quasi-static crack growth with the extended finite element method, Part II: Numerical applications. *International Journal of Solids and Structures*, **40**, 7539–7552.

Hughes, T.J.R. (2000) *The Finite Element Method*: Linear Static and Dynamic Finite Element Analysis, Dover Publications.

Hughes, T.J.R., Cottrell, J.A. and Bazilevs, Y. (2005) Isogeometric analysis: CAD, finite elements, NURBS, exact geometry and mesh refinement. *Computer Methods in Applied Mechanics and Engineering*, **194**, 4135–4195.

Hussain, M., Pu, S. and Underwood, J. (1974) Strain energy release rate for a crack under combined mode I and mode II. *ASTM, STP*, **560**, 2–28.

Hutchinson, J. (1968) Singular behavior at the end of a tensile crack tip in a power-law hardening material. *Journal of Mechanics and Physics of Solids*, **16**, 13–31.

Hutchinson, J.W., Mear, M. and Rice, J.R. (1987) Crack paralleling in between dissimilar materials. *Journal of Applied Mechanics*, **54**, 828–832.

Hwu, C. (1993a) Explicit solutions for co-linear crack problems. *International Journal of Solids and Structures*, **3**, 301–312.

Hwu, C. (1993b) Fracture parameters for the orthotropic bimaterial cracks. *Engineering Fracture Mechanics*, **45**, 89–97.

Ilschner, B. and Cherradi, N. (eds) (1995) *Proceedings of the Third International Symposium on Structural and Functional Gradient Materials*, Presses Polytechniques et Universitaires Romands, Lausanne, Switzerland.

Inglis, C.E. (1913) Stresses in a plate due to the presence of cracks and sharp corners. *Transactions of Institute of Naval Architects*, **55**, 219–241.

Irwin, G.R. (1948) Fracture Dynamics fracturing of metals. *American Society for Metals, Cleveland*, 147–166.

Irwin, G.R. (1957) Analysis of stresses and strains near the end of a crack transversing a plate. *Journal of Applied Mechanics, Transactions ASME*, **24**, 361–364.

Irwin, G.R. (1958) Fracture, in *Handbuch der Physik*, vol. **6** (ed. S. Flugge), Springer-Verlag, Berlin, Germany, pp. 551–590.

Irwin, G.R. (1961) Plastic zone near a crack tip and fracture toughness. Proceedings of the 7th Sagamore Conference, vol. IV, New York, USA, pp. 63–76.

Irwin, G.R., Kies, G.A. and Smith, H.L. (1958) Fracture strength relative to the onset and arrest of crack propagation. *Proceedings of ASTM*, **58**, 640–657.

Isenberg, B.C., DiMilla, P.A., Walker, M. *et al.* (2009) Vascular smooth muscle cell durotaxis depends on substrate stiffness gradient strength. *Biophysics Journal*, **97**, 1313.

Ishikawa, H. (1980) A finite element analysis of stress intensity factors for combined tensile and shear loading by only a virtual crack extension. *International Journal of Fracture* **16**(5), 243–246.

Jernkvist, L.O. (2001a) Fracture of wood under mixed mode loading I. Derivation of fracture criteria. *Engineering Fracture Mechanics*, **68**, 549–563.

Jernkvist, L.O. (2001b) Fracture of wood under mixed mode loading II. Experimental investigation of Picea abies. *Engineering Fracture Mechanics*, **68**, 565–576.

Jernkevist, L.O. and Thuvander, F. (2001) Experimental determination of stiffness variation across growth rings in Picea abies. *Holzforschung*, **55**, 309–317.

Jin, C., Shao, J., Zhu, Q. and He, Q. (2011) XFEM modeling of interface damage in heterogeneous geomaterials, in *International Conference on Extended Finite Element Methods - XFEM 2011* (eds S. Bordas, B. Karihaloo and P. Kerfriden), Cardiff, United Kingdom.

Jirásek, M. (2002) Numerical modeling of strong discontinuities. *Revue Française de Génie Civil*, **6**(6) 1133–1146.

Jirásek, M. and Zimmermann, T. (2001a) Embedded crack model. Part I: Basic formulation. *International Journal for Numerical Methods in Engineering*, **50**, 1269–1290.

Jirásek, M. and Zimmermann, T. (2001b) Embedded crack model. Part II: Combination with smeared cracks. *International Journal for Numerical Methods in Engineering*, **50**, 1291–1305.

Kabiri, M.M.R. (2009) XFEM Analysis of Wave Propagation in Cracked Media. M.Sc. Thesis, University of Tehran, Iran.

Kalthoff, J.F. (1987), The shadow optical method of caustics, in *Static and Dynamic Photoelasticity and Caustics* (ed. A. Lagarde), Springer-Verlag, New York, pp. 407–522.

Kalthoff, J.F. (2000) Modes of dynamic shear failure in solids. *International Journal of Fracture*, **101**, 1–31.

Kalthoff, J.F. and Winkler, S. (1987) Failure mode transmission at high rates of shear loading. International Conference on Impact Loading and Dynamic Behavior of Materials, Vol.**1**, 185–195.

Kame, M. and Yamashita, T. (1999) Simulation of the spontaneous growth of a dynamic crack without constraints on the crack tip path. *Geophysical Journal International*, **139**, 345–358.

Kanninen, M. (1984) Application of fracture mechanics to fiber composite materials and adhesive joint, a review. Third International Conference on Numerical Methods in Fracture Mechanics, Swansea, UK.

Kanninen, M. and Popelar, C.H. (1985) *Advanced Fracture Mechanics*, Oxford University Press, Oxford.

Karihaloo, B.L. and Xiao, Q.Z. (2003) Modelling of stationary and growing cracks in FE framework without remeshing: a state-of-the-art review. *Computers and Structures*, **81**, 119–129.

Karlsson, A. and Backlund, J. (1978) J-integral at loaded crack surfaces. *International Journal of Fracture*, **14**(6), R311–R314.

Kasmalkar, M. (1996) Surface and internal crack problems in a homogeneous substrate coated by a graded layer. Ph.D. Dissertation, Department of Mechanical Engineering and Mechanics, Lehigh University, Bethlehem, PA, U.S.A.

Kaysser, W.A. and Ilschner, B. (1995) FGM research activities in Europe. *MRS Bulletin* **20**(1), 22–26.

Keesecker, A.L., Dávila, C.G., Johnson, E.R. and Starnes Jr., J.H. (2003) Crack path bifurcation at a tear strap in a pressurized shell. *Computers and Structures*, **81**, 1633–1642.

Kerkhof, F. (1970) *Bruchvorgange in Gldsern*, Verlag der Deutschen Glastechnischen Gesellschaft, Frankfurt, 13–22.

Kerkhof, F. (1973) Wave fractographic investigations of brittle fracture dynamics, in *Dynamic Crack Propagation* (eds G.C. Sih), Noordhoff, Leyden, pp. 3–35.

Khoei, A.R. and Nikbakht, M. (2006) Contact friction modeling with the extended finite element method (X-FEM). *Journal of Materials Processing Technology*, **177**, 58–62.

Khoei, A.R. and Nikbakht, M. (2007) An enriched finite element algorithm for numerical computation of contact friction problems. *International Journal of Mechanical Sciences*, **49**, 183–199.

Khoei, A.R., Shamloo, A. and Azami, A.R. (2006) Extended finite element method in plasticity forming of powder compaction with contact friction. *International Journal of Solids and Structures*, **43**(18–19), 5421–5448.

Khoei, A.R., Shamloo, A., Anahid, M. and Shahim, K. (2006) The extended finite element method (X-FEM) for powder forming problems. *Journal of Materials Processing Technology*, **177**, 53–57.

Khoei, A.R., Anahid, M. and Shahim, K. (2008) An extended arbitrary Lagrangian–Eulerian finite element method for large deformation of solid mechanics. *Finite Elements in Analysis and Design*, **44**, 401–416.

Khoei, A.R., Anahid, M., Shahim, K. and DorMohammadi, H. (2008) Arbitrary Lagrangian-Eulerian method in plasticity of pressure-sensitive material with reference to powder forming process. *Computational Mechanics*, **42**, 13–38.

Khoei, S.R., Biabanaki, S.O.R. and Anahid, M. (2008) Extended finite element method for three–dimensional large plasticity deformations on arbitrary interfaces. *Computer Methods in Applied Mechanics and Engineering*, **197**, 1100–1114.

Kiendl, J., Bletzinger, K.-U., Linhard, J. and Wüchner, R. (2009) Isogeometric shell analysis with Kirchhoff-Love Elements. *Computer Methods in Applied Mechanics and Engineering*, **198**, 3902–3914.

Kiendl, J., Bazilevs, Y., Hsu, M.-C., Wüchner, R. and Bletzinger, K.-U. (2010) The bending strip method for isogeometric analysis of Kirchhoff-Love shell structures comprised of multiple patches. *Computer Methods in Applied Mechanics and Engineering*, **199**, 2403–2416.

Kim, J.-H., Paulino, G.H. (2002a) Finite element evaluation of mixed-mode stress intensity factors in functionally graded materials. *International Journal for Numerical Methods in Engineering*, **53**(8), 1903–1935.

Kim, J.-H. and Paulino, G.H. (2002b) Isoparametric graded finite elements for nonhomogeneous isotropic and orthotropic materials. *ASME Journal of Applied Mechanics*, **69**(4), 502–514.

Kim, J.-H. and Paulino, G.H. (2002c) Mixed-mode fracture of orthotropic functionally graded materials using finite elements and the modified crack closure method. *Engineering Fracture Mechanics*, **69**, 1557–1586.

Kim, J.-H., Paulino, G.H. (2003a) Mixed-mode J-integral formulation and implementation using graded finite elements for fracture analysis of nonhomogeneous orthotropic materials. *Mechanics of Materials*, **35**(1–2), 107–128.

Kim, J.-H., Paulino, G.H. (2003b) An accurate scheme for mixed-mode fracture analysis of functionally graded materials using the interaction integral and micromechanics models. *International Journal for Numerical Methods in Engineering*, **58**, 1457–1497.

Kim, J.-H., Paulino, G.H. (2003c) T-stress, mixed-mode stress intensity factors, and crack initiation angles in functionally graded materials: a unified approach using the interaction integral method. *Computer Methods in Applied Mechanics and Engineering*, **192**(11–12), 1463–1494.

Kim, J.H. and Paulino, G.H. (2003d) The interaction integral for fracture of orthotropic functionally graded materials: evaluation of stress intensity factors. *International Journal of Solids and Structures*, **40**, 3967–4001.

Kim, J.-H. and Paulino, G. H. (2004) T-stress in orthotropic functionally graded materials: Lekhnitskii and Stroh formalisms. *International Journal of Fracture*, **126**(4), 345–389.

Kim J.-H. and Paulino, G. H. (2005) Consistent formulation of the interaction integral method for fracture of functionally graded materials. *ASME Journal of Applied Mechanics*, **72**, 351–364.

Kim J.-H. and Paulino, G. H. (2007) On fracture criteria for mixed-mode crack propagation in functionally graded materials. *Mechanics of Advanced Materials and Structures*, **14**, 227–244.

Kirkaldy, D. (1863) *Experiments on Wrought Iron and Steel*, 2nd edn, Scribner, New York.

Kirsh, G. (1898) Infinite plate containing a circular hole (Die theorie der elastizitat und die bedurfnisse der festigkeitslehre). *Zeitschrift der Vereines Deutscher Ingenieure*, **42**, 797–807.

Knein, M. (1972) Zur Theorie des Druckvershchs. *Abhandlungen aus dem Aerodynamischen Institut an der T.H. Aachen*, **7**, 62.

Kolosov, G.V. (1909) On an application of complex function theory to a plane problem of the mathematical theory of elasticity. Infinite Plate Containing An Elliptical Hole, Yuriev.

Kolsky, H. (1953) *Stress Waves in Solids*, Dover, Mineola, NY.

Konda, N., Erdogan, F. (1994) The mixed mode crack problem in a nonhomogeneous elastic medium. *Engineering Fracture Mechanics*, **47**, 533–545.

Krysl, P. and Belytschko, T. (1999) The element free Galerkin method for dynamic propagation of arbitrary 3-D cracks. *International Journal for Numerical Methods in Engineering*. **44**, 767–800.

Kudryavtsev, B.A., Parton, V.Z. and Rakitin, V.I. (1975a) A rectilinear tunnel crack in the piezoelectric conductor interface. *Prikladnaya Matematika I Mekhanika*, **39**, 145–159 (in Russian).

Kudryavtsev, B.A., Parton, V.Z. and Rakitin, V.I. (1975b) Fracture mechanics of piezoelectric materials - An axisymmetric crack in the interface with a conductor. *Prikladnaya Matemika I Mekhanika*, **39**, 352–362 (in Russian).

Kuo, M.C. and Bogy, D.B. (1974) Plane solutions for the displacement and traction-displacement problem for anisotropic elastic wedges. *Journal of Applied Mechanics*, **41**, 197–203.

Kurihara, K., Sasaki, K. and Kawarada, M. (1990) Adhesion Improvement of Diamond Films, in FGM-90, Proceedings of the First International Symposium on Functionally Graded Materials, FGM Forum, Tokoyo, Japan (eds M. Yamanouchi, T. Hirai, M. Koizumi, and I. Shiota).

Kysar, J.W., Gan, Y.X. and Mendez-Arzuza, G. (2005) Cylindrical void in a rigid-ideally plastic single crystal. Part I: Anisotropic slip line theory solution for face-centered cubic crystals. *International Journal of Plasticity*, **21**, 1481–1520.

Laborde, P., Pommier, J., Renard, Y. and Salaün, M. (2005) High-order extended finite element method for cracked domains. *International Journal for Numerical Methods in Engineering*, **64**, 354–381.

Larsson, R. and Fagerström, M. (2005) A framework for fracture modelling based on the material forces concept with XFEM kinematics. *International Journal for Numerical Methods in Engineering*, **62**, 1763–1788.

Lau, K.-T. and Zhou, L.-M. (2001) Mechanical performance of composite-strengthened concrete structures. *Composites Part B*, **32**, 21–31.

Lecampion, B. (2009) An extended finite element method for hydraulic fracture problems. *Communications in Numerical Methods in Engineering*, **25**(2), 121–133.

Lee, D. and Barber, J.R. (2006) An automated procedure for determining asymptotic elastic fields at singular points. *Journal of Strain Analysis*, **41**, 287–295.

Lee, K.H. (2000) Stress and displacement fields for propagating the crack along the interface of dissimilar orthotropic materials under dynamic mode I and II load. *Journal of Applied Mechanics*, **67**, 223–228.

Lee, K.H., Hawong, J.S. and Choi, S.H. (1996) Dynamic stress intensity factors KI, KII and dynamic crack propagation characteristics of orthotropic material. *Engineering Fracture Mechanics*, **53**(1), 119–140.

Lee, S.H., Song, J.H., Yoon, Y.C. *et al.* (2004) Combined extended and superimposed finite element method for cracks. *International Journal for Numerical Methods in Engineering*, **59**, 1119–1136.

Lee, Y.D. and Erdogan, E. (1995) Residual stresses in FGM and laminated thermal barrier coatings. *International Journal of Fracture*, **69**, 145–165.

Legay, A., Wang, H.W. and Belytschko, T. (2005) Strong and weak arbitrary discontinuities in spectral finite elements. *International Journal for Numerical Methods in Engineering*, **64**, 991–1008.

Legrain, G., Moës, N. and Verron, E. (2005) Stress analysis around crack tips in finite strain problems using the extended finite element method. *International Journal for Numerical Methods in Engineering*, **63**, 290–314.

Lekhnitskii, S.G. (1968) *Anisotropic Plates*, Gordon and Breach Science Publishers, New York, USA.

Li, F.Z., Shih, C.F. and Needleman, A. (1985) A comparison of methods for calculating energy release rates. *Engineering Fracture Mechanics*, **21**(2), 405–421.

Li, X., Yao, D. and Lewis, R.W. (2003) A discontinuous Galerkin finite element method for dynamic and wave propagation problems in nonlinear solids and saturated porous media. *International Journal for Numerical Methods in Engineering*, **57**, 1775–1800.

Lim, W.K., Choi, S.Y. and Sankar, B.V. (2001) Biaxial load effects on crack extension in anisotropic solids. *Engineering Fracture Mechanics*, **68**, 403–416.

Liu, W.K., Chen, Y., Jun, S. *et al.* (1996) Overview and applications of the reproducing kernel particle methods. Archives of Computational Methods in Engineering; *State of the Art Reviews*, **3**, 3–80.

Liu, X.Y., Xiao, Q.Z. and Karihaloo, B.L. (2004) XFEM for direct evaluation of mixed mode SIFs in homogeneous and bi-materials. *International Journal for Numerical Methods in Engineering*, **59**, 1103–1118.

Liu, Sh. (1994) Quasi-impact damage initiation and growth of thick-section and toughened composite materials. *International Journal of Solids and Structures*, **31**(22), 3079–3098.

Liu, M., Hsia, K.J. (2003) Interfacial cracks between piezoelectric and elastic materials under in-plane electric loading. *Journal of the Mechanics and Physics of Solids*, **51**, 921–944.

Liu, W.K., Liu, Y., Farrell, D. *et al.* (2006) Immersed finite element method and applications to biological systems. *Computer Methods in Applied Mechanics and Engineering*, **195**, 1722–1749.

Liu, Y., Liu, W.K., Belytschko, T. *et al.* (2007) Immersed electrokinetic finite element method. *International Journal for Numerical Methods in Engineering*, **71**, 379–405.

Liu, G.R. and Trung, N.T. (2010) *Smoothed Finite Element Methods*. C.R.C. Press.

Lu, J. (2011) isogeometric contact analysis: geometric basis and formulation for frictionless contact. *Computer Methods in Applied Mechanics and Engineering*, **200**, 726–741.

Lu, Y.Y., Belytschko, T. and Tabbara, M. (1995) Element-free Galerkin method for wave propagation and dynamic fracture. *Computational Methods in Applied Mechanics and Engineering*, **126**, 131–53.

Lu, X.Z., Teng, J.G., Ye, L.P. and Jiang, J.J. (2005) Bond-slip models for FRP sheets/plates bonded to concrete. *Engineering Structures*, **27**(6), 920–37.

Lu, X.Z., Jiang, J.J., Teng, J.G. and Ye, L.P. (2006) Finite element simulation of debonding in FRP-to-concrete bonded joints. *Construction and Building Materials*, **20**, 412–424.

Macneal, R.H. (1994) *Finite Elements: Their Design and Performance*, Marcell Dekker, Inc., USA.

Madani, S.A. and Mohammadi, S. (2011) Simulation of underwater shockwave by corrective smoothed particle method (CSPM). International Conference on Particles (Particles 2011), Barcelona, Spain.

Mahdavi, A. and Mohammadi, S. (2012) An extended finite element method for pressurized cohesive crack analysis in concrete gravity dams. 6th European Congress on Computational Methods in Applied Sciences and Engineering (ECCOMAS 2012), Vienna, Austria.

Mahin, S.A. (1997) Lessons from steel buildings damaged during the Northridge earthquake, Tech. rep., *National Information Service for Earthquake Engineering*, University of California, Berkeley, http://nisee.berkeley.edu/northridge/mahin.html.

Mahmoudi, S.N. (2009) Crack Analysis in FGM Composites by the Extended Finite Element Method. M.Sc. Thesis, University of Tehran, Iran.

Maigre, H. and Rittel, D. (1993) Mixed-mode quantification for dynamic fracture initiation: application to the compact compression specimen. *International Journal Solids Structures*, **30**(23), 3233–3244.

Malekafzali, S. (2010) An Extended Finite Element Method for Dislocation Dynamics in Nano Scale Anisotropic Material. M.Sc. Thesis, University of Tehran (In Persian).

Malnic, E. and Meyer, R. (1988) Aging jet had troubled past, records show. *Los Angeles Times*, Section 1, Page 1.

Malysev, B.M. and Salganik, R.L. (1965) The strength of adhesive joints using the theory of cracks. *International Journal of Fracture*, **1**, 114–127.

Manogg, P. (1966) Investigation of the rupture of a plexiglas plate by means of an optical method involving high-speed filming of the shadows originating around holes drilled in the plate. *International Journal of Fracture*, **2**(4), 604–613.

Mariani, S. and Perego, U. (2003) Extended finite element method for quasi-brittle fracture. *International Journal for Numerical Methods in Engineering*, **58**, 103–126.

McMeeking, R.M. (1989) Electrostructive forces near crack-like flaws. *International Journal of Engineering Science*, **40**, 615–627.

McMeeking, R.M. (2004) The energy release rate for a Griffith crack in a piezoelectric material. *Engineering Fracture Mechanics*, **71**, 1149–1163.

Meguid, S.A. (1989) *Engineering Fracture Mechanics*, Elsevier Applied Science, UK.

Meier, U. (1995) Strengthening of structures using carbon fibre/epoxy composites. *Construction and Building Materials*, **9**(6), 341–51.

Meier, S.M., Nissley, D.M. and Sheffler, K.D. (1991) Thermal barrier coating life prediction model development. NASA Contractor Report 189111.

Melenk, J.M. and Babuska, I. (1996) The Partition of Unity Finite Element Method: Basic Theory and Applications. Seminar fur Angewandte Mathematik, Eidgenossische Technische Hochschule, Research Report No. 96-01, January, CH-8092 Zurich, Switzerland.

Menouillard, T., Réthoré, J., Combescure A. and Bung, H. (2006) Efficient explicit time stepping for the extended finite element method (X-FEM). *International Journal for Numerical Methods in Engineering*, **68**(9) 911–939.

Menouillard, T., Réthoré, J., Moes, N., Combescure A. and Bung, A. (2008) Mass lumping strategies for X-FEM explicit dynamics: application to crack propagation. *International Journal for Numerical Methods in Engineering*, **74**(3) 447–474.

Menouillard, T., Song, J.H., Duan, Q. and Belytschko, T. (2010) Time dependent crack tip enrichment for dynamic crack propagation. *International Journal of Fracture*, **162**, 33–49.

Mergheim, J., Kuh, E. and Steinmann, P. (2005) A finite element method for the computational modelling of cohesive cracks. *International Journal for Numerical Methods in Engineering*, **63**, 276–289.

Mergheim, J., Kuh, E. and Steinmann, P. (2006) Towards the algorithmic treatment of 3D strong discontinuities. *Communications in Numerical Methods in Engineering*, **23**(2) 97–108.

Mi, Y. and Crisfield, M.A. (1996) Analytical derivation of load/displacement relationship for the DCB and MMb and proof of the FEA formulation. *IC-AERO Report 97-02*, Department of Aeronautics, Imperial College, London, UK.

Mi, Y., Crisfield, M.A., Davies, G.A.O. and Hellweg, H.-B. (1998) Progressive delamination using interface elements. *Journal of Composite Materials*, **32**(14), 1246–1272.

Mirzaei, M. (2008) Failure analysis of an exploded gas cylinder. *Engineering Failure Analysis*, **15**(7) 820–834.

Mirzaei, M., Razmjoo, A. and Pourkamali, A. (2001) Failure analysis of the girth gear of an industrial ball mill. Proceedings of the 10th International Congress of Fracture (ICF10), USA.

Moës, N. (2011) The thick level method (TLS) to model damage growth and transition to fracture, in *International Conference on Extended Finite Element Methods - XFEM 2011* Cardiff, United Kingdom, (eds S. Bordas, B. Karihaloo and P. Kerfriden).

Moës, N. and Belytschko, T. (2002a) Extended finite element method for cohesive crack growth. *Engineering Fracture Mechanics*, **69**, 813–833.

Moës, N. and Belytschko, T. (2002b) X-FEM: De nouvelles frontières pour les éléments finis. *Revue Européenne des Eléments Finis*, **11**, 305–318.

Moës, N., Dolbow, J. and Belytschko, T. (1999) A finite element method for crack growth without remeshing. *International Journal for Numerical Methods in Engineering*, **46**, 131–150.

Moës, N., Gravouil, A. and Belytschko, T. (2002) Non-planar 3D crack growth by the extended finite element and level sets–Part I: Mechanical model. *International Journal for Numerical Methods in Engineering*, **53**, 2549–2568.

Moës, N., Cloirec, M., Cartraud, P. and Remacle, J.F. (2003) A computational approach to handle complex microstructure geometries. *Computer Methods in Applied Mechanics and Engineering*, **192**, 3163–3177.

Moës, N., Stolz, C., Bernard, P.-E. and Chevaugeon, N. (2011) A level set based model for damage growth: The thick level set approach. *International Journal for Numerical Methods in Engineering*, **86**(3), 358–380.

Moghaddam, A.S., Ghajar, R. and Alfano, M. (2011) Finite element evaluation of stress intensity factors in curved non-planar cracks in FGMs. *Mechanics Research Communications*, **38**, 17–23.

Mohammadi, S. (2003a) *Discontinuum Mechanics by Combined Finite/Discrete Elements*, WIT Press, UK.

Mohammadi, S. (2003b) Impact resistance of composite structures. International Conference on FRP Composites in Civil Engineering (CICE 2001), Hong Kong, pp. 1479–1486.

Mohammadi, S. (2008) *Extended Finite Element Method for Fracture Analysis of Structures*, Wiley-Blackwell, UK.

Mohammadi, S. (2012) Multiscale Analysis. Lecture notes, University of Tehran, Tehran, Iran.

Mohammadi, S. and Forouzan-Sepehr, S. (2003) 3D adaptive multi fracture analysis of composites. *Materials Science Forum, Part B*, **440–441**, 145–152.

Mohammadi, S. and Asadpoure, A. (2006) A novel approach to analyze a crack with XFEM in composite meida. International Conference on Computational Methods in Engineering (ICOME), Hefei, China.

Mohammadi, S. and Moosavi, A.A. (2007) 3D multi delamination/fracture analysis of composites subjected to impact loadings. *Journal of Composite Materials, Part B*, **41**(12), 1459–1475.

Mohammadi, S. and Malekafzali, S. (2011) Application of Disclocation Dynamics for Analysis of New Orthotropic Materials in Nano-Scale. Report HRDC-22139, Iran Nano Technology Initiative Council.

Mohammadi, S. and Bayesteh, H. (2011) *Fracture Analysis of Cracked Pressurizdd Shells*. University of Tehran, Report: 8102051-1-1.

Mohammadi, S. and Bayesteh, H. (2012) XFEM Fracture *Analysis of Orthotropic Functionally Graded Materials*. Iran National Sience Foundation, Report 90003421-1

Mohammadi, S. and Esna Ashari, S. (2012) *Delamination Analysis of FRP Strengthened Structures*. University of Tehran, Report: 8102051-1-2.

Mohammadi, S. and Motamedi, D. (2012) *Dynaimc XFEM Fracture Analysis of Orthotropic Composites*. Iran National Sience Foundation, Report 90003421-2.

Mohammadi, S., Owen, D.R.J. and Peric, D. (1997) Delamination analysis of composites by discrete element method. *Computational Plasticity*, Barcelona, Spain, pp. 1206–1213.

Mohammadi S., Owen, D.R.J. and Peric, D. (1998) A combined finite/discrete element algorithm for delamination analysis of composites. *Finite Elements in Analysis and Design*, **28**, 321–336.

Mohammadi, S., Owen D.R.J. and Peric D. (1999) Progressive fracture analysis of layered composites by a combined finite/discrete element algorithm. *Scientia Iranica*, **6**(3-4), 225–232.

Mohammadi, S., Forouzan-Sepehr, S. and Asadollahi, A. (2002) Contact based delamination and fracture analysis of composites. *Thin Walled Structures*, **40**, 595–609.

Moosavi, A.A. and Mohammadi, S. (2007) Modelling dynamic response of concrete beams strengthened by FRP composites. 1st National Civil Engineering Conference, Tehran, Iran, p. 119 (In Persian).

Moran, B. and Shih, C.F. (1987) Crack tip and associated integrals from momentum and energy balance. *Engineering Fracture Mechanics*, **17**(6) 615–642.

Motamedi, D. (2008) Dynamic Fracture Analysis of Composites by XFEM. M.Sc. Thesis, University of Tehran, Tehran, Iran.

Motamedi, D. and Mohammadi, S. (2010a) Dynamic analysis of fixed cracks in composites by the extended finite element method. *Engineering Fracture Mechanics*, **77**, 3373–3393.

Motamedi, D. and Mohammadi, S. (2010b) Dynamic crack propagation analysis of orthotropic media by the extended finite element method. *International Journal of Fracture*, **161**, 21–39.

Motamedi, D. and Mohammadi, S. (2012) Fracture Analysis of Composites by Time Independent Moving-Crack Orthotropic XFEM. *International Journal of Mechanical Sciences*, **54**, 20–37.

Murrel, S.A.F. (1964) The theory of the propagation of elliptical Grifith cracks under various conditions of plane strain or plane stress: Part 1. *British Journal of Applied Physics*, **15**(10), 119–1210.

Muskhelishvili, N.I. (1953) *Some Basic Problems on the Mathematical Theory of Elasticity*. Translated by J.R.M. Radok, Noordhoof, Groningen, The Netherlands.

Nadeau, J. and Ferrari,M. (1999) Microstructural optimization of a functionally graded transversely isotropic layer, *Mechanics of Materials*, **31**, 637–651.

Nagashima, T. and Suemasu, H. (2004) Application of extended finite element method to fracture of composite materials. European Congress on Computational Methods in Applied Sciences and Engineering (ECCOMAS), Jyväskylä, Finland.

Nagashima, T. and Suemasu, H. (2006) Stress analysis of composite laminates with delamination using XFEM. *International Journal of Computational Methods*, **3**, 521–543.

Nagashima, T., Omoto, Y. and Tani, S. (2003) Stress intensity factor analysis of interface cracks using X-FEM. *International Journal for Numerical Methods in Engineering*, **56**, 1151–1173.

Natarajan, S., Baiz, P.M., Bordas, S. *et al.* (2011) Natural frequencies of cracked functionally graded material plates by the extended finite element method. *Composite Structures*, article in press, doi: 10.1016/j.compstruct.2011.04.007

Neto, E. de S., Peric, D. and Owen, D.R.J. (2008) *Computational Methods for Plasticity - Theory and Applications*, John Wiley & Sons, Chichester.

Nielsen, P.N., Gersborg, A.R., Gravesen, J. and Pedersen, N.L. (2011) Discretizations in Isogeometric Analysis of Navier-Stokes Flow. *Computer Methods in Applied Mechanics and Engineering*, **200**(45–46), 3242–3253.

Niino, M. and Maeda, S. (1990) Recent development status of functionally graded materials. *The Iron and Steel Institute of Japan*, **30**, 699–703.

Nikishkov, G.P. and Atluri, S.N. (1987) Calculation of fracture mechanics parameters for an arbitrary three-dimensional crack, by the 'equivalent domain integral' method. *International Journal for Numerical Methods in Engineering*, **24**, 1801–1821.

Nilson, R.H., Proffer, W.J. and Duff, R.E. (1985) Modeling of gas-driven fracture induced by propellant combustion within a borehole. *International Journal of Rock Mechanics and Mining Science & Geomechics*, **22**(1), 3–19.

Nilsson, F. (1974) A note on the stress singularity at a nonuniformly moving crack tip. *Journal of Elasticity*, **4**, 73.

Nishioka, T. and Atluri, S.N. (1980a) Numerical modeling of dynamic crack propagation in finite bodies, by moving singular elements, Part 1: Formulation. *Journal of Applied Mechanics*, **47**, 570–576.

Nishioka, T. and Atluri, S.N. (1980b) Numerical modeling of dynamic crack propagation in finite bodies, by moving singular elements, Part 2: Results. *Journal of Applied Mechanics*, **47**, 577–583.

Nishioka, T. and Atluri, S.N. (1983) Path-independent integrals, energy release rates and general solutions of near tip fields in mixed-mode dynamic fracture mechanics. *Engineering Fracture Mechanics*, **18**, 1–22.

Nishioka, T. and Atluri, S.N. (1984) On the computation of mixed-mode K-factors for a dynamically propagating crack using path-independent integrals J'k. *Engineering Fracture Mechanics*, **20**(2) 193–208.

Nishioka, T., Tokudome, H. and Kinoshita, M. (2001) Dynamic fracture-path prediction in impact fracture phenomena using moving finite element method based on Delaunay automatic mesh generation. *International Journal of Solids and Structures*, **38**, 5273–5301.

Nistor, I., Pantale, O. and Caperaa, S. (2008) Numerical implementation of the extended finite element method for dynamic crack analysis. *International Journal of Advances in Engineering Software*, **39**, 573–587.

Nobile, L. and Carloni, C. (2005) Fracture analysis for orthotropic cracked plates. *Composite Structures*, **68**(33) 285–293.

O'Brien, K. (1985) Analysis of local delaminations and their influence on composite laminate behaviour, in *Delamination and Debonding of Materials* (ed. W.S. Johnson), ASTM STP 876, pp. 282–297.

Oliver, J., Huespe, A.E., Pulido, M.D.G. and Samaniego, E. (2003) On the strong discontinuity approach in finite deformation settings. *International Journal for Numerical Methods in Engineering*, **56**, 1051–1082.

Oliver, J., Huespe, A.E., Blanco, S. and Linero, D.L. (2006) Stability and robustness issues in numerical modeling of material failure with the strong discontinuity approach. *Computer Methods in Applied Mechanics and Engineering*, **195**(52), 7093–7114.

Onate, E., Idelsohn, S., Fischer, T. and Zienkiewicz, O.C. (1995) A finite point method for analysis of fluid flow problems. Ninth International Conference on Finite Elements in Fluids, Venice, Italy, pp. 15–21.

Orowan, E. (1948) Fracture and strength of solids. *Reports on Progress in Physics*, **XII**, 185–232.

Osher, S. and Sethian, J.A. (1988) Fronts propagating with curvature-dependent speed: algorithms based on Hamilton–Jacobi formulations. *Journal of Computational Physics*, **79**(1), 12–49.

Ostad, H. and Mohammadi, S. (2012) A stabilized particle method for large deformation dynamic analysis of structures. *International Journal of Structural Stability and Dynamics*, In press.

Ou, Z.C. and Chen, Y.H. (2004) Near-tip stress fields and intensity factors for an interface crack in metal/piezoelectric biomaterials. *International Journal of Engineering Science*, **42**, 1407–1438.

Owen, D.R.J. and Fawkes, A. (1983) *Engineering Fracture Mechanics: Numerical Methods and Applications*, Pineridge Press Ltd, Swansea, UK.

Ozturk, M., Erdogan, F. (1997) Mode I crack problem in an inhomogeneous orthotropic medium. *International Journal of Engineering Science*, **35**(9), 869–883.

Pagano, N.J. (1974) On the Calculation of Interlaminar Normal Stress in Composite Laminate. *Journal of Composite Materials*, **8**, 65–81.

Pak, Y.E. and Tobin, A. (1993) On electric field effects in fracture of piezoelectric materials, in *Mechanics of Electromagnetic Materials and Structures* (eds J. S.Lee and G.A. Maugin), ASME, AMD-**161**, MD-42, 51–62.

Parchei, M. (2012) Application of XFEM for Simulation of Fault Sliding. M.Sc. Thesis, University of Tehran, Iran.

Parchei, M., Mohammadi, S. and Zafarani, H. (2011) Two-dimensional dynamic extended finite element method for simulation of seismic fault rupture, *International Conference on Extended Finite Element Methods - XFEM 2011* Cardiff, United Kingdom, (eds S. Bordas, B. Karihaloo, and P. Kerfriden).

Parchei, M., Mohammadi, S. and Zafarani, H. (2012) Extended finite element method for simulation of dynamic fault rupture. *Journal of Seismology and Earthquake Engineering*, IIESS, Iran. In press.

Paris, P.C., Gomez, M.P. and Anderson, W.E. (1961) A rational analytic theory of fatigue. *The Trend in Engineering*, **13**(1), 9–14.

Parker, A.P. (1981) *The Mechanics of Fracture and Fatigue, An Introduction*, E. & F.N. SPON Ltd.

Parks, D. (1974) A stiffness derivative finite element technique for determination of crack tip stress intensity factors. *International Journal of Fracture*, **10**(4) 487–502.

Parton, V.Z. (1976) Fracture mechanics of piezoelectric materials. *Acta Astronautica*, **3**, 671–683.

Patrikalakis, N.M. (2003) Computational Geometry. Lecture Notes, Massachusetts Institute of Technology, USA.

Patzak, B. and Jirásek, M. (2003) Process zone resolution by extended finite elements. *Engineering Fracture Mechanics*, **70**, 957–977.

Peerlings, R.H.J., de Borst, R., Brekelmans, W.A.M. and Geers, M.G.D. (2002) Localisation issues in local and nonlocal continuum approaches to fracture. *European Journal of Mechanics – A: Solids*, **21**, 175–189.

Pesic, N. and Pilakoutas, K. (2003) Concrete beams with externally bonded flexural FRP-reinforcement: analytical investigation of debonding failure. *Composites: Part B*, **34**, 327–338.

Peters, M. and Hack, K. (2005) Numerical aspects of the extended finite element method. *Proceedings of Applied Mathematics and Mechanics*, **5**, 355–356.

Pichugin, A.V., Askes, H. and Tyas, A. (2008) The equivalence of asymptotic homogenisation procedures and fine-tuning of continuum theories. *Journal of Sound and Vibration*, **313**(3–5), 858–874.

Piegl, L. and Tiller, W. (1997) *The NURBS book (Monographs in Visual Communication)*, 2nd edn, Springer-Verlag, New York.

Pipes, R.B. and Pagano, N.J. (1970) Interlaminar stresses in composite laminates under uniform axial extensions. *Journal of Composite Materials*, **4**, 538–548.

Pipes, R.B. and Pagano, N.J. (1974) Interlaminar stresses in composite laminates – an approximate elasticity solution. *Journal of Applied Mechanics*, **41**, 668–672.

Piva, A. and Viola, E. (1988) Crack propagation in an orthotropic medium. *Engineering Fracture Mechanics*. **29**, 535–548.

Piva, A., Viola, E. and Tornabene, F. (2005) Crack propagation in an orthotropic medium with coupled elastodynamic properties. *Mechanics Research Communications*, **32**, 153–159.

Pommier, S., Gravouil, A., Combescure, A. and Moes, N. (2011) *Extended Finite Element Method for Crack Propagation*, Wiley-ISTE.

Prabel, B., Marie, S. and Combescure, A. (2008) Using the X-FEM method to model the dynamic propagation and arrest of cleavage cracks in ferritic steel. *Engineering Fracture Mechanics*, **75**(10), 2984–3009.

Qian, W., Sun, C.T. (1998) Methods for calculating stress intensity factors for interfacial cracks between two orthotropic solids. *International Journal of Solids and Structures*, **35**, 3317–3330.

QingWen, J. L., YuWen, D. and TianTang, Y.U. (2009) Numerical modeling of concrete hydraulic fracturing with extended finite element method. *Science in China Series E: Technological Science*, **52**(3), 559–565.

Qu, J. and Bassani, J.L. (1993) Interfacial fracture mechanics for anisotropic biomaterials. *Journal of Applied Mechanics*, **60**, 422–431.

Qun, L. and Yiheng, C. (2007) Analysis of crack-tip singularities for an interfacial permeable crack in metal/piezoelectric biomaterials. *Acta Mechanica Solida Sinica*, **20**(3). doi: 10.1007/s10338-007-0729-6

Rabczuk, T. and Wall, W.A. (2006) Extended finite element and meshfree methods. Technical University of Munich, Germany, WS200/2007.

Rabinovitch, O. (2004) Fracture-mechanics failure criteria for RC beams strengthened with FRP strips – a simplified approach. *Composite Structures*, **64**(3–4), 479–492.

Rabinovitch, O. (2008) Debonding analysis of fiber-reinforced-polymer strengthened beams: Cohesive zone modeling versus a linear elastic fracture mechanics approach. *Engineering Fracture Mechanics*, **75**, 2842–2859.

Rabinovitch, O. and Frosting, Y. (2001) Delamination failure of RC beams strengthened with FRP strips: a closed form high order fracture mechanics approach. *Journal of Engineering Mechanics*, **127**(8), 852–861.

Rao, B.N. and Rahman, S. (2003) Mesh-free analysis of cracks in isotropic functionally graded materials. *Engineering Fracture Mechanics*, **70**(1), 1–27.

Rashetnia, R. (2012) Finite Strain Fracture Analysis of Structures using the Extended Finite Element Method. M.Sc. Thesis, University of Tehran, Iran.

Remmers, J.J.C., Wells, G.N. and de Borst, R. (2003) A solid-like shell element allowing for arbitrary delaminations. *International Journal for Numerical Methods in Engineering*, **58**, 2013–2040.

Rethore, J., Gravouil, A. and Combescure, A. (2004) A stable numerical scheme for the finite element simulation of dynamic crack propagation with remeshing. *Computer Methods in Applied Mechanics and Engineering*, **193**(42–44), 4493–4510

Réthoré, J., Gravouil, A. and Combescure, A. (2005a) An energy-conserving scheme for dynamic crack growth using the extended finite element method. *International Journal for Numerical Methods in Engineering*, **63**, 631–659.

Réthoré, J., Gravouil, A. and Combescure, A. (2005b) A combined space–time extended finite element method. *International Journal for Numerical Methods in Engineering*, **63**, 631–659.

Réthoré, J., Gravouil, A., Morestin, F. and Combescure, A. (2005) Estimation of mixed-mode stress intensity factors using digital image correlation and an interaction integral. *International Journal of Fracture*, **132**, 65–79.

Rezaei, S.N. (2010) Development of an Extended Finite Element Method for Simulation of Space-Time Singularities. M.Sc. Thesis, University of Tehran, Iran.

Rice, J.R. (1968a) Path-independent integral and the approximate analysis of strain concentration by notches and cracks. *Journal of Applied Mechanics, Transactions ASME*, **35**(2), 379–386.

Rice, J.R. (1968b) Mathematical analysis in the mechanics of fracture, in *Fracture* (ed. H. Leibowitzm), Academic Press, p. 191.

Rice, J.R. (1987) Tensile crack tip fields in elastic-ideally plastic crystals. *Mechanics of Materials*, **6**, 317–335.

Rice, J.R. (1988) Elastic fracture mechanics concepts for interfacial cracks. *Journal of Applied Mechanics, Transactions ASME*, **55**, 98–103.

Rice, J.R. and Levy, N. (1972) The part-through surface crack in an elastic plate. *Journal of Applied Mechanics, Transactions ASME*, **39**, 185–194.

Rice, J.R. and Rosengren, G.F. (1968) Plane strain deformation near a crack tip in a power-law hardening material. *Journal of Mechanics and Physics of Solids*, **16**, 1–12.

Rice, J.R. and Sih, G.C. (1965) Plane problems of cracks in dissimilar media. *Journal Of Applied Mechanics*, **32**, 418–423.

Rice, J.R., Hawk, D.E. and Asaro, R.J. (1990) Crack tip fields in ductile crystals. *International Journal of Fracture*, **42**, 301–321.

Rizkalla, S., Hassan, T. and Hassan, N. (2003) Design recommendations for the use of FRP for reinforcement and strengthening of concrete structures. *Progress in Structural Engineering and Materials*, **5**(1), 16–28.

Roberts, T.M. (1989) Shear and normal stresses in adhesive joints. *Journal of Engineering Mechanics*, ASCE **115**(11), 2460–2476.

Roh, H.Y. and Cho, M. (2004) The application of geometrically exact shell elements to B-spline surfaces. *Computer Methods in Applied Mechanics and Engineering*, **193**, 2261–2299.

Rojas-Dıaz, R., Sukumar, N., Saez, A. and Garcıa-Sanchez, F. (2011) Fracture in magnetoelectroelastic materials using the extended finite element method. *International Journal for Numerical Methods in Engineering*, **88**(12), 1238–1259.

Romanowicz, M. and Seweryn, A. (2008) Verification of a non-local stress criterion for mixed mode fracture in wood. *Engineering Fracture Mechanics*, **75**, 3141–3160.

Rosakis, A.J. and Freund, L.B. (1982) Optical measurement of the plastic strain concentration at a crack tip in a ductile steel plate. *Journal of Engineering Materialsand Technology*, **104**, 115–120.

Rousseau, C.-E. and Tippur, H.V. (2000) Compositionally graded materials with cracks normal to the elastic gradient. *Acta Materialia*, **48**(16), 4021–4033.

Rowlands, R.E. (1985) Strength (failure) theories and their experimental correlation. in *Handbook of Composites*, vol. **3**—*Failure Mechanics of Composites* (eds G.C. Sih and A.M. Skudra), Elsevier, Amsterdam, Ch. **2**, pp. 71–125.

Rubio-Gonzales, C., Mason, J.J. (1998) Closed form solutions for dynamic stress intensity factors at the tip of uniformly loaded semi-infinite cracks in orthotropic materials. *Journal of Mechanics and Physics of Solids*, **48**, 899–925.

Sadd, M.H. (2005) *Elasticity: Theory, Applications, and Numeric*, Elsevier.

Sadeghirad, A. and Mohammadi, S. (2007) Equilibrium on the line method (ELM) for imposition of Neumann boundary conditions in the finite point method (FPM). *International Journal for Numerical Methods in Engineering*, **69**, 60–86.

Samaniego, E. and Belytschko, T. (2005) Continuum–discontinuum modelling of shear bands. *International Journal for Numerical Methods in Engineering*, **62**, 1857–1872.

Sampath, S., Herman, H., Shimoda, N. and Saito, T. (1995) Thermal spray processing of FGMs. *MRS Bulletin*, **20**(1), 27–31.

Saouma, V.E. (2000) Fracture mechanics. Lecture notes CVEN-6831, Univesity of Colorado, USA.

Saouma, V.E. and Sikiotis, E.S. (1986) Stress intensity factors in anisotropic bodies using singular isoparametric elements. *Engineering Fracture Mechanics*, **25**, 115–121.

Saouma, V.E., Ayari, M. and Leavell, D. (1987) Mixed mode crack propagation in homogeneous anisotropic solids. *Engineering Fracture Mechanics*, **27**(2), 171–184.

Schardin, H. (1959) Velocity effects in fracture, in *Fracture* (eds B.L. Averbach *et al.*), MIT Press, Cambridge, MA, pp. 297–330.

Sethi, M., Gupta, K.C., Kakar, R. and Gupta, M.P. (2011) Propagation of love waves in a non-homogeneous orthotropic layer under compression 'p' overlying semi-infinite non-homogeneous medium. *International Journal of Applied Mathematics and Mechanics*, **7**(10), 97–110.

Sethian, J.A. (1987) Numerical methods for propagating fronts, in *Variational Methods for Free Surface Interfaces* (eds P. Concus and R. Finn), Springer, New York, USA, pp. 155–164.

Sethian, J.A. (1996) A marching level set method for monotonically advancing fronts. *Proceedings of National Academy of Science*, **93**(4), 1591–1595.

Sethian, J.A. (1999a) Fast marching methods. *SIAM Review*, **41**(2), 199–235.

Sethian, J.A. (1999b) *Level Set Methods and Fast Marching Methods*, Cambridge University Press, Cambridge, UK.

Sethian, J.A. (2001) Evolution, implementation, and application of level set and fast marching methods for advancing fronts. *Journal of Computational Physics*, **169**(2), 503–555.

Sevilla, R., Fernández-Méndez, S. and Huerta, A. (2008) NURBS-enhanced finite element method (NEFEM). *International Journal for Numerical Methods in Engineering*, **76**(1), 56–83.

Sha, G. (1984) On the virtual crack extension technique for stress intensity factors and energy release rate calculation for mixed fracture modes. *International Journal of Fracture*, **25**(2) 33–42.

Shahi, S. and Mohammadi, S. (2012) Heart Simulation: A Biomechanical Multiscale Approach. University of Tehran, Report: 8102051.

Shamloo, A., Azami, A.R. and Khoei, A.R. (2005) Modeling of pressure-sensitive materials using a cap plasticity theory in extended finite element method. *Journal of Materials Processing Technology*, 164–165, 1248–1257.

Sharifabadi, H. (1990) Theory of Elasticity. Lecture Notes, University of Tehran, Iran.

Sharma, K., Biu, T.Q., Zhang, Ch. and Bhargava, R.R. (2012) Analysis of subinterface crack in piezoelectric bimaterials with the extended finite element method. *Engineering Fracture Mechanics*, In press.

Shih, C., de Lorenzi, H. and German, M. (1976) Crack extension modelling with singular quadratic isoparametric elements. *International Journal of Fracture*, **12**, 647–651.

Shih, C.F. (1991) Cracks on bimaterial interfaces: Elasticity and plasticity aspects. *Material Science and Engineering A*, **143**, 77–90.

Shih, W.M. and Lin, C. (2010) Knitting complex weaves with DNA origami. *Current Opinion in Structural Biology*, **20**(3), 276–282.

Shindo, Y. and Hiroaki, H. (1990) Impact response of symmetric edge cracks in an orthotropic strip. *Fracture and Strength*, 436–441.

Shindo, Y.Z., Murakami, H., Horiguchi, K. and Narita, F. (2002) Evaluation of electric fracture properties of piezoelectric ceramics using the finite element and single-edge pre-cracked-beam methods. *Journal of Americam Ceramics Society*, **85**, 1243–1248.

Shishvan, S.S., Mohammadi, S., Rahimian, M. and Van der Giessen, E. (2011) Plane-strain discrete dislocation plasticity incorporating anisotropic elasticity. *International Journal of Solids and Structures*, **48**, 374–387.

Shukla, A., Chalivendra, V.B., Parameswran, V. and Lee, K.H. (2003) Photoelastic investigation of fracture between orthotropic and isotropic materials. *Optic and Lasers in Engineering*, **40**, 307–324.

Sih, G.C. (1968) Some elastodynamic problems of cracks. *International Journal of Fracture*, **4**, 51–68.

Sih, G.C. (1973) *Handbook of Stress Intensity Factors for Researchers and Engineers*, Lehigh University, Bethlehem, PA, USA.

Sih, G.C. (1974) Strain energy density factors applied to mixed mode crack problems. *International Journal of Fracture*, **10**, 305–321.

Sih, G.C. and Irwin, G.R. (1970) Dynamic analysis for two-dimensional multiple crack division. *Engineering Fracture Mechanics*, **1**, 603–614.

Sih, G.C. and Chen, E.P. (1980) Normal and shear impact of layered composite with a crack: dynamic stress intensification. *Journal of Applied Mechanics*, **47**, 351–358.

Sih, G.C., Paris, P. and Irwin, G. (1965) On cracks in rectilinearly anisotropic bodies. *International Journal of Fracture Mechanics*, **1**(3) 189–203.

Soleymani Shishvan, S., Mohammadi, S. and Rahimian, M. (2008) A dislocation dynamics based derivation of the Frank-Read sources characteristics for discrete dislocation plasticity. *Modelling and Simulation in Materials Science and Engineering*, **16**, 075002.

Song, J.H., Areias, P.M.A. and Belytschko, T. (2006) A method for dynamic crack and shear band propagation with phantom nodes. *International Journal for Numerical Methods in Engineering*, **67**, 868–893.

Sosa, H. and Khutoryansky, N. (1996) New Development Concerning Piezoelectric Materials with Defect. *International Journal of Solids and Structures*, **33**, 3399–3414.

Sou, Z. (1990) Singularities, interfaces and cracks in dissimilar anisotropic media. *Proceedings of Royal Society of London*, **427**, 331–358.

Sprenger, W., Gruttmann, F. and Wagner, W. (2000) Delamination growth analysis in laminated structures with continuum-based 3D-shell elements and a viscoplastic softening model. *Computer Methods in Applied Mechanics and Engineering*, **185**, 123–139.

Stazi, F.L., Budyn, E., Chessa, J. and Belytschko, T. (2003) An extended finite element method with higher-order elements for curved cracks. *Computational Mechanics*, **31**, 38–48.

Stolarska, M. and Chopp, D.L. (2003) Modeling thermal fatigue cracking in integrated circuits by level sets and the extended finite element method. *International Journal of Engineering Science*, **41**, 2381–2410.

Stolarska, M., Chopp, D.L., Moës, N. and Belytschko, T. (2001) Modelling crack growth by level sets in the extended finite element method. *International Journal for Numerical Methods in Engineering*, **51**, 943–960.

Stroh, A.N. (1958) Dislocations and cracks in anisotropic elasticity. *Philosophical Magazine*, **3**, 625–646.

Sukumar, N., Chopp, D.L., Moës, N. and Belytschko, T. (2001) Modeling holes and inclusions by level sets in the extended finite-element method. *Computer Methods in Applied Mechanics and Engineering*, **190**, 6183–6200.

Sukumar, N., Chopp, D.L. and Moran, B. (2003) Extended finite element method and fast marching method for three-dimensional fatigue crack propagation. *Engineering Fracture Mechanics*, **70**, 29–48.

Sukumar, N., Huang, Z., Prevost, J.H. and Suo, Z. (2004) Partition of unity enrichment for bimaterial interface cracks. *International Journal for Numerical Methods in Engineering*, **59**, 1075–1102.

Sukumar, N., Moës, N., Moran, B. and Belytschko, T. (2000) Extended finite element method for three-dimensional crack modeling. *International Journal for Numerical Methods in Engineering*, **48**, 1549–1570.

Sukumar, N. and Prevost, J.H. (2003) Modeling quasi-static crack growth with the extended finite element method, Part I: Computer implementation. *International Journal of Solids and Structures*, **40**, 7513–7537.

Sukumar, N., Srolovitz, D.J., Baker, T.J. and Prevost, J.H. (2003) Brittle fracture in polycrystalline microstructures with the extended finite element method. *International Journal for Numerical Methods in Engineering*, **56**, 2015–2037.

Sumi, Y. and Kagohashi, Y. (1983) A fundamental research on the growth pattern of cracks (second report). *Journal of the Society of Naval Architects of Japan*, **152**, 397–404 (in Japanese).

Sun, C.T. and Jih, C.J. (1987) On strain energy release rate for interfacial cracks in bimaterial media. *Engineering Fracture Mechanics*, **28**, 13–20.

Sun, C.T. and Manoharan, M.G. (1989) Strain energy release rate of an interfacial crack between two orthotropic solids. *Journal of Composite Materials*, **23**, 460–478.

Sun, W. and Sacks, M.S. (2005) Finite element implementation of a generalized Fung-elastic constitutive model for planar soft tissues. *Biomechanics and Modeling in Mechanobiology*, **4**(2–3), 190–199.

Suo, Z. (1990) Singularities, interfaces and cracks in dissimilar anisotropic media. *Proceedings Royal Society of London Series A*, **427**(1873), 331–358.

Suo, X.Z. and Combescure, A. (1992) On the application of G(O) method and its comparison with De Lorenzi's approach. *Nuclear Engineering and Design*, **135**, 207–224.

Svensson, A.S. (2012) Comet 1 worlds first jetliner. http://w1.901.telia.com/u90113819/archives/comet.htm (accessed 14 April 2012).

Swenson, D. (1986) Derivation of the near-tip stress and displacement fields for a moving crack without using complex functions. *Engineering Fracture Mechanics*, **24**(2) 315–321.

Swenson, D. and Ingraffea, A. (1988) Modeling mixed mode dynamic crack propagation using finite elements: Theory and applications. *Computational Mechanics*, **3**, 381–397.

Szabo, B.A. and Babuska, I. (1991) *Finite Element Analysis*, John Wiley & Sons, New York.

Tada, T. and Yamashita, T. (1997) Non-hypersingular boundary integral equations for two-dimensional non-planar crack analysis. *Geophysical Journal International*, **130**(2), 269–282.

Takahashi, H., Ishikawa, T., Okugawa, D. and Hashida, T. (1993) Laser and plasma-arc thermal shock/fatigue fracture evaluation procedure for functionally gradient materials, in (eds G. Schneider and G. Petzow ), *Thermal Shock and Thermal Fatigue Behavior of Advanced Ceramics*, Kluwer Academic Publishers, Dordrecht, pp. 543–554.

Taljsten, B. (1997) Strengthening of beams by plate bonding. *Journal of Materials in Civil Engineering*, **9**(4), 206–12.

Tarancon, J.E., Vercher, A., Giner, E. and Fuenmayor, F.J. (2009) Enhanced blending elements for XFEMapplied to linear elastic fracture mechanics. *International Journal for Numerical Methods in Engineering*, **77**, 126–148.

Temizer, I., Wriggers, P. and Hughes, T.J.R. (2011) Contact treatment in isogeometric analysis with NURBS. *Computer Methods in Applied Mechanics and Engineering*, **200**, 1100–1112.

Teng, J.G., Chen, J.F., Smith, S.T. and Lam, L. (2002) *FRP-Strengthened RC Structures*, John Wiley & Sons, Chichester.

Teng, J.G., Yuan, H. and Chen, J.F. (2006) FRP-to-concrete interfaces between two adjacent cracks. Theoretical model for debonding failure. *International Journal of Solids and Structures*, **43**(18–19), 5750–5778.

Tewary, V.K., Wagoner, R.H. and Hirth, J.P. (1989) Elastic Green's function for a composite solid with a planar interface. *Journal of Materials Research*, **4**, 113–136.

TianTang, Y. and QingWen, R. (2011) Modeling crack in viscoelastic media using the extended finite element method. *Science China*, **54**(6), 1599–1606.

Ting, T.C.T. (1986) Explicit solution and invariance of the singularities at a crack in anisotropic composites. *International Journal of Solids and Structures*, **22**, 965–983.

Tipper, C.F. (1962) *The Brittle Fracture Story*, Cambridge University Press, Cambridge.

Tupholme, G.E. (1974) A study of cracks in orthotropic crystals using dislocations layers. *Journal of Engineering and Mathematics*, **8**, 57–69.

Tvergaard, V. (1990) Effect of fibre debonding in a whisker-reinforced metal. *Materials Science and Engineering A*, **125**, 203–213.

Vaidya, N.K., Huang, H. and Takagi, S. (2010) Coarse grained molecular dynamics simulation of interaction between hemagglutinin fusion peptides and lipid bilayer membranes. *Advances in Applied Mathematics and Mechanics*, **2**(4), 430–450.

Ventura, G. (2006) On the elimination of quadrature subcells for discontinuous functions in the extended finite-element method. *International Journal for Numerical Methods in Engineering*, **66**, 761–795.

Ventura, G., Budyn, E. and Belytschko, T. (2003) Vector level sets for description of propagating cracks in finite elements. *International Journal for Numerical Methods in Engineering*, **58**, 1571–1592.

Ventura, G., Gracie, R. and Belytschko, T. (2009) Fast integration and weight function blending in the extended finite element method. *International Journal for Numerical Methods in Engineering*, **77**, 1–29.

Ventura, G., Moran, B. and Belytschko, T. (2005) Dislocations by partition of unity. *International Journal for Numerical Methods in Engineering*, **62**, 1463–1487.

Verhoosel, C.V., Scott, M.A., Hughes, T.J.R. and de Borst, R (2010a) An isogeometric analysis approach to gradient damage models. *International Journal for Numerical Methods in Engineering*, **86**(1), 115–134.

Verhoosel, C.V., Scott, M.A., de Borst, R. and Hughes, T.J.R. (2010b) An isogeometric approach to cohesive zone modeling. *International Journal for Numerical Methods in Engineering*, doi: 10.1002/nme.3061

Viola, E., Piva, A. and Radi, E. (1989) Crack propagation in an orthotropic medium under general loading. *Engineering Fracture Mechanics*, **34**(5), 1155–1174.

Wang, J. (2006) Debonding of FRP plated reinforced concrete beam, a bond-slip analysis. I: Theoretical formulation. *International Journal of Solids and Structures*, **43**(21), 6649–6664.

Wang, S.S. and Choi, I. (1983a) The crack between dissimilar anisotropic materials. *Journal of Applied Mechanics*, **50**, 169–178.

Wang, S.S. and Choi, I. (1983b) The crack between dissimilar anisotropic composites under mixed-mode loading. *Journal of Applied Mechanics*, **50**, 179–183.

Wang, S.S., Yau, J.F. and Corten, H.T. (1980) A mixed mode crack analysis of rectilinear anisotropic solids using conservation laws of elasticity. *International Journal of Fracture*, **16**, 247–259.

Wang, X. and Shen, Y.-P (2002) Exact solution for mixed boundary value problems at anisotropic piezoelectric bimaterial interface and unification of various interface defects. *International Journal of Solids and Structures*, **39**, 1591–1619.

Watson, G.N. (1995) *A Treatise on the Theory of Bessel Functions*, 2nd edn, Cambridge University Press, Cambridge.

Wells, A. (1963) Application of fracture mechanics at and beyond general yielding. British Welding Research Association, Report M13, UK.

Wells, A.A. (1955) The condition of fast fracture in aluminium alloys with particular reference to comet failures. British Welding Research Association Report.

Wells, A.A. (1961) Unstable crack propagation in metals: cleavage and fast fracture. *Proceedings of the Crack Propagation Symposium*, vol. **1**, Paper 84, Cranfield, UK.

Wells, A.A. and Post, D. (1958) The dynamic stress distribution surrounding a running crack - a photoelastic analysis. *Proceedings of the SESA*, **16**, 69–92.

Westergaard, H. (1939) Bearing pressures and cracks. *ASME Journal of Applied Mechanics*, **6**, 49–53.

Williams, M.L. (1952) Stress singularities resulting from various boundary conditions in angular corners of plates in extension. *Journal of Applied Mechanics, Transactions ASME*, **19**, 526–528.

Williams, M.L. (1957) On the stress distribution at the base of a stationary crack. *Journal of Applied Mechanics*, **24**, 109–114.

Williams, M.L. (1959) The stress around a fault or crack in dissimilar media. *Bulletin of Seismology Society of America*, **49**, 199–204.

Willis, J.R. (1971) Fracture mechanics of interfacial cracks. *Journal of the Mechanics and Physics of Solids*, **19**, 353–368.

Winnie, D.H.J. and Wundt, B.M. (1958) Application of the Griffith-Irwin theory of crack propagation to the bursting behaviour of disks, including analytical and experimental studies. *Transactions of ASME*, **80**, 1643–1658.

Wong, R.S.Y. and Vecchio, F.J. (2003) Towards modeling of reinforced concrete members with externally bonded fiber-reinforced polymer composite. *ACI Structures Journal*, **100**(1), 47–55.

Wu, K.C. (1990) Stress intensity factor and energy release rate for interfacial cracks between dissimilar anisotropic materials. *Journal of Applied Mechanics*, **57**, 882–886.

Wu, K.C. (2000) Dynamic crack growth in anisotropic material. *International Journal of Fracture*, **106**, 1–12.

Wu, Z. and Liu, Y. (2010) Singular stress field near interface edge in orthotropic/isotropic bi-materials. *International Journal of Solids and Structures*, **47**(17), 2328–2335.

Wu, Z.S., Yuan, H. and Niu, H.D. (2002) Stress transfer and fracture propagation in different kinds of adhesive joints. *Journal of Engineering Mechanics*, ASCE, **128**(5), 562–573.

Wu, Z.S. and Yin, J. (2003) Fracture behaviors of FRP-strengthened concrete structures. *Engineering Fracture Mechanics*, **70**, 1339–1355.

Xiao, Q.Z. and Karihaloo, B.L. (2006) Improving the accuracy of XFEM crack tip fields using higher order quadrature and statically admissible stress recovery. *International Journal for Numerical Methods in Engineering*, **66**, 1378–1410.

Xiao, Z.M. and Zhao, J.F. (2004) Electro-elastic stress analysis for a Zener–Stroh crack at the metal/piezoelectric bi-material interface. *International Journal of Solids and Structures*, **41**, 2501–2519.

Yamanouchi, M., Koizumi, M., Hirai, T. and Shiota, I. (eds) (1990) Proceedings of the First International Symposium on Functionally Graded Materials, FGM-90, FGM Forum. Tokyo, Japan.

Yang, Q.S., Peng, X.R., Kwan, A.K.H. (2006) Strain energy release rate for interfacial cracks in hybrid beams. *Mechanical Research Communications*, **33**(6), 796–803.

Yang, W., Sou, Z. and Shih, C.F. (1991) Mechanics of dynamic debonding. *Proceedings of Royal Society of London*, **33**, 679–697.

Yau, J.F. (1979) Mixed Mode Crack Analysis Using a Conservation Integral. PhD thesis, Dept. of Theoretical and Applied Mechanics, University of Illinois.

Yau, J.F., Wang, S.S. and Corten, H.T. (1980) A mixed-mode crack analysi of isotropic solids using conservation laws of elasticity. *ASME Journal of Applied Mechanics*, **47**(2), 335–341.

Ye, Z. and Ayari, M.L. (1994) Prediction of crack propagation in anisotropic solids. *Engineering Fracture Mechanics*, **49**(6), 797–808.

Ying, L. (1982) A note on the singularity and the strain energy of singular elements. *International Journal for Numerical Methods in Engineering*, **18**, 31–39.

Yoffe, E.H. (1951) The moving Griffith crack. *Philiosiphical Magazine*, **42**(7), 739.

Zhang, C.Y. (2006) *Viscoelastic Fracture Mechanics*, Science Press, Beijing.

Zhang, Ch., Cui, M., Wang, J. *et al.* (2011) 3D crack analysis in functionally graded materials. *Engineering Fracture Mechanics*, **78**, 585–604.

Zhang, H.H., Rong, G. and Li, L.X. (2010) Numerical study on deformations in a cracked viscoelastic body with the extended finite element method. *Engineering Analysis with Boundary Elements*, **34**, 619–624.

Zhang, T.Y. and Gao, C.F. (2004) Fracture behaviors of piezoelectric materials. *Theoretical and Applied Fracture Mechanics*, **41**, 339–379.

Zhang, T.Y., Qian, C.F. and Tong, P. (1998) Linear electro-elastic analysis of a cavity or a crack in a piezoelectric material. *International Journal of Solids and Structures*, **35**, 2121–2149.

Zhang, Y., Bazilevs, Y., Goswami, S. *et al.* (2007) Patient-Specific Vascular NURBS Modeling for Isogeometric Analysis of Blood Flow. *Computer Methods in Applied Mechanics and Engineering*, **196**(29–30), 2943–2959.

Zehnder, A.T. and Viz, M.J. (2005) Fracture mechanics of thin plates and shells under combined membrane, bending, and twisting loads. *Applied Mechanics Reviews*, **58**(1–6), 37–48.

Zi, G. and Belytschko, T. (2003) New crack-tip elements for XFEM and applications to cohesive cracks. *International Journal for Numerical Methods in Engineering*, **57**, 2221–2240.

Zi, G., Song, J.H., Budyn, E. *et al.* (2004) A method for growing multiple cracks without remeshing and its application to fatigue crack growth. *Modeling and Simulations for Matererial Science and Engineering*, **12**, 901–915.

Zi, G., Chen, H., Xu, J. and Belytschko, T. (2005) The extended finite element method for dynamic fractures. *Shock and Vibration*, **12**, 9–23.

Zienkiewicz, O.C., Taylor, R.L. and Zhu, J.Z. (2005) *The Finite Element Method*, 6th edn, Elsevier, USA.

# Index

---

*XFEM Fracture Analysis of Composites*, First Edition. Soheil Mohammadi.
© 2012 John Wiley & Sons, Ltd. Published 2012 by John Wiley & Sons, Ltd.